Damage Models and Algorithms for Assessment of Structures under Operating Conditions

Structures and Infrastructures Series

ISSN 1747-7735

Book Series Editor:

Dan M. Frangopol

Professor of Civil Engineering and
Fazlur R. Khan Endowed Chair of Structural Engineering and Architecture
Department of Civil and Environmental Engineering
Center for Advanced Technology for Large Structural Systems (ATLSS Center)
Lehigh University
Bethlehem, PA, USA

Volume 5

Damage Models and Algorithms for Assessment of Structures under Operating Conditions

Siu-Seong Law[1] and Xin-Qun Zhu[2]

[1]Civil and Structural Engineering Department, Hong Kong Polytechnic University, Kowloon, Hong Kong
[2]School of Engineering, University of Western Sydney, Australia

CRC Press
Taylor & Francis Group
Boca Raton   London   New York   Leiden

CRC Press is an imprint of the
Taylor & Francis Group, an **informa** business

A BALKEMA BOOK

Colophon

Book Series Editor:
Dan M. Frangopol

Volume Authors:
Siu-Seong Law & Xin-Qun Zhu

Cover illustration:
Spatial mathematical model of the Tsing Ma Suspension Bridge Deck.

Taylor & Francis is an imprint of the Taylor & Francis Group,
an informa business

© 2009 Taylor & Francis Group, London, UK

Typeset by Charon Tec Ltd (A Macmillan company), Chennai, India
Printed and bound in Great Britain by Antony Rowe (a CPI Group company),
Chippenham, Wiltshire

British Library Cataloguing in Publication Data
A catalogue record for this book is available from the British Library

Library of Congress Cataloging-in-Publication Data

Law, S. S.
  Damage models and algorithms for assessment of structures under
operating conditions / S.S. Law and X.Q. Zhu.
    p. cm. — (Structures and infrastructures series, ISSN 1747-7735 ; v. 5)
  Includes bibliographical references.
  ISBN 978-0-415-42195-9 (hardcover : alk. paper) — ISBN 978-0-203-87087-7
  (e-book : alk. paper) 1. Structural failures—Mathematical models. 2. Buildings—
Evaluation—Mathematics. I. Zhu, X. Q. II. Title. III. Series.
TA656.L39 2009
624.1'71015118—dc22

                                                            2009017893

Published by: CRC Press/Balkema
               P.O. Box 447, 2300 AK Leiden, The Netherlands
               e-mail: Pub.NL@taylorandfrancis.com
               www.crcpress.com – www.taylorandfrancis.co.uk – www.balkema.nl

ISBN13   978-0-415-42195-9(Hbk)
ISBN13   978-0-203-87087-7(eBook)
Structures and Infrastructures Series: ISSN 1747-7735
Volume 5

Table of Contents

Editorial   XIII
About the Book Series Editor   XV
Preface   XVII
Dedication   XXI
Acknowledgements   XXIII
About the Authors   XXV

Chapter 1   Introduction   1

1.1   Condition monitoring of civil infrastructures   1
       1.1.1   Background to the book   1
       1.1.2   What information should be obtained from the structural health
                monitoring system?   1
1.2   General requirements of a structural condition assessment algorithm   3
1.3   Special requirements for concrete structures   3
1.4   Other considerations   4
       1.4.1   Sensor requirements   4
       1.4.2   The problem of a structure with a large number of
                degrees-of-freedom   4
       1.4.3   Dynamic approach versus static approach   5
       1.4.4   Time-domain approach versus frequency-domain approach   5
       1.4.5   The operation loading and the environmental effects   5
       1.4.6   The uncertainties   6
1.5   The ideal algorithm/strategy of condition assessment   6

Chapter 2   Mathematical concepts for discrete inverse problems   9

2.1   Introduction   9
2.2   Discrete inverse problems   9
       2.2.1   Mathematical concepts   9
       2.2.2   The ill-posedness of the inverse problem   10

2.3   General inversion by singular value decomposition   11
       2.3.1   Singular value decomposition   11
       2.3.2   The generalized singular value decomposition   12
       2.3.3   The discrete Picard condition and filter factors   13
2.4   Solution by optimization   15
       2.4.1   Gradient-based approach   15
       2.4.2   Genetic algorithm   17
       2.4.3   Simulated annealing   19
2.5   Tikhonov regularization   19
       2.5.1   Truncated singular value decomposition   20
       2.5.2   Generalized cross-validation   21
       2.5.3   The L-curve   23
2.6   General optimization procedure for the inverse problem   25
2.7   The criteria of convergence   25
2.8   Summary   25

Chapter 3   Damage description and modelling   27

3.1   Introduction   27
       3.1.1   Damage models   27
       3.1.2   Model on pre-stress   27
3.2   Damage-detection-oriented model   28
       3.2.1   Beam element with end flexibilities   28
                3.2.1.1   Hybrid beam with shear flexibility   29
                3.2.1.2   Hybrid beam with both shear and flexural
                           flexibilities   42
       3.2.2   Decomposition of system matrices   46
                3.2.2.1   The generic element   47
                3.2.2.2   The eigen-decomposition   52
       3.2.3   Super-element   69
                3.2.3.1   Beam element with semi-rigid joints   70
                3.2.3.2   The Tsing Ma bridge deck   73
       3.2.4   Concrete beam with flexural crack and debonding at the steel
                and concrete interface   78
       3.2.5   Beam with unbonded pre-stress tendon   86
       3.2.6   Pre-stressed concrete box-girder with bonded tendon 92
       3.2.7   Models with thin plate   97
                3.2.7.1   Anisotropic model of elliptical crack with strain
                           energy equivalence   97
                3.2.7.2   Thin plates with anisotropic crack from dynamic
                           characteristic equivalence   100
       3.2.8   Model with thick plate   107
                3.2.8.1   Thick plate with anisotropic crack model 107
       3.2.9   Model of thick plate reinforced with Fibre-Reinforced-Plastic 113
                3.2.9.1   Damage-detection-oriented model of delamination
                           of fibre-reinforced plastic and thick plate 116
3.3   Conclusions   128

Chapter 4   Model reduction   131

4.1   Introduction   131
4.2   Static condensation   131
4.3   Dynamic condensation   132
4.4   Iterative condensation   135
4.5   Moving force identification using the improved reduced system   135
    4.5.1   Theory of moving force identification   135
    4.5.2   Numerical example   137
4.6   Structural damage detection using incomplete modal data   139
    4.6.1   Mode shape expansion   139
    4.6.2   Application   140
4.7   Remarks on more recent developments   144

Chapter 5   Damage detection from static measurement   145

5.1   Introduction   145
5.2   Constrained minimization   145
    5.2.1   Output error function   146
        5.2.1.1   Displacement output error function   146
        5.2.1.2   Strain output error function   147
    5.2.2   Damage detection from the static response changes   148
    5.2.3   Damage detection from combined static and dynamic
        measurements   150
5.3   Variation of static deflection profile with damage   152
    5.3.1   The static deflection profile   152
    5.3.2   Spatial wavelet transform   154
5.4   Application   154
    5.4.1   Damage assessment of concrete beams   154
        5.4.1.1   Effect of measurement noise   154
        5.4.1.2   Damage identification   156
        5.4.1.3   Damage evolution under load   158
        5.4.1.4   Damage identification – Simulating practical
            assessment   159
    5.4.2   Assessment of bonding condition in reinforced concrete beams   160
        5.4.2.1   Local beam damage identification   160
        5.4.2.2   Identification of local bonding   161
        5.4.2.3   Simultaneous identification of local bonding and
            beam damages   163
5.5   Limitations with static measurements   164
5.6   Conclusions   165

Chapter 6   Damage detection in the frequency domain   167

6.1   Introduction   167
6.2   Spatial distributed system   167

6.3   The eigenvalue problem   168
   6.3.1   Sensitivity of eigenvalues and eigenvectors   168
   6.3.2   System with close or repeated eigenvalues   170
6.4   Localization and quantification of damage   172
6.5   Finite element model updating   172
6.6   Higher order modal parameters and their sensitivity   173
   6.6.1   Elemental modal strain energy   173
      6.6.1.1   Model strain energy change sensitivity   174
   6.6.2   Modal flexibility   176
      6.6.2.1   Model flexibility sensitivity   177
   6.6.3   Unit load surface   180
6.7   The curvatures   180
   6.7.1   Mode shape curvature   181
   6.7.2   Modal flexibility curvature   181
   6.7.3   Unit load surface curvature   181
   6.7.4   Chebyshev polynomial approximation   182
   6.7.5   The gap-smoothing technique   184
      6.7.5.1   The uniform load surface curvature sensitivity 185
   6.7.6   Numerical examples of damage localization   188
      6.7.6.1   Simply supported plate   189
      6.7.6.2   Study on truncation effect   189
      6.7.6.3   Comparison of curvature methods   191
      6.7.6.4   Resolution of damage localization   192
      6.7.6.5   Cantilever plate   192
      6.7.6.6   Effect of sensor sparsity   192
      6.7.6.7   Effect of measurement noise   195
      6.7.6.8   When the damage changes the boundary condition
         of the structure   196
6.8   Conclusions   197

Chapter 7   System identification based on response sensitivity   199

7.1   Time-domain methods   199
7.2   The response sensitivity   199
   7.2.1   The computational approach   199
   7.2.2   The analytical formulation   200
   7.2.3   Main features of the response sensitivity   201
7.3   Applications in system identification   205
   7.3.1   Excitation force identification   205
      7.3.1.1   The response sensitivity   207
      7.3.1.2   Experimental verification   207
   7.3.2   Condition assessment from output only   209
      7.3.2.1   Algorithm of iteration   209
      7.3.2.2   Experimental verification   211
   7.3.3   Removal of the temperature effect   212
   7.3.4   Identification with coupled system parameters   215
   7.3.5   Condition assessment of structural parameters having
      a wide range of sensitivities   216

7.4   Condition assessment of load resistance of isotropic structural
           components   216
           7.4.1   Dynamic test for model updating   218
           7.4.2   Damage scenarios   219
           7.4.3   Dynamic test for damage detection   219
                      Damage scenario E1   219
                      Damage scenario E2   221
                      Damage scenario E3   221
                      The false positives in the identified results   222
7.5   System identification under operational loads   223
           7.5.1   Existing approaches   223
                      7.5.1.1   The equation of motion   223
                      7.5.1.2   Damage detection from displacement measurement   224
           7.5.2   The generalized orthogonal function expansion   226
           7.5.3   Application to a bridge-vehicle system   227
                      7.5.3.1   The vehicle and bridge system   227
                      7.5.3.2   The residual pre-stress identification   229
7.6   Conclusions   230

Chapter 8   System identification with wavelet   231

8.1   Introduction   231
           8.1.1   The wavelets   231
           8.1.2   The wavelet packets   233
8.2   Identification of crack in beam under operating load   235
           8.2.1   Dynamic behaviour of the cracked beam subject to moving load   236
           8.2.2   The crack model   237
           8.2.3   Crack identification using continuous wavelet transform   239
           8.2.4   Numerical study   240
           8.2.5   Experimental verification   243
8.3   The sensitivity approach   245
           8.3.1   The wavelet packet component energy sensitivity and the
                      solution algorithm   246
                      The solution algorithm   247
           8.3.2   The wavelet sensitivity and the solution algorithm   247
                      Analytical approach   248
                      Computational approach   249
                      The solution algorithm   250
           8.3.3   The wavelet packet transform sensitivity   251
           8.3.4   Damage information from different wavelet bandwidths   252
                      Damage scenarios and their detection   255
                      Effect of measurement noise and model error   256
           8.3.5   Damage information from different wavelet coefficients   257
           8.3.6   Frequency and energy content of wavelet coefficients   258
                      Comparison with response sensitivity   258
                      Damage identification   259
                      Effect of model error   260
           8.3.7   Noise effect   263

8.4 Approaches that are independent of input excitation 264
    8.4.1 The unit impulse response function sensitivity 264
        8.4.1.1 Wavelet-based unit impulse response 265
        8.4.1.2 Impulse response function via discrete wavelet transform 267
        8.4.1.3 Solution algorithm 268
        8.4.1.4 Simulation study 269
                Damage identification with model error and noise effect 269
        8.4.1.5 Discussions 270
    8.4.2 The covariance sensitivity 271
        8.4.2.1 Covariance of measured responses 271
        8.4.2.2 When under single random excitation 272
        8.4.2.3 When under multiple random excitations 273
        8.4.2.4 Sensitivity of the cross-correlation function 275
8.5 Condition assessment including the load environment 275
    8.5.1 Sources of external excitation 275
    8.5.2 Under earthquake loading or ground-borne excitation 275
        8.5.2.1 Simulation studies 275
        8.5.2.2 The sensitivities 276
        8.5.2.3 Damage identification from WPT sensitivity and response sensitivity 277
                Effect of model error and noise 279
                Performance from a subset of the measured response 280
    8.5.3 Under normal random support excitation 280
        8.5.3.1 Damage localization based on mode shape changes 281
        8.5.3.2 Laboratory experiment 282
                Modelling of the structure 282
                Ambient vibration test for damage detection 283
                Damage scenarios 284
                Model improvement for damage detection 285
8.6 Conclusions 286

Chapter 9 Uncertainty analysis 287

9.1 Introduction 287
9.2 System uncertainties 288
    9.2.1 Modelling uncertainty 288
    9.2.2 Parameter uncertainty 289
    9.2.3 Measurement and environmental uncertainty 289
9.3 System identification with parameter uncertainty 290
    9.3.1 Monte Carlo simulation 291
    9.3.2 Integrated perturbed and Bayesian method 291
9.4 Modelling the uncertainty 294
9.5 Propagation of uncertainties in the condition assessment process 295
    9.5.1 Theoretical formulation 295
        9.5.1.1 Uncertainties of the system 295

	9.5.1.2	Derivatives of local damage with respect to the uncertainties	296
	9.5.1.3	Uncertainty in the system parameter	296
	9.5.1.4	Uncertainty in the exciting force	297
	9.5.1.5	Uncertainty in the structural response	298
	9.5.1.6	Statistical characteristics of the damage vector	299
	9.5.1.7	Statistical analysis in damage identification	301
9.5.2	Numerical example		301
	9.5.2.1	The structure	301
	9.5.2.2	Uncertainty with the mass density	302
	9.5.2.3	Uncertainty with the elastic modulus of material	304
	9.5.2.4	Uncertainty with the excitation force and measured response	305
9.5.3	Discussions		306
9.6	Integration of system uncertainties with the reliability analysis of a box-section bridge deck structure		306
9.6.1	Numerical example		307
9.6.2	Condition assessment		308
9.6.3	Reliability analysis		310
9.7	Conclusions		313

References	315
Subject Index	329
Structures and Infrastructures Series	333

Editorial

Welcome to the Book Series *Structures and Infrastructures*.

Our knowledge to model, analyze, design, maintain, manage and predict the life-cycle performance of structures and infrastructures is continually growing. However, the complexity of these systems continues to increase and an integrated approach is necessary to understand the effect of technological, environmental, economical, social and political interactions on the life-cycle performance of engineering structures and infrastructures. In order to accomplish this, methods have to be developed to systematically analyze structure and infrastructure systems, and models have to be formulated for evaluating and comparing the risks and benefits associated with various alternatives. We must maximize the life-cycle benefits of these systems to serve the needs of our society by selecting the best balance of the safety, economy and sustainability requirements despite imperfect information and knowledge.

In recognition of the need for such methods and models, the aim of this Book Series is to present research, developments, and applications written by experts on the most advanced technologies for analyzing, predicting and optimizing the performance of structures and infrastructures such as buildings, bridges, dams, underground construction, offshore platforms, pipelines, naval vessels, ocean structures, nuclear power plants, and also airplanes, aerospace and automotive structures.

The scope of this Book Series covers the entire spectrum of structures and infrastructures. Thus it includes, but is not restricted to, mathematical modeling, computer and experimental methods, practical applications in the areas of assessment and evaluation, construction and design for durability, decision making, deterioration modeling and aging, failure analysis, field testing, structural health monitoring, financial planning, inspection and diagnostics, life-cycle analysis and prediction, loads, maintenance strategies, management systems, nondestructive testing, optimization of maintenance and management, specifications and codes, structural safety and reliability, system analysis, time-dependent performance, rehabilitation, repair, replacement, reliability and risk management, service life prediction, strengthening and whole life costing.

This Book Series is intended for an audience of researchers, practitioners, and students world-wide with a background in civil, aerospace, mechanical, marine and automotive engineering, as well as people working in infrastructure maintenance, monitoring, management and cost analysis of structures and infrastructures. Some volumes are monographs defining the current state of the art and/or practice in the field, and some are textbooks to be used in undergraduate (mostly seniors), graduate and postgraduate courses. This Book Series is affiliated to *Structure and Infrastructure Engineering* (http://www.informaworld.com/sie), an international peer-reviewed journal which is included in the Science Citation Index.

It is now up to you, authors, editors, and readers, to make *Structures and Infrastructures* a success.

Dan M. Frangopol
Book Series Editor

About the Book Series Editor

Dr. Dan M. Frangopol is the first holder of the Fazlur R. Khan Endowed Chair of Structural Engineering and Architecture at Lehigh University, Bethlehem, Pennsylvania, USA, and a Professor in the Department of Civil and Environmental Engineering at Lehigh University. He is also an Emeritus Professor of Civil Engineering at the University of Colorado at Boulder, USA, where he taught for more than two decades (1983–2006). Before joining the University of Colorado, he worked for four years (1979–1983) in structural design with A. Lipski Consulting Engineers in Brussels, Belgium. In 1976, he received his doctorate in Applied Sciences from the University of Liège, Belgium, and holds two honorary doctorates (Doctor Honoris Causa) from the Technical University of Civil Engineering in Bucharest, Romania, and the University of Liège, Belgium. He is a Fellow of the American Society of Civil Engineers (ASCE), American Concrete Institute (ACI), and International Association for Bridge and Structural Engineering (IABSE). He is also an Honorary Member of both the Romanian Academy of Technical Sciences and the Portuguese Association for Bridge Maintenance and Safety. He is the initiator and organizer of the Fazlur R. Khan Lecture Series (www.lehigh.edu/frkseries) at Lehigh University.

Dan Frangopol is an experienced researcher and consultant to industry and government agencies, both nationally and abroad. His main areas of expertise are structural reliability, structural optimization, bridge engineering, and life-cycle analysis, design, maintenance, monitoring, and management of structures and infrastructures. He is the Founding President of the International Association for Bridge Maintenance and Safety (IABMAS, www.iabmas.org) and of the International Association for Life-Cycle Civil Engineering (IALCCE, www.ialcce.org), and Past Director of the Consortium on Advanced Life-Cycle Engineering for Sustainable Civil Environments (COALESCE). He is also the Chair of the Executive Board of the International Association for Structural Safety and Reliability (IASSAR, www.columbia.edu/cu/civileng/iassar) and the Vice-President of the International Society for Health Monitoring of Intelligent Infrastructures (ISHMII, www.ishmii.org). Dan Frangopol is the recipient of several prestigious awards including the 2008 IALCCE Senior Award, the 2007 ASCE Ernest Howard Award, the 2006 IABSE OPAC Award, the 2006 Elsevier Munro Prize, the 2006 T. Y. Lin Medal, the 2005 ASCE Nathan M. Newmark Medal, the 2004 Kajima

Research Award, the 2003 ASCE Moisseiff Award, the 2002 JSPS Fellowship Award for Research in Japan, the 2001 ASCE J. James R. Croes Medal, the 2001 IASSAR Research Prize, the 1998 and 2004 ASCE State-of-the-Art of Civil Engineering Award, and the 1996 Distinguished Probabilistic Methods Educator Award of the Society of Automotive Engineers (SAE).

Dan Frangopol is the Founding Editor-in-Chief of *Structure and Infrastructure Engineering* (Taylor & Francis, www.informaworld.com/sie) an international peer-reviewed journal, which is included in the Science Citation Index. This journal is dedicated to recent advances in maintenance, management, and life-cycle performance of a wide range of structures and infrastructures. He is the author or co-author of over 400 refereed publications, and co-author, editor or co-editor of more than 20 books published by ASCE, Balkema, CIMNE, CRC Press, Elsevier, McGraw-Hill, Taylor & Francis, and Thomas Telford and an editorial board member of several international journals. Additionally, he has chaired and organized several national and international structural engineering conferences and workshops. Dan Frangopol has supervised over 70 Ph.D. and M.Sc. students. Many of his former students are professors at major universities in the United States, Asia, Europe, and South America, and several are prominent in professional practice and research laboratories.

For additional information on Dan M. Frangopol's activities, please visit www.lehigh.edu/~dmf206/

Preface

The economy of the world has undergone a significant leap in the last few decades. Many major infrastructures were constructed to meet the growing demand of rapid and heavy inter-city passages and freight transportation. These structures are important to the economy, and significant losses will be incurred if they are out of service. Some sort of health monitoring system has been included in some of these infrastructures as part of the management system to ensure the smooth operation of these civil structures. Operational data has been collected over many years and yet there is no analysis algorithm which can give the exact working state of the structure on-line. The maintenance engineer would like to know the exact location and damage state of the structural components involved in a damage scenario after an earthquake, a major intentional attack or an unintentional accident to the structure. This knowledge is required in a matter of hours for life saving or any necessary military action. Also, the client would like to have a rapid diagnosis of the structure to make a decision on any necessary remedial work.

Existing methods that use the modal parameters for a diagnosis of the structure are not feasible, as they demand a large number of measurement stations and full or partial closure of the structure. Also, measurement with operation loads on top would not give an accurate estimate of the modal parameters. Other existing methods that use time-response histories do not include the operating load in the analysis. This book is devoted to the condition assessment problem with the structure under operating loads, with many illustrations related to a bridge deck under a group of moving vehicular loads. The loading environment under which the structure is exposed serves as the excitation. It may be a group of vehicular loads, earthquake excitation or ambient random excitation at the supports. It may be the wind loads acting at the deck level in a flexible cable-supported bridge deck. Different algorithms based on these excitations are discussed. These excitation forces are used directly in the equation of motion of the structure for the estimation of local changes in the structure, and different time-domain approaches, including those developed by the authors, are discussed in detail.

These algorithms enable real-time identification with deterministic results on the state of the structure. This meets the needs of maintenance engineers, who would like to know the damage state of the structural components, so that they can judge the suitability of remedial measures and estimate the effect on the performance of the structure. This also matches the current practice of deterministic design of the infrastructure, thus giving the maintenance engineer a good feeling about the safety of the structure.

A description of the type of damage is an essential component of structural condition assessment. However, damage models are scare and are mostly limited to crack(s) in a beam. This book covers a group of Damage-Detection-Oriented Models, including a new decomposition of the elemental matrices of the beam element and plate element that automatically differentiates changes in the different load resisting stiffnesses of the structural component. These models give a more precise description of the damage component than ordinary models, which usually treat the damage as an average reduction in the elastic modulus of the material, and hence they are more suitable for detecting damage.

The owner of the infrastructure is also concerned with the safety and reliability of the structure in the form of a statistical estimate of its remaining life. A method that can extend the deterministic condition assessment to provide statistical information is also included in this book.

The group of methods and algorithms described in this book can be implemented for on-line condition assessment of a structure through model updating during the course of an earthquake, when under normal ambient excitation or operation excitation from passing vehicles. These capabilities are demonstrated with examples of the condition assessment of different structures supplemented with major references.

Chapter 1 gives the background to structural condition assessment and its main components. The requirements of an ideal and practical structural condition assessment algorithm are discussed. Chapter 2 gives a summary of the mathematical techniques that are needed to solve the inverse problem with the condition assessment algorithms presented in this book. The Tikhonov regularization and other optimization methods are noted to be frequently used with the algorithms.

Chapter 3 summarizes the more recently developed models on damage in frame and plate elements. These include, the crack size and orientation in a thin and thick plate; the delamination of Fibre Reinforced Plastic from a concrete plate; the super-element model of the Tsing Ma Bridge deck; a general-purpose joint model with both rotational and transverse flexibility; and the pre-stressing effect of a concrete member. The stiffness matrix of a rectangular shell element can be decomposed analytically into its macro-stiffnesses and the corresponding natural modes. This pair of parameters has been shown to be associated with the axial, bending, shear and torsional capacities of the element. The pattern of the decomposed parameters in a structure has been shown to be capable of indicating the load path and possible failure associated with the different load-carrying capabilities of the structural element. The modelling of damping in the concrete–steel interface of a concrete beam is also included. It is known that the accuracy of the condition assessment result depends on the correctness of the damage model in the model-based approach. These models are different from existing models, which were originally developed for the study of the static and dynamic behaviour under load. These models are grouped under the name of Damage-Detection-Oriented Models, with parameters representing the damage state of the structural element.

Chapter 4 summarizes the formulation of model reduction methods and mode-shape expansion methods, which may be appropriate for the solution of the inverse problem with small- and medium-size structures. Remarks are given on their limitations, and a new direction is discussed whereby a large-scale structure is considered as an assembly of 'sub-structures' and the interface forces between sub-structures are treated as input to the sub-structures in the condition assessment.

Examples of condition assessment using static measurement are given in Chapter 5. Although it is not a popular approach, the examples illustrate the essential features of the inverse identification problem, including the fact that the identified local damage is a function of the load level. Chapter 6 gives the more recently developed high-order sensitive dynamic parameters in the frequency domain. The analytical relationship between the different modal parameters and the parameters of the structure are presented. The modal flexibility and unit load surface curvatures are applied in the assessment of cracks in thin and thick plates.

Chapter 7 deals with the more recent developments in the time-domain with the structure under operation load. The measured response is used directly in a sensitivity approach for both the localization and quantification of local damages. This time-domain approach provides a virtually unlimited supply of measured information from as few as one sensor. Features of these approaches are discussed, including the identification from output response only; the treatment with coupled structural parameters; the problem with a wide range of sensitivity in the inverse analysis; and whether the operation load and system parameters can be identified separately or simultaneously. The temperature effect in the different measurements can also be accounted for with these techniques. Chapter 8 further develops the time-domain sensitivity approach with wavelet and wavelet packet representations, where the information from different bandwidths of the measured responses for the condition assessment can be explored. The unit-impulse response function sensitivity and covariance sensitivity are formulated to remove the dependence of the problem on input excitation. The different types of load environments, such as, earthquake excitation, vehicular excitation and random white noise support excitation, are included in the condition assessment.

Chapter 9 summarizes the different uncertainties involved in the structural condition assessment and the more common methods for the reliability analysis of a structure. An example is given on how the system uncertainties in the inverse problem are integrated into the condition assessment process, resulting in propagation of these uncertainties from the system model into the final identified results. The statistics of the basic variables of the system are altered, resulting in an updated set of reliability indices for the structure. A box-section bridge deck is taken as an example to explain the integration of these uncertainties in the condition assessment and the subsequent reliability analysis.

Xin-Qun Zhu
Siu-Seong Law
July 2009

Dedication

This book is dedicated to our wives Connie Lam and Yan Wang and our families for their support and patience during the preparation of this book, and also to all of our students and colleagues who over the years have contributed to our knowledge of structural damage detection and health monitoring.

Acknowledgements

A special acknowledgement to the American Society of Mechanical Engineers and the American Institute of Aeronautics and Astronautics for their permission to use some of the materials which were originally published in the following journal articles:

- Wu. D. and Law, S.S. (2005) Sensitivity of uniform load surface curvature for damage identification in plate structures. *Journal of Vibration and Acoustics*, ASME, 127(1): 84–92.
- Wu, D. and Law, S.S. (2005) Crack identification in thin plates with anisotropic damage model and vibration measurements. *Journal of Applied Mechanics*, ASME. 72(6): 852–861.
- Wu, D. and Law, S.S. (2007) Delamination detection oriented finite element model for a FRP bonded concrete plate and its application with vibration measurements. *Journal of Applied Mechanics*, ASME. 74(2): 240–248.
- Law, S.S. and Li, X.Y. (2007) Wavelet-based sensitivity analysis of the impulse response function for damage detection. *Journal of Applied Mechanics*, ASME. 74(2): 375–377.
- Li, X.Y. and Law, S.S. (2008) Damage identification of structures including system uncertainties and measurement noise. *American Institute of Aeronautics and Astronautics Journal*. 46(1): 263–276.

About the Authors

Xin-Qun Zhu – Dr. Zhu, who received his Ph.D. in civil engineering from the Hong Kong Polytechnic University (2001), is currently a lecturer in Structural Engineering at the University of Western Sydney. His research interests are primarily in structural dynamics, with emphasis on structural health monitoring and condition assessment, vehicle-bridge/road/track interaction analysis, moving load identification, damage mechanism of concrete structures and smart sensor technology. He has published over 100 refereed papers in journals and international conferences.

Siu-Seong Law – Dr. Law, is currently an Associate Professor of the Civil and Structural Engineering Department of the Hong Kong Polytechnic University. He received his doctorate in civil engineering from the University of Bristol, United Kingdom (1991). His main area of research is in the inverse analysis of force identification and condition assessment of structures with special application in bridge engineering. He has published extensively in the area of damage models and time domain approach for the inverse analysis of structure.

Chapter 1

Introduction

1.1 Condition monitoring of civil infrastructures

1.1.1 Background to the book

The world economy has undergone a significant leap in the last few decades with the construction of many large-scale infrastructures. The construction of many long span bridges is usually accompanied by the installation of a structural health monitoring system. Some of these structures have been monitored on their performances for over a decade, and yet there is no condition assessment method that can use the collected data to yield useful information for the bridge owner towards the maintenance scheduling and on the evolution of the structural conditions of the bridge. Also, many of the highways and bridges constructed in the fifties and sixties in the United States and Europe are aging with the wear from usage and poor maintenance. The failure of these highways and bridges would be disastrous for the economy of the area and for the whole country. The collapse of two major bridges in JiuJiang, in China and in Minneapolis in the United States in 2007, highlighted the urgent need for a simple and realistic approach for condition assessment integrated with the reliability rating of the bridge structure. However, the limited resources of short-term structural health monitoring of the stock of infrastructures in any country is noted.

The existing practice of condition assessment of highway bridges is based on visual inspections or theoretical/numerical models and is typically oriented towards the detection of local anomalies, localization and identification. Other technical approaches that use low load level static and dynamic tests, underestimate the local anomalies which are often functions of the load level.

1.1.2 What information should be obtained from the structural health monitoring system?

The most important information required by the owner of the infrastructure is that which helps the engineer to decide on the maintenance schedule and to prepare an emergency plan in case of an accident. The basic requirements are: Is there any damage to the structure? Where is the damage? How bad is the damage scenario? and, how will the damage affect the remaining useful life of the structure? They are generally referred to as the Level 1, Level 2, Level 3 and Level 4 problems. Answers to the first three problems are usually provided by the Structural Health Monitoring (SHM) system, which involves the observation of a structure over time using periodically sampled response measurements from an array of sensors, the extraction of damage features from these measurements and their analysis to determine the current state of the structure. For long-term structural health monitoring, this process is periodically repeated with updated information on the performance of the structure. This process is usually referred to as Condition Monitoring. The answer to the Level 4 problem is usually provided through the process of Damage Prognosis, which is the estimation of the performance of the structure via predictive models, including the past and present condition of the structure; the environmental influence; and the original design assumptions regarding the loading and operational environments. This question is difficult to answer and the problem will not readily be solved in the next few years. While there are many methods, particularly non-model-based methods, that can handle the Level 1 problem, and, to a certain extent, the Level 2 problem, the answer to the Level 3 problem needs a correlation of the damage features with the load resisting model of the structure, which in turn requires a damage model to represent the extent of damage. However, only a few damage models can be found in the literature. This book aims to provide the answer mainly to the Level 3 problem with information provided by either short-term or long-term structural health monitoring.

Maintenance engineers would like to know the exact location and damage state of the structural components involved with a damage scenario, so that they can judge the suitability of remedial measures and estimate the effect on the performance of the structure. These requirements are consistent with the existing practice of deterministic design of the infrastructures.

The interpretation of the assessment results must be related to some basic parameters of the structure to have physical meaning. In a discretized model of a structure, such parameters are usually averaged over the entire element with no details on the state of damage in the element and its relation to the state in adjacent elements. Damage models are scarce and the types are limited. The identification of equivalent changes in the stiffnesses of a large number of discretized finite elements of a structure to define its performance in the limit and serviceability states would be meaningful only to the structure of isotropic homogeneous materials and is constrained by the capability of optimization algorithms with many unknowns. There is unfortunately no direct link between the stiffness change and the load-carrying capacity of a structure.

Promising types of vibration-based methods (Doebling et al., 1998b) for structural health monitoring include primarily model-based and non-model-based statistical pattern recognition methods. The first group of methods updates the required structural parameters of the damaged structure with respect to the model of the intact structure, and the parameters can be interpreted to locate and evaluate the damage, as has been done by Abdel Wahab et al. (1999) with their reinforced concrete beams. The key is to find and use features that are sensitive to damage. Most commonly used features in vibration-based damage identification are model-based linear features, such as modal frequencies, mode shapes, mode shape derivatives, modal macro-strain vectors, modal flexibility/stiffness and load-dependent Ritz vectors. These features can be applied to either linear or nonlinear response data, but are based on linear concepts. The parameters of linear (physics-based) finite element models of structures are also used as features for damage identification purposes. The use of these parameters needs 'data-mining' through flexible software to manipulate the basic measured data, and it is not discussed in this book.

1.2 General requirements of a structural condition assessment algorithm

An effective structural condition assessment method should consist of the following components: a strategy of measurement; a selection of parameters to be updated; the updating algorithm; and a library of damage models plus on-line assessment from short duration measurements. The set of measured information should be sensitive to the physical parameters to be identified, and the measured locations should be determined using different criteria (Kammer, 1997). The set of parameters to be updated should be considered from an engineering perspective, and they, as a whole, should be able to give a full description of the condition of the structure. These parameters, when linked with the measured location, should enable an optimum selection of the parameters with sufficient sensitivity. Existing damage models are not universal and therefore it is necessary to repeat the identification for a best match with different damage models of the structure. Some of these models are discussed in Chapter 3. The updating algorithm should be iterative to take account of nonlinearities in the anomaly. The uniqueness of the solution is not guaranteed in all existing updating algorithms, but this is constrained by the capability of the minimization algorithm not falling into local minima. The ill-conditioned solution will need to be improved with regularization (Law et al., 2001b). This is discussed further in Chapter 2. It is clear from the above discussions that the assessment method depends on the target structure. Also, with the uncertainties involved and the difficulty of fully eliminating these errors in the assessment process, probabilistic estimation methods have also been developed (Farrar et al., 1999). The assessment result usually contains some statistical characteristics.

While there are numerous problems associated with the condition assessment of a structure, the following are the major problems that need to be solved for any practical application:

- How to minimize the required measured information? Which are the best sensor locations?
- How to incorporate the operational loading into the algorithm?
- How to assess a structure with many structural components?
- How to include or exclude the effect of environmental parameters?
- How to set up the threshold value that triggers an alarm?

1.3 Special requirements for concrete structures

The problem of condition assessment for a pre-stressed concrete bridge deck lies in the fact that, the definition of the damage state of the structure in terms of the EI, GJ, etc. with an isotropic homogeneous material, does not have the same physical interpretation as the non-homogeneous reinforced concrete member. The damage zone in a beam has been assessed using a three-parameter model with dynamic loads (Maeck et al., 2000; Law and Zhu, 2004). The load-carrying capacity of a pre-stressed concrete structural component is largely determined by the pre-stressing force in the cables, and in most cases, the cracks are closed under the pre-stress. Therefore, a damage model on the bonding effect (Limkatanyu and Spacone, 2002; Zhu and Law, 2007b) is required with the pre-stress as an identifiable parameter. The damage model may be incorporated into a finite element model or a finite strip model of the structure. The finite strip approach has been developed with fewer unknowns in the system identification than the former, to take account of the continuous structure with a non-uniform profile under the action of point loads. The pre-stress force is modelled as equivalent forces at the strip nodal points (Choi et al., 2002), and at the nodal points (Figueiras and Póvoas, 1994) in the case of modelling using the finite element.

1.4 Other considerations

1.4.1 Sensor requirements

Different types of sensors, ranging from strain gauges, linear displacement transducers and accelerometers to GPS and laser vibrometers, can be used to collect an array of deformations and stresses of the structure under load. Their locations should be optimized before the sensor installation, so that the effectiveness and sensitivity of the measured information are maximized. Kammer (1991) proposed the sensor placement to maximize the Fisher information and to provide linearly independent mode shapes. Hemez and Farhat (1994) modified Kammer's Effective Independence method according to the strain energy distribution of the structure with the sensors placed near the load paths, so that any structural change becomes more observable. Shi et al. (2000c) developed the sensor placement method for structural damage detection, whereby sensors are placed at locations most sensitive to structural changes of the structure. Chapters 7 and 8 also show that different types of measured information have different sensitivities with respect to the local damages under study. While other sensors are collecting information on the temperature, wind conditions, humidity, etc., they calibrate the health monitoring system with respect to the different variables of the system. The data 'fusion' (Jiang et al., 2005; Guo, 2006; Smyth and Wu, 2007), created by combining groups of sensor information into new virtual sensors to produce hybrid information taking advantage of their spatial relationship, can also be achieved through flexible, enabling software for dynamically establishing and managing the sensor groups with commercially available software packages.

1.4.2 The problem of a structure with a large number of degrees-of-freedom

Existing condition assessment techniques based on measurement make use of the *global* response of the structure for the assessment of *local* anomalies, which are subject to measurement error and errors in the analytical model in a model-based approach. These errors distribute throughout the set of results masking those identified for the local anomalies. As a result of this deficiency, the identification of a large structure with many structural components does not give the correct estimation on the condition of the structure.

Many attempts have made to reduce the structure into sub-structures with fewer degrees-of-freedom (DOFs) or to expand the measured information into the full set of DOFs of the structural system, and they are subject to the error distribution in the final set of identified results. However, these methods are very useful for solving medium- and small-size problems, and more details of the different formulations are presented in Chapter 4.

1.4.3 Dynamic approach versus static approach

The static approach uses the responses from the operational, or close to the operational, load for the assessment. This is important as most types of damages do not show up under a small load level and they are difficult to detect. These damages affect the gradient of the load-deformation curve when under operational load, and hence are closely associated with the reserve load-carrying capacity of the structure. However, the information obtained from static tests is limited and it is expensive to repeat the test to get more sets of data for the condition assessment. The dynamic approach, however, can provide a large amount of dynamic data in both the frequency and time domains but with the limitation of measuring at a low load level, so some of the local damages may not be detected with this approach. This book discusses one way to overcome this limitation, by including the effect of the operational load in the condition assessment of the structure as shown in Chapters 5, 7 and 8.

1.4.4 Time-domain approach versus frequency-domain approach

The dynamic approach in the frequency domain, though more flexible than the static approach in terms of data collection, still has the disadvantage of a limitation of the measured data in terms of the number of modal frequencies of the structure and the number of measured points to define the mode shapes. The investment in the number of sensors and the data collection system would be limited. Also, the Fourier transformation that converts the measured time series into a spectrum, suffers from a loss of information which is of the same order of the information from the local damages, while the time-domain approach makes direct use of the measured time series in the condition assessment. The time measurement can be collected continuously with time and the experiment can be repeated easily with only a limited number of sensors. When the measured time series is decomposed into wavelets, the damage detection can further be performed with damage information contained in different frequency bandwidths of the response. Also, the wavelet decomposition does not have the data corruption as the Fourier transform. It retains all the information from the local damages in the decomposition. These two groups of methods are discussed with examples in Chapters 7 and 8.

1.4.5 The operation loading and the environmental effects

A structure is subjected to different types of loading during its life span. They may be the operational load, seismic load, wind load and ground tremor. All these loads generate sufficient vibrational response in the structure to reveal some of the hidden damages which would otherwise be impossible to detect with the lack of sufficiently large energy for the artificial excitation. The inclusion of all these loads would be an advantage for a practical damage detection algorithm. Chapters 7 and 8 show some of the works towards this end, including one algorithm using white noise random excitation for the condition assessment. The environmental effects in terms of the temperature, humidity and wind conditions should be treated as random variables of the measured system and they will be handled as random processes in the identification in Chapter 9. A rudimentary treatment of the temperature effect on the identification is also presented in Chapter 7.

1.4.6 *The uncertainties*

Each of the system parameters is treated as a random variable with a mean and a variance. When they go through the condition assessment process, their statistics change and affect the statistics of the identified results. This fact has not been considered in existing condition assessment procedures, leading to incorrect indices in the subsequent reliability analysis. This is elaborated further in Chapter 9, with remarks on how these random variables could be integrated into the structural condition assessment resulting in an updated reliability index of the structure.

1.5 The ideal algorithm/strategy of condition assessment

This book includes analysis methods for evaluating, calibrating and applying deterministic approaches for detecting structural changes or anomalies in a structure and quantifying their effects in a form for the engineer to make a decision. Other approaches, e.g. the non-parametric methods, such as neural networks; statistical pattern recognition; integration of non-destructive damage identification method with reliability and risk analysis (Stubbs et al., 1998); and the use of probabilistic networks and computational decision theory (Pearl, 1988), to integrate system uncertainties and derive rational decision policies are not discussed in this book.

A promising model-based condition assessment method consists of updating the parameters of a physics-based nonlinear finite element model of the bridge deck using response measurement (Lu and Law, 2007a) or its wavelet decomposition (Law et al., 2005; Law et al., 2006) with possibly the input data. The solution is based on the response or wavelet sensitivity with respect to the different system parameters. The environmental temperature, the pre-stress force (Law and Lu, 2005; Lu and Law, 2006b) and load environment (Lu and Law, 2005; Zhu and Law, 2007) of the operating structure can be considered, while the effect of the modelling error can be alleviated, particularly with the wavelet approach (Law et al., 2005; Law et al., 2006) where the parameter identification can be conducted in different bandwidths of the response measurement. The response and wavelet sensitivity approaches are linear, but when used iteratively with regularization of the solution, they give accurate estimates of the nonlinear anomalies. A study of the distribution of the model error effect in the bandwidth of the measured response is also required, so that the error can be avoided by not using that particular bandwidth of wavelet coefficients (Law et al., 2006). The best sensor location and the best wavelet coefficients/packets with respect to the configuration of the structural system are studied with experiences gained in previous studies (Law et al., 2005; Law et al., 2006). A research challenge in performing the parameter updating is the propagation of uncertainties from the data and the model into the identified parameters of a nonlinear finite element model. This is included, taking advantage of the recent formulation of the uncertainty sensitivities (Xia et al., 2002; Li and Law, 2008). The local anomalies in the bridge deck modelled explicitly with an existing damage model can be identified in the structural condition assessment using the moving vehicle technique (Law and Zhu, 2004).

A new sub-structuring method will be developed taking the *local* dynamic forces at the interfacing DOFs between the sub-structure as the criteria of acceptance of the accuracy of the reduced model. This is different from the existing practice of taking the modal parameters of the structure as the criteria, which are *global* responses. The adjacent sub-structures can be replaced by a substitute set of known forces (Devriendt and Fontul, 2005; Law et al., 2008) at the same coupling coordinates. Thus, the sub-structural analysis technique can be integrated with visual inspection where part of the structure, which has been checked to contain minimal model errors and local anomalies, can be represented by the set of interfacing forces of the sub-structure, while other parts, which are prone to local anomalies and model errors or contain critical components, are monitored closely.

The finite element model of the sub-structure consists of a fraction of the number of DOFs of the whole structure, and the system identification is more effective and accurate compared with existing methods with measurement from a few selected accelerometers on the structure (Kammer, 1991; Hemez and Farhat, 1994; Shi et al., 2000).

Chapter 2

Mathematical concepts for discrete inverse problems

2.1 Introduction

Inverse problems can be found in many areas of engineering mechanics (Tanaka and Bui, 1992; Bui, 1994; Zabaras et al., 1993; Friswell and Mottershead, 1996; Trujillo and Busby, 1997; Tanaka and Dulikravich, 1998; Friswell et al., 1999; Tanaka and Dulikravich, 2000). A successful solution of the inverse problems covers damage detection (Ge and Soong, 1998), model updating (Fregolent et al., 1996; Ahmadian et al., 1998), load identification (Lee and Park, 1995), image or signal reconstruction (Mammone, 1992) and inverse heat conduction problems (Trujillo and Busby, 1997). Generally, the inverse problem is concerned with the determination of the input and the characteristics of a system given certain information on its output. Mathematically, such problems are ill-posed and have to be overcome through the development of new computational schemes, regularization techniques, objective functions and experimental procedures.

This chapter gives a brief description of the basic knowledge of ill-conditioned matrices. Discussions on the Singular Value Decomposition (SVD) and the discrete Picard condition give insight into the discrete ill-posed problem. Section 2.4 gives three optimization algorithms for the solution of the inverse problem. Section 2.5 describes some of the techniques to obtain a regularized solution. Finally, criteria for convergence of the solution are discussed in Section 2.7.

Information in this chapter forms the basis for understanding the solution process of system identification in the following chapters, apart from Chapter three that deals with damage models of a structure.

2.2 Discrete inverse problems

2.2.1 Mathematical concepts

In general, the inverse problem centres on the equation

$$Ax = b \qquad (2.1)$$

where $A \in \Re^{m \times n}, b \in \Re^{m \times 1}, x \in \Re^{n \times 1}$, and x is the vector of required parameters or the input. In the inverse problem, vector b is measured with the aim of estimating the unknown vector, x. This is a linear least-squares problem, as

$$\min_{x}\|Ax - b\|_2 \tag{2.2}$$

It is well known that the least-squares solution is unique and unbiased when $m > n$ provided that rank $(A) = n$. Matrix A becomes unstable or ill-conditioned when A is close to being rank deficient. The inverse problem is a discrete ill-posed problem if it satisfies the following criteria (Hansen, 1994):

(1)   the singular values of A decay gradually to zero;
(2)   the ratio between the largest and the smallest nonzero singular values is large.

Criterion (1) implies that there is no nearby problem with a well-conditioned coefficient matrix and with a well-determined numerical rank. Criterion (2) implies that the matrix A is ill-conditioned, i.e. the solution is potentially very sensitive to perturbations. Singular values are discussed in detail in Section 2.3.1.

2.2.2   The ill-posedness of the inverse problem

There is an interesting and important feature of the discrete ill-posed problem. The ill-conditioning of the problem does not mean that a meaningful approximate solution cannot be computed. Rather, the ill-conditioning implies that standard methods in numerical linear algebra for solving Equations (2.1) and (2.2), cannot be used directly to compute such a solution. More sophisticated methods must be applied instead to ensure the computation of a meaningful solution. The regularization methods have been developed with the aim of achieving this goal.

The primary difficulty with the discrete ill-posed problem is that it is essentially under-determined due to the existence of the group of small singular values of A. Hence, it is necessary to incorporate further information about the desired solution in order to stabilize the problem and to single out a useful and stable solution. This is how the regularization works.

Among the various types of available methods, the more popular approach to regulate the ill-posed problem is to have the second-norm or an appropriate semi-norm of the solution to be small. An estimate, x^*, of the solution may also be included in a side constraint. The most common and well-known form of regularization is the one known as Tikhonov Regularization (Tikhonov, 1963; Morozov, 1984). The idea is to define the regularized solution, x_λ, as the optimal solution of the following weight combination of the residual norm and the smoothing norm

$$x_\lambda = \arg\min\{\|Ax - b\|_2^2 + \lambda\|L(x - x^*)\|_2^2\} \tag{2.3}$$

where the regularization parameter, λ, controls the weight given to minimize the side constraint relative to the minimization of the residual norm. The matrix $L \in \Re^{m \times n}$ is typically either the identity matrix I_n or a $(p \times n)$ discrete approximation of the $(n - p)$th derivative operator, in which case L is a banded matrix with full row rank. The optimal solution is sought that provides a balance between minimizing the smoothing norm and the residual norm. The basic idea behind Equation (2.3) is that a regularized solution with a small semi-norm and a suitable small residual norm is not too far from the desired and unknown solution of the unperturbed problem underlying the given problem. Clearly, a large λ favours a small smoothed semi-norm at the cost of a large residual norm, while a small λ has the opposite effect. If $\lambda = 0$, we return to the least-squares problem and the unregularized solution is computed. The regularization parameter, λ, controls the degree with which the sought regularized solution should fit to the data in b.

The use of Equation (2.3) in regularizing an ill-posed problem has the assumption that the errors on the right-hand-side of the equation are unbiased and that their covariance matrix is proportional to the identity matrix. If the second condition is not satisfied, then the problem should be scaled as suggested by Hansen (1994). Besides Tikhonov regularization, there are many other regularization methods with properties that make them better suited to specific types of problems (Hansen, 1994).

2.3 General inversion by singular value decomposition

2.3.1 Singular value decomposition

Let $A \in \Re^{m \times n}$ be a rectangular matrix with $m \geq n$. The singular value decomposition (SVD) of A is a decomposition of the form (Golub, 1996)

$$A = U \Sigma V^T = \sum_{i=1}^{n} u_i \sigma_i v_i^T \tag{2.4}$$

where $U = (u_1, u_2, \cdots, u_m)$ and $V = (v_1, v_2, \cdots, v_n)$ are matrices with orthonormal columns, with $U^T U = I_m, V^T V = I_n$ and $\Sigma = diag(\sigma_1, \sigma_2, \cdots, \sigma_n)$ has non-negative diagonal elements appearing in descending order such that

$$\sigma_1 \geq \sigma_2 \geq \cdots \geq \sigma_n \geq 0 \tag{2.5}$$

The terms σ_i are the singular values of A, while the vectors u_i and v_i are the left and right singular vectors of A, respectively.

It is noted from the relationships $A^T A = V \Sigma^2 V^T$ and $AA^T = U \Sigma^2 U^T$ that the SVD of A is strongly linked to the eigenvalue decompositions of the symmetric positive semi-definite matrices $A^T A$ and AA^T. This shows that the SVD is unique for a given matrix A, except for singular vectors associated with multiple singular values.

Two characteristic features of the SVD of A are very often found in connection with a discrete ill-posed problem.

- The singular values, σ_i, decay gradually to zero with no zero value and with no particular gap in the spectrum. An increase in the dimensions of A increase the number of small singular values.
- The left and right singular vectors, u_i and v_i, tend to have more sign changes in their elements as the index i increases, i.e. the vectors become more oscillatory when σ_i decreases.

Although these features are found in many discrete ill-posed problems arising in practical applications, they are unfortunately very difficult or perhaps impossible to prove in general.

To have more understanding on the ill-conditioning of matrix A, the following relations, which follow directly from Equation (2.4), are studied:

$$\begin{cases} Av_i = \sigma_i u_i \\ \|Av_i\|_2 = \sigma_i \end{cases} \quad i = 1, 2, \cdots, n \qquad (2.6)$$

It is noted that a small singular value, σ_i, compared to $\|Av_1\|_2 = \sigma_1$, means that there exists a certain linear combination of the columns of A, characterized by the elements of the right singular vector, v_i, such that $\|Av_i\|_2 = \sigma_i$ is small. In other words, one or more small σ_i implies that A is nearly rank deficient (with near zero singular values), and the vector, v_i, associated with the small σ_i are numerical null-vectors of A. From this characteristic feature of A, it can be concluded that the matrix in a discrete ill-posed problem is always highly ill-conditioned and its numerical null-space is spanned by vectors with many sign changes. The null-space is the subset of matrix A corresponding to the unknowns, x, that are mapped onto $b = 0$.

The SVD also gives an important insight into another aspect of the discrete ill-posed problems, namely the smoothing effect typically associated with a square integrable kernel. Notice that as σ_i decreases, the singular vectors u_i and v_i become increasingly oscillatory. With the mapping Ax of an arbitrary vector x using the SVD,

$$x = \sum_{i=1}^{n} (v_i^T x) v_i \text{ and } Ax = \sum_{i=1}^{n} \sigma_i (v_i^T x) u_i \qquad (2.7)$$

This clearly shows that, due to the multiplication with σ_i, the high-frequency components of x are more damped in Ax than the low-frequency components. Moreover, the inverse problem, namely that of computing x from $Ax = b$ or $\min \|Ax - b\|_2$, must have the opposite effect, i.e. it amplifies the high-frequency oscillations in the right-hand-side of vector b.

2.3.2 The generalized singular value decomposition

The generalized singular value decomposition (GSVD) of the matrix pair (A, L) is a generalization of the SVD of A in the sense that the generalized singular values of (A, L) are the square roots of the generalized eigenvalues of the matrix pair $(A^T A, L^T L)$. The dimensions of $A \in \Re^{m \times n}$ and $L \in \Re^{p \times n}$ are assumed to satisfy $m \geq n \geq p$, which is always the case with a discrete ill-posed problem. Then the GSVD is a decomposition of A and L in the form (Hansen, 1994)

$$A = U \begin{pmatrix} \Sigma & 0 \\ 0 & I_{n-p} \end{pmatrix} X^{-1}, \quad L = V(M, 0) X^{-1} \qquad (2.8)$$

where the columns of $U \in \Re^{m \times n}$ and $V \in \Re^{p \times p}$ are orthonormal; $X \in \Re^{n \times n}$ is non-singular; and Σ and M are $(p \times p)$ diagonal matrices, i.e. $\Sigma = diag(\sigma_1, \cdots, \sigma_p)$, $M = diag(u_1, \cdots, u_p)$. Moreover, the diagonal entries of Σ and M are non-negative and ordered such that

$$0 \leq \sigma_1 \leq \sigma_2 \leq \cdots \leq \sigma_p \leq 1, \qquad 1 \geq u_1 \geq \cdots \geq u_p > 0$$

and they are normalized such that

$$\sigma_i^2 + u_i^2 = 1, \qquad i = 1, \cdots, p$$

Then the generalized singular values γ_i of (A, L) are defined as the ratios

$$\gamma_i = \sigma_i / u_i \qquad i = 1, \cdots, p \qquad (2.9)$$

and they obviously appear in ascending order, which is opposite to the ordering of the ordinary singular values of A.

For $p < n$ the matrix $L \in \Re^{p \times n}$ always has a non-trivial null-space $N(L)$. For example, if L is an approximation to the second derivative operator on a regular mesh, i.e. $L = tridiag(1, -2, 1)$, then $N(L)$ is spanned by the two vectors $(1, 1, \cdots, 1)^T$ and $(1, 2, \cdots, n)^T$. In the GSVD, the last $(n - p)$ columns, x_i, of the non-singular matrix X satisfy

$$Lx_i = 0, \qquad i = p + 1, \cdots, n. \qquad (2.10)$$

and they are therefore basis vectors for the null-space $N(L)$.

There is a slight notational problem here because the matrices U, Σ and V in the GSVD of (A, L) are different from the matrices with the same symbols in the SVD of A. However, in this chapter it will always be clear from the context which decomposition is used. When L is the identity matrix, I_n, then the U and V of the GSVD are identical to the U and V of the SVD, and the generalized singular values of (A, I_n) are identical to the singular values of A, except for the ordering of the singular values and vectors.

2.3.3 The discrete Picard condition and filter factors

There is, strictly speaking, no Picard condition for a discrete ill-posed problem because the norm of the solution is always bounded. Nevertheless, a discrete Picard condition could be implemented in a real-world application. The measurement vector b is usually contaminated with various types of error, such as measurement error, approximation error and rounding error. Hence, b can be written as

$$b = \bar{b} + e \qquad (2.11)$$

where e is a vector of the errors and $\bar{b}$ is the unperturbed right-hand-side. Both $\bar{b}$ and the corresponding unperturbed solution, $\bar{x}$, represent the underlying unperturbed and unknown problem. Now, to compute a regularized solution, x_{reg}, from the given vector b, such that x_{reg} approximates the exact solution $\bar{x}$, the corresponding right-hand-side vector $\bar{b}$ must satisfy the following criterion.

The unperturbed vector $\bar{b}$ in a discrete ill-posed problem with regularization matrix L satisfies the discrete Picard condition if the Fourier coefficients $|u_i^T \bar{b}|$ on average decay to zero faster than the generalized singular values, γ_i (Hansen, 1990). The fulfilment of this condition implies that the exact, unknown solution can be approximated by a regularized solution.

Consider Equations (2.1) and (2.2), and assume for simplicity that A has no exact zero singular values. It is easy to show with SVD that the solutions to both systems are given by the same equation:

$$x_{LSQ} = \sum_{i=1}^{n} \frac{u_i^T b}{\sigma_i} v_i \tag{2.12}$$

Since the Fourier coefficients, $|u_i^T b|$, corresponding to the small singular values, σ_i, do not decay as fast as the singular values, but rather tend to level off due to contamination. The solution, x_{LSQ}, is dominated by the terms in the sum corresponding to the small σ_i. Consequently, the solution x_{LSQ} has many sign changes and thus appears completely random.

Figure 2.1 shows the Picard plot by Visser (2001) in near-field acoustic source identification. Figure 2.1(a) gives the discrete Picard condition for the unperturbed data vector, $\bar{b}$. The 'average' decay of the SVD coefficients (crosses) is clearly steeper than that of the singular values. This ensures that a meaningful regularized solution can be obtained. The circles in the figure show the participation of each mode to the solution. The solution is noted to be determined by the first few modes with no dominance of the higher modes.

Figure 2.1(b) gives the Picard plot when the data vector, b, is contaminated with Gaussian noise at a signal to noise ratio of 20dB. The first few SVD coefficients fall off more steeply than the singular values and it is still possible to reconstruct a meaningful solution. But it is also noted that the coefficients (crosses) level off at the noise level. The location of the circles for the higher modes clearly shows their dominating contribution with respect to the first few lower modes, and this phenomenon is important in the physically meaningful solution. This shows the disastrous influence of noise in ill-conditioned problems.

The purpose of a regularization method is to dampen or filter out the contributions to the solution corresponding to the small, generalized singular values. Hence the regularized solution, x_{reg}, which, for $x^* = 0$, can be written as

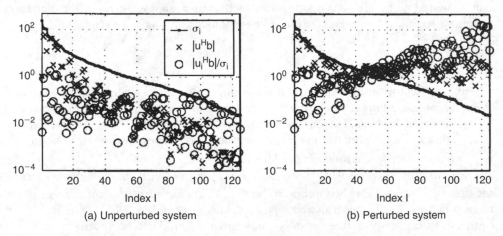

(a) Unperturbed system   (b) Perturbed system

Figure 2.1 Picard plot

$$x_{reg} = \sum_{i=1}^{n} f_i \frac{u_i^T b}{\sigma_i} v_i \qquad \text{if } L = I_n \tag{2.13}$$

$$x_{reg} = \sum_{i=1}^{p} f_i \frac{u_i^T b}{\sigma_i} x_i + \sum_{i=p+1}^{n} (u_i^T b) x_i \qquad \text{if } L \neq I_n \tag{2.14}$$

Here, the terms f_i are the filter factors for the particular regularization method. The filter factors have the important property that as σ_i decreases, the corresponding f_i tends to zero in such a way that the contributions $(u_i^T b/\sigma_i) x_i$ to the solution from the smaller σ_i are effectively filtered out. The difference between the various regularization methods lies essentially in the way these filter factors, f_i, are defined. Hence, the filter factors play an important role in regularization theory, and it is worthwhile characterizing the filter factors for the various regularization methods that are presented below.

For Tikhonov regularization, which plays a central role in regularization theory, the filter factors are either $f_i = \sigma_i^2/(\sigma_i^2 + \lambda^2)$ (for $L = I_n$) or $f_i = \gamma_i^2/(\gamma_i^2 + \lambda^2)$ (for $L \neq I_n$) and the filtering effectively sets in for $\sigma_i < \lambda$ and $\gamma_i < \lambda$, respectively. This shows that the discrete ill-posed problems are essentially un-regularized by Tikhonov's method for $\lambda < \sigma_n$ and $\lambda < \gamma_p$, respectively.

2.4 Solution by optimization

The detection and identification of structural damage is formulated as an optimization problem. The mathematical model of a physical structural system is established to fit the behaviour of a real system through minimizing the discrepancy between the computed and measured responses. Many methods have been developed to solve the optimization problem. Three of these methods are discussed here: gradient-based approach, genetic algorithm (GA) and simulated annealing.

2.4.1 Gradient-based approach

Many excellent and comprehensive texts on mathematical optimization have been written, particularly in gradient-based algorithms (Snyman, 2005). Gradient-based optimization strategies iteratively search a minimum of an n-dimensional objective function $f(x)$. For the function $f(x) \in C^2$, a vector of first-order partial derivatives, or a gradient vector can be computed at any point x, such that

$$\nabla f(x) = \begin{bmatrix} \dfrac{\partial f(x)}{\partial x_1} \\ \dfrac{\partial f(x)}{\partial x_2} \\ \vdots \\ \dfrac{\partial f(x)}{\partial x_n} \end{bmatrix} = g(x) \tag{2.15}$$

where $x = [x_1, x_2, \cdots, x_n]^T \in \mathfrak{R}^n$. The actual optimization can be performed iteratively, and details of the iteration of the optimization problem by a gradient search technique are given below (Snyman, 2005):

(1) Given starting points x^0 and positive tolerances ε_1, ε_2 and ε_3, set $i = 1$.
(2) Select a descent direction, p^i.
(3) Perform a linear search in direction p^i to give the step size, λ_i.
(4) Set $x^i = x^{i-1} + \lambda_i \cdot p^i$ and compute the objective function, $f(x^i)$.
(5) Check the convergence criterion of $f(x^i)$. The algorithm is terminated if a convergence criterion is satisfied. Termination is usually enforced at iteration i if one, or a combination, of the following criteria is met:

$$\text{a) } \|x^i - x^{i-1}\| < \varepsilon_1; \quad \text{b) } \|\nabla f(x^i)\| < \varepsilon_2; \quad \text{c) } \|f(x_i) - f(x_{i-1})\| < \varepsilon_3.$$

(6) Set $i = i + 1$ and go back to Step 2.

To compute the step direction, p^i, a linear (first-order) approximation of the objective function can be used:

$$f(x^i + \lambda_i p^i) \approx f(x^i) + (\nabla f(x^i))^T p^i \tag{2.16}$$

which results in the step direction:

$$p^i = -\nabla f(x^i) \tag{2.17}$$

This is called the steepest descent method. A second-order approach uses a quadratic approximation:

$$\nabla f(x^i) + \nabla^2 f(x^i) p^i = 0 \tag{2.18}$$

and this is referred to as the Newton's direction method.

For an analytical objective function, the first and second derivatives can be directly transferred to a computer program. However if no explicit formula can be defined, the objective function is computed numerically by means of a simulation where approximations for the derivatives are necessary. The finite difference approximation can be applied for each dimension for a multivariate objective function. The gradient vector can be approximated by the forward finite differences as

$$\frac{\partial f(x)}{\partial x_j} \cong \frac{\Delta f(x)}{\delta_j} = \frac{f(x + \delta_j) - f(x)}{\delta_j} \tag{2.19}$$

where $\delta_j = \{0, 0, \cdots, \delta_j, 0, \cdots, 0\}^T$, $\delta_j > 0$ at the j-th position. Better approximations may be obtained using central finite differences.

The performance of a gradient-based method strongly depends on the available initial values. Several optimization runs with different initial values might be necessary if no a priori knowledge (e.g. the result of a process simulation) on the function to be optimized is available.

2.4.2 *Genetic algorithm*

Genetic Algorithm (GA) is based on the principles of evolutionary theory, which are natural selection and evolution. The GA is a 'non-traditional' search or optimization method that simulates the phenomenon of natural evolution according to Darwin's theory. This technique was developed originally to operate on an initial population of randomly generated candidate solutions, encoded as chromosomes, and applied to produce increasingly better approximations to a solution (Figure 2.2) with the principle of survival of the fittest (Holland, 1975). A new set of approximations in each generation is created by the process of selecting individuals according to their level of fitness in the problem domain and breeding them together using operators adopted from natural genetics. This process leads to the evolution of populations of individuals that are better suited to their environment than the individuals that they were created from, just as in natural adaptation. Within the chromosome are separate genes that represent the independent variables of the problem under study. To obtain better-fit chromosomes, three basic randomized operators, the selection, crossover and mutation are used in the evolution.

Chromosomes are selected based on their fitness for the reproduction of future populations. Selection is a very important step within a GA, as the quality of an individual is measured by its fitness value. If selection involves only the fittest chromosomes, the solution space may be very limited due to the lack of diversity. However, a random selection does not guarantee that future generations will increase in fitness.

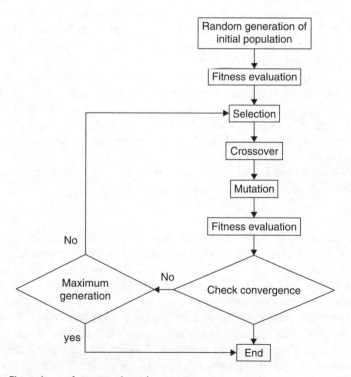

Figure 2.2 Flow chart of genetic algorithm

Crossover is the most important operator in a GA. This operator takes the chromosomes of two parents which are randomly selected, and then exchanges part of their genes resulting in two new chromosomes for the child generation. Therefore, the crossover does not create new material within the population; it simply inter-mixes the existing population. The usual schemes to generate new chromosomes are the single-point crossover, the multipoint crossover and the uniform crossover. The probability of crossover defines the ratio of the number of offspring produced in each generation to the population size.

The mutation operator introduces a change in one or more of the chromosome's genes. New material is introduced in the population with this operator and its main goal is to prevent the population from converging to a local minimum. The probability of mutation is defined as the ratio of the number of mutated genes to the total number of genes in the population and its value is usually low, typically in the range 0.01 and 0.001. However, in some cases it can take higher values with the purpose of increasing the diversity of the population.

From the above discussion, it can be seen that GAs differ substantially from traditional gradient-based search techniques. In fact, they have several advantages that make them suitable for dealing with complex problems where traditional search techniques fail. The major advantages are:

- GAs work on a population of points in parallel in the search space, while traditional search techniques work only on a single point at a time. Because of this, the guess of the initial point has a large effect in traditional methods, since there is a possibility of converging to a local optimal point rather than the global optimal point. Therefore, GAs are more advantageous in complex, nonlinear and multimodal optimization problems.
- GAs, unlike traditional optimization techniques, do not require the evaluation of gradients or higher-order derivatives. Only the objective function and the corresponding fitness levels influence the directions of search. Therefore, GAs are applicable in problems where the objective function is not differentiable.
- GAs use probabilistic search rules, not deterministic ones like gradient-based optimization.
- GAs work on an encoding of the parameter set rather than the parameter itself (except for where real-valued individuals are used).

Due to all these advantages, GAs, although slow in execution, are best applied to problems where traditional optimization techniques do not work well, as in the case of complex problems with many local optima and where the global optimum is required. However, the main disadvantage of GAs, compared to traditional methods, is their high computational cost. However, this drawback can be overcome with faster computers or by using simple objective functions that can be quickly computed. In addition, GAs are not suitable for problems with too many variables, since the search space becomes much larger with an increase in the number of variables. In these cases, GAs appear to be relatively imprecise in performance near to the global optimum when compared with conventional optimization techniques.

2.4.3 Simulated annealing

Another important computational intelligence approach, simulated annealing (SA), is a popular stochastic method based on the physical process of annealing (Van Laarhoven and Aarts, 1987). SA is the simulation of the annealing of a physical multi-particle system for finding the global optimum solution of a large combinatorial optimization problem. If a system is in a configuration q at time t, then a new configuration r of the system at time $t+1$ is generated randomly. The configuration r is accepted according to the acceptance probability $Pro(r)$.

$$Pr\,o(r) = e^{-\left(\frac{E_r - E_q}{K_B T}\right)} \tag{2.20}$$

where K_B is the Boltzmann constant and T is referred to as control parameter (or temperature in the original SA). The energy of the original configuration q and the new configuration r is represented as Eq and Er, respectively. In addition, an annealing process is needed to obtain a lower energy configuration. The well-known cooling schedule that provides the necessary and sufficient conditions for convergence is

$$C(t) = \frac{C_0}{\log t} \quad \forall t > 0 \tag{2.21}$$

where $C(t)$ is a sequence of control parameters, C_0 is a constant and t is the time. It is noted that $C(t)$ approaches zero when t goes to infinity.

When the simulated annealing schedule is applied to an optimization problem, the energy function becomes the objective function, and the configuration becomes the solution configuration of the parameters. Also Equations (2.20) and (2.21) can be further transformed as

$$Pr\,o(r) = e^{-\left(\frac{E_r - E_q}{T(l)}\right)} \tag{2.22}$$

$$T(l) = \frac{T}{\log l} \tag{2.23}$$

where l denotes an integer step sequence, T_0 is the initial constant control parameter and $T(l)$ is a sequence of control parameter. Equations (2.22) and (2.23) then give

$$Pr\,o(l) = l^{-\left(\frac{E_r - E_q}{T_0}\right)} \tag{2.24}$$

It is noted that $Pro(l)$ equals 1.0 when $l = 1.0$.

2.5 Tikhonov regularization

Regularization techniques include direct regularization methods and iterative regularization methods. A brief review of the regularization methods for numerical treatment of discrete ill-posed problems is found in Hansen (1994). Equation. (2.3) shows that the zeroth-order regularization when $L = I$ (the identify matrix). It becomes the first-order regularization when L is a gradient operator and a second-order regularization when L is a surface Laplacian operator. Only one of the three orders of regularization is employed under most circumstances. The zeroth-order regularization biases the estimates towards zero but also greatly reduces large-magnitude oscillations in the parameter values, whereas first-order regularization biases the estimates towards a constant and reduces the tendency to fluctuate from one value to the next. Three regularization methods (truncated SVD, generalized cross-validation and L-curve) are discussed in the following sections.

2.5.1 Truncated singular value decomposition

The SVD allows the solution of singular systems by separating the components of operators belonging to its range from those belonging to its null-space (corresponding to the null singular values). If the whole set of n singular values in Equation (2.4) is nonzero, the solution becomes:

$$x = \sum_{i=1}^{n} \sigma_i^{-1}(u_i^T b)v_i \qquad (2.25)$$

The small singular values may cause solution instability as discussed in Section 2.3.1, and the terms corresponding to the smallest singular values are affected. It would be useful to consider these terms as belonging to the null-space and neglecting the corresponding singular values. In this way, the solution is deprived of some information content but becomes more regular without solution instability due to the small singular values. This is one way to treat the ill-conditioning of A to generate a new problem with a well-conditioned rank deficient coefficient matrix. The rank deficient matrix, which is the closest rank-k approximation A_k to A, is measured in the 2-norm and is obtained by truncating the SVD expansion in Equation (2.4) at k, to give

$$A_k = \sum_{i=1}^{k} u_i \sigma_i v_i^T, \quad k \le n \qquad (2.26)$$

and the solution is given by

$$x_k = \sum_{i=1}^{k} \frac{u_i^T b}{\sigma_i} v_i \qquad (2.27)$$

The number of neglected singular values should be neither too high to preserve the information content, nor too low to obtain a more stable solution. If an estimate, $\|\delta b\|$, of the amount of noise in the data is available, the summation in Equation (2.25) can be truncated when the following condition is not satisfied:

$$\sigma_k \ge \|\delta b\| \sigma_1 \qquad (2.28)$$

This means that the first k singular values can be retained when error in the data can be removed by filtering.

An alternative method to treat the problem is to use the discrete Picard condition to determine the number of terms in the summation in Equation. (2.12). Since the error-free right-hand-side of Equation (2.1) is generally unknown, the Picard condition can be expressed using the perturbed right-hand-side, b, i.e. a bounded solution of the ill-posed problems exists if the terms $|u_i^T b|$ decay faster than the corresponding singular values, σ_i. When $|u_i^T b|$ do not decay faster than the corresponding singular values, the summation should be truncated because the remaining singular values are not able to filter out the error contained in the data. This method does not require knowledge of the error in the data. However, it can be used only when the operator A is noise-free and the error is only on the right-hand-side. In fact, the error in operator A affects both the singular values and the singular vectors. The trend of the terms $|u_i^T b|$ can exhibit large oscillations and it is very difficult to decide when the Picard condition is satisfied.

The third method is based on the minimization of the output residual. If b_{pred} is the reconstructed measurements from the solution by Equation (2.1), then the non-dimensional output residual ς can be defined as

$$\varsigma = \frac{\|b_{pred} - b\|}{\|b\|} \tag{2.29}$$

The output residual is computed for different numbers of retained singular values varying from 1 to n. The truncation of the summation can be performed for a value of k such that ς is a minimum.

2.5.2 Generalized cross-validation

The idea of cross-validation (Stone, 1974) is to maximize the predictability of the model with a choice of the regularization parameter, λ. A predictability test can be arranged by omitting one measured data point, b_k $(k = 1, 2, \cdots, n)$, at a time and determining an estimate, $_k x(\lambda)$, using the remaining data points. Then for each of the estimates, the missing data is predicted and the value of λ that predicts the b_k $(k = 1, 2, \cdots, n)$ best is found. The procedure of this cross-validation is explained in the following steps.

(1)   Find the estimate, $_k x$, which minimizes

$$\sum_{\substack{i=1 \\ i \neq k}}^{n} \left(b_i - \sum_{j=1}^{m} a_{ij} x_j\right)^2 + \lambda \|L(x - x^*)\|_2^2 \tag{2.30}$$

(2)   Predict the missing data point,

$$\widetilde{b}_k(\lambda) = \sum_{j=1}^{m} a_{kj_k} x_j(\lambda) \tag{2.31}$$

(3)   Choose the value of λ which minimizes the cross-validation function,

$$V_0 = \frac{1}{n} \sum_{k=1}^{n} (b_k - \widetilde{b}_k(\lambda))^2 \tag{2.32}$$

To simplify the cross-validation, consider the identity,

$$b_k - \widetilde{b}_k(\lambda) = \frac{b_k - \sum\limits_{i=1}^{m} a_{kj_k} x_j(\lambda)}{1 - \widetilde{r}_{kk}} \tag{2.33}$$

where,

$$\widetilde{r}_{kk} = \frac{\sum\limits_{j=1}^{m} a_{kj_k} x_j(\lambda) - \widetilde{b}_k(\lambda)}{b_k - \widetilde{b}_k(\lambda)} \tag{2.34}$$

and $x_j(\lambda)$ is the j-th term in $x(\lambda)$. Since $\widetilde{b}_k = \sum\limits_{j=1}^{m} a_{kj\,k} x_j(\lambda)$ it follows that

$$\widetilde{r}_{kk} = \sum_{j=1}^{m} \frac{a_{jk}((x_j(\lambda) - {}_k x(\lambda))}{b_k - \widetilde{b}_k(\lambda)} \tag{2.35}$$

Replacing the term on the right-hand-side by a derivative, gives

$$\widetilde{r}_{kk} = \frac{\partial}{\partial b_k} \left(\sum_{j=1}^{m} a_{kj} x_j(\lambda) \right) = r_{kk}(\lambda) \tag{2.36}$$

Combining Equations. (2.32), (2.33) and (2.36) gives

$$V_0(\lambda) = \frac{1}{n} \sum_{i=1}^{n} \left(\frac{b_k - \sum\limits_{j=1}^{m} a_{kj} x_j(\lambda)}{1 - r_{kk}} \right)^2 \tag{2.37}$$

Equation (2.37) can be rewritten in the form

$$V_0 = \frac{1}{n} \| Q(\lambda)(Ax(\lambda) - b) \|_2^2 \tag{2.38}$$

where

$$Q(\lambda) = diag\left(\frac{1}{1 - r_{kk}(\lambda)} \right), \qquad (i = 1, 2, \cdots, n) \tag{2.39}$$

and r_{kk} is the kk-th element of the influence matrix, $R(\lambda)$, where

$$R(\lambda) = A(A^T A + \lambda L^T L)^{-1} A^T \tag{2.40}$$

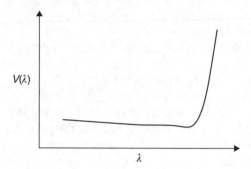

Figure 2.3 The normal GCV function

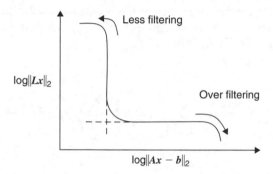

Figure 2.4 The generic form of L-curve

Golub et al. (1979) showed that the 'ordinary' cross-validation method led to solutions that were rotationally dependent. They replaced $r_{ii}(\lambda)$ in Equation (2.37) with $(1/n)trace(R(\lambda))$ to give the generalized cross-validation (GCV) function,

$$V(\lambda) = \frac{\frac{1}{n}\|Ax(\lambda) - b\|_2^2}{\left(\frac{1}{n}trace(I - R(\lambda))\right)^2} \tag{2.41}$$

Figure 2.3 shows a normal GCV function. The GCV function usually has a flat minimum and it works well in determining the optimal λ value. However, the minimum may be difficult to locate numerically under some situations.

2.5.3 The L-curve

Perhaps the most convenient graphical tool for the analysis of discrete ill-posed problems is the L-curve (Hansen, 1992), which is a plot for all valid regularization parameters of the norm, $\|Lx_{reg}\|_2$, of the regularized solution versus the corresponding residual norm, $\|Ax_{reg} - b\|_2$. The L-curve clearly displays the compromise between the minimization of these two quantities, which is the essence of any regularization method. Figure 2.4 shows the generic form of the L-curve.

For a discrete ill-posed problem, the L-curve, when plotted in the log-log scale, usually has a characteristic L-shape appearance with a distinct corner separating the vertical and the horizontal parts of the curve. It is noted that if x^* denotes the exact, un-regularized solution corresponding to the exact right-hand-side, $\bar{b}$, in Equation (2.11), then the error, $x_{reg} - x^*$, in the regularized solution consists of two components, namely, a perturbation error from the error e in the given vector b, and a regularization error due to the regularization of the error-free component, $\bar{b}$, in the right-hand-side. The vertical part of the L-curve corresponds to solutions where $\|Lx_{reg}\|_2$ is very sensitive to changes in the regularization parameter because the perturbation error e dominates x_{reg} and it does not satisfy the discrete Picard condition. The horizontal part of the L-curve corresponds to solutions where it is the residual norm, $\|Ax_{reg} - b\|_2$, that is most sensitive to the regularization parameter because x_{reg} is dominated by the regularization error, as long as $\bar{b}$ satisfies the discrete Picard condition.

This can be substantiated by means of the regularized solution, x_{reg}, expressed in terms of the filter factors. For the general form of regularization, Equation (2.14) yields the following expression for the error in x_{reg}:

$$x_{reg} - x^* = \left(\sum_{i=1}^{p} f_i \frac{u_i^T e}{\sigma_i} x_i + \sum_{i=p+1}^{n} (u_i^T e) x_i \right) + \sum_{i=1}^{p} (f_i - 1) \frac{u_i^T \bar{b}}{\sigma_i} x_i \qquad (2.42)$$

Here, the term in parentheses is the perturbation error due to the perturbation, e; and the last term is the regularization error caused by regularization of the unperturbed component, $\bar{b}$, on the right-hand-side. When only a little regularization is introduced, most of the filter factors, f_i, are approximately equal to unity and the error, $x_{reg} - x^*$, is dominated by the perturbation error. However, when most of the filter factors are small, i.e. $f_i \ll 1$, $x_{reg} - x^*$ is dominated by the regularization error.

The L-curve for the Tikhonov regularization plays a central role in the regularization methods for discrete ill-posed problems because it divides the first quadrant into two regions. It is impossible to construct any solution that corresponds to a point below the Tikhonov L-curve; any regularized solution must lie on or above this curve. The solution computed by Tikhonov regularization is therefore optimal in the sense that for a given residual norm there does not exists a solution with a smaller semi-norm than the Tikhonov solution and the same is true with the roles of the norms interchanged. A consequence of this is that other regularization methods can be compared with the Tikhonov regularization by inspecting how close the L-curve for the alternative method is to the Tikhonov L-curve.

For the regularized solution in Equation (2.14), there is obviously an optimal regularization parameter that balances the perturbation error and the regularization error in solution, x_{reg}. An essential feature of the L-curve is that this optimal regularization parameter defined above, is not far from the regularization parameter that corresponds to the L-curve's corner. In other words, by locating the corner of the L-curve, an approximation to the optimal regularization parameter can be computed and thus, in turn, a regularized solution with a good balance between the two types of errors can be computed. For continuous L-curves, a computationally convenient definition of the L-curve's corner is the point with maximum curvature in the log-log scale.

$$\max(\log\|Ax_\lambda - b\|_2, \log\|Lx_\lambda\|_2) \tag{2.43}$$

When the regularization parameter is discrete, the discrete L-curve in the log-log scale can be approximated by a two-dimensional spline curve. The point on the spline curve with maximum curvature was computed and the corner of the discrete L-curve at that point, which is closest to the corner of the spline curve, was defined.

2.6 General optimization procedure for the inverse problem

The procedure of a general iterative algorithm used to solve the nonlinear optimization problem is as follows:

1) Initially assume x_0.
2) Since the measurements of the baseline model are not available, they are calculated using the theoretical model instead of the measured information.
3) Solve the inverse problem using one of the optimization procedures.
4) Reconstruct the responses using the identified results.
5) Calculate the criteria of convergence in Equation (2.44). Convergence is achieved when the errors are less than the predefined tolerance values.
6) When the computed error does not converge, repeat Steps 3 to 5 until convergence is reached.

2.7 The criteria of convergence

The usual criteria of convergence are

$$\begin{cases} Error1 = \dfrac{\|b - b_{\text{Pr}edicted}\|}{\|U_s\|} \times 100\% \\[4mm] Error2 = \dfrac{\|x_{j+1} - x_j\|}{\|x_j\|} \times 100\% \end{cases} \tag{2.44}$$

The first criterion is based on the output residual between the measurements and the reconstructed responses. The second criterion is based on the difference between the identified results of two adjacent steps.

2.8 Summary

This chapter gives a brief description of the basic knowledge of ill-posed inverse problems and three optimization algorithms for the solution of inverse problems. The techniques to obtain the regularization solution are discussed and these techniques are used in the following chapters for the condition assessment of structures.

Chapter 3

Damage description and modelling

3.1 Introduction

3.1.1 *Damage models*

The following two questions are often asked: Where does the damage occur? And what is its magnitude? In fact, there may be many types of damage due to various reasons over the lifetime of the structure, and they affect the structural dynamic characteristics differently. The error in modelling the damage itself plays a significant role in the damage detection process. A more satisfactory set of question would be: Where does the damage occur? What kind of damage it is? And what is its magnitude? Many damage detection methods take the local damage as an overall reduction in the local stiffness of an element. Efforts have been made to build damage models but most of them are related to cracks in a beam. More comprehensive damage models with cracks in a plate have also been formulated based on the constitutive relations. If the initial model is not provided with the description of a particular type of damage, then the subsequent model improvement cannot yield the correct description of the damage in the finite element. A good example is the model error of a Timoshenko beam, where the initial model is based on that of an Euler–Bernoulli beam. Condition assessment based on an incorrect description of the damage yields system errors that spread throughout the identified results, and they account for a significant portion of the total error in the result.

3.1.2 *Model on pre-stress*

The load-carrying capacity of a pre-stressed structure depends mainly on the level of pre-stress in the structure. A significant reduction in the pre-stress jeopardizes the safety level of the structure. The modelling of pre-stress at the design stage is based on an equilibrating force system from which the internal stresses are calculated. Such a model is useful when static measured data are used for the system identification of the structure. But when dynamic data are used, the restoring forces from the equilibrating force system always tend to zero with no indication of the level of pre-stress. Two models of the pre-stress in a structure are given in this chapter with the pre-stressing effect modelled as physical parameters in the structural system, and they are specially developed for structural damage detection.

3.2 Damage-detection-oriented model

Many damage models can be found in the literature. However, most of them are explicitly developed for studying the changes in static and dynamic responses of the structure due to the damage, which is a forward problem mathematically. Thus, the use of these models is inconvenient or even impossible for detecting damage in structures from measurements, which is usually an inverse problem.

A type of model of the damages, the 'damage-detection-oriented model', specially developed for the inverse problem is needed. The requirements that need to be fulfilled with a damage-detection-oriented model are defined as:

- The analytical model can represent the real structure accurately, so that the modal properties predicted by the model are well correlated with the measured data from the undamaged structure, and modal uncertainty due to model errors is far less than the modal parameter changes caused by structural damage.
- The size of the analytical model is adequate such that the number of degrees-of-freedom (DOFs) is not much larger than that of the measured DOFs.
- The analytical model is detailed enough not to mask the damage occurring in an individual structural member.

3.2.1 Beam element with end flexibilities

Conventional approaches to the design and analysis of steel frames commonly assume the structural components to be connected either by rigid or pinned joints. In practice, all joints are semi-rigid, in which the slope of the lateral deflected shape and rotation are discontinuous between two connected members. The rotational properties of semi-rigid connections have been studied based on the complementary energy principle by Shi and Atluri (1989) with the weak formulation of the dynamic and large deflection response of semi-rigid jointed steel frames. They have also modelled the hysteretic damping at the joint (Shi and Atluri, 1992) with a bilinear Coulomb model to account for the damping effect due to the slip at the joint. Gao and Haldar (1995) developed a numerical procedure for the tangent stiffness formulation using a stress-based finite element method; whilst Chan and Zhou (1998) proposed a robust displacement-based finite element for a detailed and accurate analysis of semi-rigid jointed steel frames. There are also many reported works on the seismic and dynamic analysis of semi-rigid connected frames (Chan, 1994; Chen et al., 1998; Chui and Chan, 1997; Kukreti and Abolmaali, 1999). A comprehensive review of the work on semi-rigid connections has been given by Chan and Chui (2000).

Beam-column connection behaviour has also been experimentally investigated by many researchers, such as Nethercot (1985) and Popov (1983). The experimental studies demonstrated that the connections are semi-rigid and behave inelastically under severe loading. Theoretical studies have been conducted to calculate the moment-rotation stiffness of the semi-rigid connections using nonlinear or piecewise linear mathematical models.

However, most of the research so far has only studied the rotational flexibility of semi-rigid joints in steel frames. The dynamic behaviour of steel structures with ordinary bolted frictional joints that have flexibility in the tangential direction has received limited attention.

3.2.1.1   *Hybrid beam with shear flexibility*

The slotted, bolted, connection element incorporated at intersections or joints of a steel frame is allowed for in the design standard (BS5950, 1990). It is used to mitigate the dynamic response by providing hysteretic damping and nonlinear stiffness to the structure. A slotted, bolted, connection element (SBCE) is a semi-rigid bolted connection with a nonlinear relationship between the shear force and the relative slippage at the interfaces. High-strength friction grip bolts are used in an SBCE to connect two parts of the steel frame with slotted holes in the components, and the frictional interface may consist of sandwiched plates as shown in Figure 3.1. A pre-stressed force is applied to the interface in the normal direction to provide the SBCE with sufficient initial stiffness before significant slip occurs. It behaves nearly rigidly when subject to a small, in-plane shear force. Large slippage may occur between the surfaces of the element when under severe excitation, and energy is dissipated with the accompanying hysteretic damping. To protect the pre-stressed bolts from transverse shear, the maximum relative displacement is limited by a stopper, as shown in Figure 3.1.

Behaviour under cyclic loading

With the assumption of an exponential distribution of peak height of spherical contact elements, Shoukry (1985) developed a micro-slip element to model the frictional behaviour between two metallic interfaces, by using Mindlin's spherical contact element (Mindlin, 1949) as the basic element. The force–deformation relationship of the frictional interface is given by:

$$Q = \mu N \left[1 - \exp\left(\frac{-\gamma}{\sigma} v \right) \right] \tag{3.1}$$

where μ and N are the friction coefficient and normal force acting on the interface, respectively; v is the lateral deformation; σ is the standard deviation of the peak height distribution of the contact element; and γ is a constant equal to $2(1 - v_p)/[\mu(2 - v_p)]$, where v_p is the Poisson ratio.

A model is presented in this section with the damping mechanism based on the micro-slip at the interfaces only. The shear-slip model consists of three parameters defining the shear-load deformation $(Q - v)$ relation as

$$Q = Q_s \frac{K_0 v}{(Q_s^m + (K_0 v)^m)^{\frac{1}{m}}} \tag{3.2a}$$

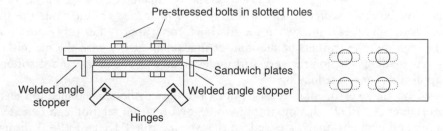

Figure 3.1   Configuration of SBCE

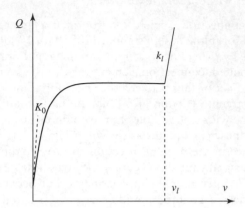

Figure 3.2 Loading curve of frictional joint

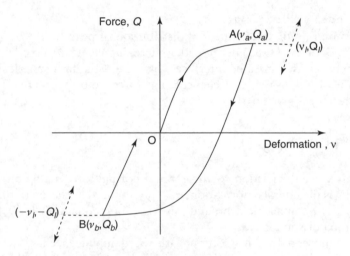

Figure 3.3 Connection under cyclic loading

where Q_s is the ultimate friction capacity of connection which is equivalent to μN in Shoukry's expression; N is the axial pre-stress force; K_0 is the initial connection stiffness, i.e. $K_0 = \frac{dQ}{dv}\big|_{v=0}$; and m is the shape parameter of the $Q-v$ curve. The connection becomes stiffer when the value of m increases. In addition, the model possesses no shear resistance when the frictional force approaches Q_s (Figure 3.2). It can be seen that the stiffness of the joint changes nonlinearly with the magnitude of the shear loading. The joint is very stiff under small shear loading, and it softens on the application of severe loading.

Based on the experimental observations of a slotted, bolted, connection element (Grigorian et al., 1992), this monotonic force–deformation relation can be extended to cyclic and dynamic analysis based on the Masing rule (Herrera, 1965). Equation (3.2a) is then used to represent the virgin loading curve of the frictional joint in the tangential direction. The corresponding instantaneous connection stiffness in the virgin curve is given by

$$r_v = \frac{\mathrm{d}Q}{\mathrm{d}v_c} = K_0 \left[1 - \frac{Q^m}{Q_s^m} \right]^{\frac{m+1}{m}}$$

$$= K_0 \left[1 - \frac{(K_0 v)^m}{(K_0 v)^m + Q_s^m} \right]^{\frac{m+1}{m}} \tag{3.2b}$$

Equation (3.2a) can also be generalized for both the unloading and reloading curves, based on the Masing rule as

$$\left| \frac{Q - Q^*}{2} \right| = Q_s \frac{K_0 \left| \frac{v - v^*}{2} \right|}{\left(Q_s^m + \left(K_0 \left| \frac{v - v^*}{2} \right| \right)^m \right)^{\frac{1}{m}}} \tag{3.3a}$$

where (v^*, Q^*) is the point at which the load reversal occurs at point A in Figure 3.3. Therefore, the corresponding instantaneous connection stiffness of both the unloading and reloading curves can be expressed as

$$r_v = \frac{\mathrm{d}Q}{\mathrm{d}v} = K_0 \left[1 - \frac{\left| \frac{Q - Q^*}{2} \right|^m}{Q_s^m} \right]^{\frac{m+1}{m}} = K_0 \left[1 - \frac{\left(K_0 \left| \frac{v - v^*}{2} \right| \right)^m}{\left(K_0 \left| \frac{v - v^*}{2} \right| \right)^m + Q_s^m} \right]^{\frac{m+1}{m}} \tag{3.3b}$$

in which, $v^* = v_a^*$ when the slippage deformation happens at the reversal point A in Figure 3.3. If point B is the next reversal point, the reloading stiffness is obtained by replacing v_a^* with v_b^* in Equation (3.3b).

Equations (3.2a and 3.2b) are physically important because the joint parameters in the loading curve, such as the initial stiffness and the frictional resistance force, are related to other physical parameters, such as the normal pressure and friction coefficient. Accordingly, the joint properties can be predicted from other physical parameters, and direct testing for the force–deformation characteristics may not be necessary.

In the following formulation, Equation (3.2b) is used to represent the virgin loading curve of the frictional joint in the tangential direction. The corresponding instantaneous connection stiffness in the virgin curve is given by

$$\begin{cases} r_v = \dfrac{\mathrm{d}Q}{\mathrm{d}v_c} = K_0 \exp\left[-\dfrac{K_0}{Q_s} v_c \right], & \text{if } |v_c| \le v_{cl}; \\[2mm] r_v = \dfrac{\mathrm{d}Q}{\mathrm{d}v_c} = k_l, & \text{if } |v_c| > v_{cl}; \end{cases} \tag{3.4}$$

Equation (3.2a) can also be generalized for both the unloading and reloading curves, based on the Masing rule as

$$\begin{cases} \left| \dfrac{Q - Q^*}{2} \right| = Q_s \left[1 - \exp\left(-\dfrac{K_0}{Q_s} \left| \dfrac{v_c - v_c^*}{2} \right| \right) \right], & \text{if } |v_c| \le v_{cl}; \\[2mm] Q = Q^* + k_l (v_c - v_c^*), & \text{if } |v_c| > v_{cl}; \end{cases} \tag{3.5}$$

where (v_c^*, Q^*) is the point at which load reversal occurs in the range of $|v_c| \leq v_{cl}$. But, when $|v_c| > v_{cl}$, the loading and reloading process is on a straight line with slope k_l. Therefore, the corresponding instantaneous connection stiffness of both the unloading and reloading curves can be expressed as

$$
\begin{cases}
r_v = \dfrac{dQ}{dv_c} = K_0 \exp\left(-\dfrac{K_0}{Q_s}\left|\dfrac{v_c - v_c^*}{2}\right|\right), & \text{if } |v_c| \leq v_{cl}; \\[4mm]
r_v = \dfrac{dQ}{dv_c} = k_l, & \text{if } |v_c| > v_{cl};
\end{cases}
\tag{3.6}
$$

in which, $v_c^* = v_{ca}^*$ when the slippage deformation happens after the reversal point A in Figure 3.3. If point B is the next reversal point, the reloading stiffness is obtained by replacing v_{ca}^* with v_{cb}^* in Equation (3.6).

Connection spring element

This semi-rigid joint can be modelled as a virtual connection spring element inserted at the intersection point between a beam and a column. A typical connection spring element in the tangential direction shown in Figure 3.4 is used. The virtual connection spring in other directions can be described in a similar manner. The connection spring elements are located at the member ends, and they are assumed infinitely small. The complete hybrid element includes a member and the corresponding connections. Nodes I and J are external nodes, and nodes i and j are internal nodes. The beam-column element is between nodes i and j. The connection spring elements are between nodes I and i and between nodes J and j, respectively, as shown in Figure 3.5.

In the derivation of the spring stiffness matrix, the three basic governing conditions, i.e. the compatibility, the equilibrium and the constitutive relations for an element, are

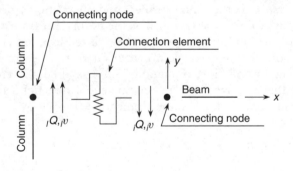

Figure 3.4 Connection spring element

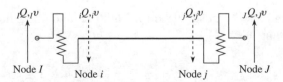

Figure 3.5 Beam-column element with end connection springs

considered. Considering a spring with stiffness, r_v; two end displacements, $_Iv$ and $_iv$ on either side; and the corresponding lateral shear forces, $_IQ$ and $_iQ$, the equilibrium condition requires

$$_IQ + _iQ = 0 \tag{3.7}$$

Assuming v_c is the relative displacement between the joint interfaces, the compatibility condition requires

$$v_c = _Iv - _iv \tag{3.8}$$

The constitutive relationship on the shearing action is

$$r_v = \frac{_IQ}{v_c} = \frac{_iQ}{v_c} \tag{3.9}$$

Substituting Equation (3.7) into Equation (3.9), gives

$$\begin{Bmatrix} _IQ \\ _iQ \end{Bmatrix} = \begin{bmatrix} r_v & -r_v \\ -r_v & r_v \end{bmatrix} \begin{Bmatrix} _Iv \\ _iv \end{Bmatrix} \tag{3.10}$$

Therefore, the stiffness matrix of the connection element is given by $\begin{bmatrix} r_v & -r_v \\ -r_v & r_v \end{bmatrix}$.

Displacement function of the hybrid element

By using the cubic Hermitian function, the lateral deflection, v, in the y-direction at a location x along the centre-line of a straight element can be written as

$$v = [(3 - 2\rho_1)\rho_1^2 \quad (3 - 2\rho_2)\rho_2^2] \begin{bmatrix} v_i \\ v_j \end{bmatrix} + [\rho_1^2\rho_2 L \quad -\rho_1\rho_2^2 L] \begin{bmatrix} \theta_i \\ \theta_j \end{bmatrix} \tag{3.11}$$

in which

$$\rho_1 = 1 - \frac{x}{L}, \quad \rho_2 = \frac{x}{L} \tag{3.12}$$

The lateral deflection, v, in Equation (3.11) has not yet accounted for the effect of connection flexibility at the ends of the beam-column element.

The elemental stiffness matrix for the complete hybrid element is assembled from those of the end springs and the beam element. The loads Q_I, M_I and Q_J, M_J are assumed to be applied only at the global nodes I and J, and hence the shear forces Q_i and Q_j are equal to zero. Thus,

$$\begin{Bmatrix} v_i \\ v_j \end{Bmatrix} = \begin{bmatrix} r_{vi} + K_{11} & K_{13} \\ K_{31} & r_{vj} + K_{33} \end{bmatrix}^{-1} \begin{bmatrix} r_{vi} & 0 \\ 0 & r_{vj} \end{bmatrix} \begin{Bmatrix} v_I \\ v_J \end{Bmatrix}$$

$$- \begin{bmatrix} r_{vi} + K_{11} & K_{13} \\ K_{31} & r_{vj} + K_{33} \end{bmatrix}^{-1} \begin{bmatrix} K_{12} & K_{14} \\ K_{32} & K_{34} \end{bmatrix} \begin{Bmatrix} \theta_i \\ \theta_j \end{Bmatrix} \tag{3.13}$$

Using the conditions $M_i = M_I$, $M_j = M_J$, $\theta_i = \theta_I$ and $\theta_j = \theta_J$, and eliminating the internal DOFs, the relationship between the external nodal force and nodal displacements/rotations of the element can be obtained.

In assembling the elemental stiffness matrices of a structure, the element stiffness matrix is rewritten in the global coordinate system for geometrical compatibility. Substituting Equation (3.13) into Equation (3.11) and transforming the external nodal rotations, θ_I and θ_J, about the local axis at the ends of the element to the global nodal rotations $\overline{\theta}_i$ and $\overline{\theta}_j$, the displacement function v can finally be written as

$$
\begin{aligned}
v = {}& [(3 - 2\rho_1)\rho_1^2 \quad (3 - 2\rho_2)\rho_2^2] \begin{bmatrix} r_{vi} + K_{11} & K_{13} \\ K_{31} & r_{vj} + K_{33} \end{bmatrix}^{-1} \begin{bmatrix} r_{vi} & 0 \\ 0 & r_{vj} \end{bmatrix} \begin{Bmatrix} v_I \\ v_J \end{Bmatrix} \\
& - [(3 - 2\rho_1)\rho_1^2 \quad (3 - 2\rho_2)\rho_2^2] \begin{bmatrix} r_{vi} + K_{11} & K_{13} \\ K_{31} & r_{vj} + K_{33} \end{bmatrix}^{-1} \begin{bmatrix} K_{12} & K_{14} \\ K_{32} & K_{34} \end{bmatrix} \begin{Bmatrix} \theta_I \\ \theta_J \end{Bmatrix} \\
& + [\rho_1^2 \rho_2 L \quad -\rho_1 \rho_2^2 L] \begin{Bmatrix} \theta_I \\ \theta_J \end{Bmatrix} \\
= {}& [(3 - 2\rho_1)\rho_1^2 \quad (3 - 2\rho_2)\rho_2^2] \begin{bmatrix} r_{vi} + K_{11} & K_{13} \\ K_{31} & r_{vj} + K_{33} \end{bmatrix}^{-1} \begin{bmatrix} r_{vi} & 0 \\ 0 & r_{vj} \end{bmatrix} \begin{Bmatrix} \overline{v}_i \\ \overline{v}_j \end{Bmatrix} \\
& + \Bigg([\rho_1^2 \rho_2 L \quad -\rho_1 \rho_2^2 L] - [(3 - 2\rho_1)\rho_1^2 \quad (3 - 2\rho_2)\rho_2^2] \\
& \quad \times \begin{bmatrix} r_{vi} + K_{11} & K_{13} \\ K_{31} & r_{vj} + K_{33} \end{bmatrix}^{-1} \begin{bmatrix} K_{12} & K_{14} \\ K_{32} & K_{34} \end{bmatrix} \Bigg) \begin{bmatrix} 1/L & 1 & -1/L & 0 \\ 1/L & 0 & -1/L & 1 \end{bmatrix} \begin{Bmatrix} \overline{v}_i \\ \overline{\theta}_i \\ \overline{v}_j \\ \overline{\theta}_j \end{Bmatrix}
\end{aligned} \tag{3.14}
$$

in which $\overline{v}_i$, $\overline{v}_j$, $\overline{\theta}_i$ and $\overline{\theta}_j$ are the nodal displacements and rotations of the element referring to the global axis.

The displacement function, w, in the z-direction can also be expressed similar to Equation (3.14) when under the shear load and bending moments at the global DOFs at the ends of the member.

Linear elemental stiffness matrix

The linear stiffness matrix, $[K_L]$, for a beam-column element in the x- and y-planes can be obtained based on the variational principle and the energy theorem,

$$
[K_L] = \int_0^L \left(\frac{\partial^2 v}{\partial x^2} \right)^T EI \left(\frac{\partial^2 v}{\partial x^2} \right) dx \tag{3.15}
$$

The nonzero terms of the upper triangle of the linear stiffness matrix (in local coordinates) of the three-dimensional element are

$$
\begin{aligned}
& K_{L1,1} = EA/L, && K_{L1,7} = -EA/L, \\
& K_{L2,2} = 48EI_z \xi_y^{-1} (r_{vi}^y)^2 (r_{vj}^y)^2, && K_{L2,6} = 12EI_z L \xi_y^{-1} (r_{vi}^y)^2 (r_{vj}^y)^2, \\
& K_{L2,8} = -48EI_z \xi_y^{-1} (r_{vi}^y)^2 (r_{vj}^y)^2, && K_{L2,12} = 12EI_z L \xi_y^{-1} (r_{vi}^y)^2 (r_{vj}^y)^2, \\
& K_{L3,3} = 48EI_y \xi_z^{-1} (r_{vi}^z)^2 (r_{vj}^z)^2, && K_{L3,5} = 12EI_y L \xi_z^{-1} (r_{vi}^z)^2 (r_{vj}^z)^2,
\end{aligned}
$$

$$K_{L3,9} = -48EI_y\xi_z^{-1}(r_{vi}^z)^2(r_{vj}^z)^2, \quad K_{L3,11} = 12EI_yL\xi_z^{-1}(r_{vi}^z)^2(r_{vj}^z)^2,$$

$$K_{L4,4} = GJ/L, \qquad\qquad K_{L4,10} = -GJ/L,$$

$$K_{L5,5} = L^2EI_y\xi_z^{-1}\{(K_{11}^z)^2(r_{vi}^z)^2 + 2(K_{11}^z)^2r_{vi}^zr_{vj}^z + 2K_{11}^z(r_{vi}^z)^2r_{vj}^z$$
$$+ (K_{11}^z)^2(r_{vj}^z)^2 + 2K_{11}^zr_{vi}^z(r_{vj}^z)^2 + 4(r_{vi}^z)^2(r_{vj}^z)^2\},$$

$$K_{L5,9} = -12EI_yL\xi_z^{-1}(r_{vi}^z)^2(r_{vj}^z)^2,$$

$$K_{L5,11} = -L^2EI_y\xi_z^{-1}\{(K_{11}^z)^2(r_{vi}^z)^2 + 2(K_{11}^z)^2r_{vi}^zr_{vj}^z + 2K_{11}^z(r_{vi}^z)^2r_{vj}^z$$
$$+ (K_{11}^z)^2(r_{vj}^z)^2 + 2K_{11}^zr_{vi}^z(r_{vj}^z)^2 - 2(r_{vi}^z)^2(r_{vj}^z)^2\},$$

$$K_{L6,6} = L^2EI_z\xi_y^{-1}\{(K_{11}^y)^2(r_{vi}^y)^2 + 2(K_{11}^y)^2r_{vi}^yr_{vj}^y$$
$$+ 2K_{11}^y(r_{vi}^y)^2r_{vj}^y + (K_{11}^y)^2(r_{vj}^y)^2 + 2K_{11}^yr_{vi}^y(r_{vj}^y)^2 + 4(r_{vi}^y)^2(r_{vj}^y)^2\},$$

$$K_{L6,8} = -12EI_zL\xi_y^{-1}(r_{vi}^y)^2(r_{vj}^y)^2,$$

$$K_{L6,12} = -L^2EI_z\xi_y^{-1}\{(K_{11}^y)^2(r_{vi}^y)^2 + 2(K_{11}^y)^2r_{vi}^yr_{vj}^y$$
$$+ 2K_{11}^y(r_{vi}^y)^2r_{vj}^y + (K_{11}^y)^2(r_{vj}^y)^2 + 2K_{11}^yr_{vi}^y(r_{vj}^y)^2 - 2(r_{vi}^y)^2(r_{vj}^y)^2\},$$

$$K_{L7,7} = EA/L, \qquad\qquad K_{L8,8} = 48EI_z\xi_y^{-1}(r_{vi}^y)^2(r_{vj}^y)^2,$$

$$K_{L8,12} = -12EI_zL\xi_y^{-1}(r_{vi}^y)^2(r_{vj}^y)^2, \quad K_{L9,9} = 48EI_y\xi_z^{-1}(r_{vi}^z)^2(r_{vj}^z)^2,$$

$$K_{L9,11} = -12EI_yL\xi_z^{-1}(r_{vi}^z)^2(r_{vj}^z)^2, \quad K_{L10,10} = GJ/L$$

$$K_{L11,11} = L^2EI_y\xi_z^{-1}\{(K_{11}^z)^2(r_{vi}^z)^2 + 2(K_{11}^z)^2r_{vi}^zr_{vj}^z$$
$$+ 2K_{11}^z(r_{vi}^z)^2r_{vj}^z + (K_{11}^z)^2(r_{vj}^z)^2 + 2K_{11}^zr_{vi}^z(r_{vj}^z)^2 + 4(r_{vi}^z)^2(r_{vj}^z)^2\},$$

$$K_{L12,12} = L^2EI_z\xi_y^{-1}\{(K_{11}^y)^2(r_{vi}^y)^2 + 2(K_{11}^y)^2r_{vi}^yr_{vj}^y$$
$$+ 2K_{11}^y(r_{vi}^y)^2r_{vj}^y + (K_{11}^y)^2(r_{vj}^y)^2 + 2K_{11}^yr_{vi}^y(r_{vj}^y)^2 + 4(r_{vi}^y)^2(r_{vj}^y)^2\}$$

$$(3.16)$$

where

$$K_{11}^z = 12EI_y/L^3, \qquad\qquad K_{11}^y = 12EI_z/L^3,$$
$$\xi_z = L^3(K_{11}^zr_{vi}^z + K_{11}^zr_{vj}^z + r_{vi}^zr_{vj}^z)^2, \qquad \xi_y = L^3(K_{11}^yr_{vi}^y + K_{11}^yr_{vj}^y + r_{vi}^yr_{vj}^y)^2.$$

Elemental geometric stiffness matrix

To consider the instability effect due to axial load [P] in the element, the geometric stiffness matrix, $[K_G]$, for a beam-column element in the x- and y-planes is obtained from

$$[K_G] = \frac{P}{2}\int_0^L \left(\frac{\partial v}{\partial x}\right)^T \left(\frac{\partial v}{\partial x}\right)\mathrm{d}x \qquad (3.17)$$

The nonzero terms in the upper triangle of the matrix (in local coordinates) of the three-dimensional element are

$$K_{G1,1} = 0,$$

$$K_{G2,2} = P\zeta_y^{-1}\{30(K_{11}^y)^2(r_{vi}^y)^2 + 60(K_{11}^y)^2 r_{vi}^y r_{vj}^y$$
$$- 60K_{11}^y(r_{vi}^y)^2 r_{vj}^y + 30(K_{11}^y)^2(r_{vj}^y)^2 - 60K_{11}^y r_{vi}^y(r_{vj}^y)^2 + 54(r_{vi}^y)^2(r_{vj}^y)^2\},$$

$$K_{G2,6} = PL\zeta_y^{-1}\{15(K_{11}^y)^2(r_{vi}^y)^2 + 30(K_{11}^y)^2 r_{vi}^y r_{vj}^y$$
$$- 15K_{11}^y(r_{vi}^y)^2 r_{vj}^y + 15(K_{11}^y)^2(r_{vj}^y)^2 - 15K_{11}^y r_{vi}^y(r_{vj}^y)^2 + 6(r_{vi}^y)^2(r_{vj}^y)^2\},$$

$$K_{G2,8} = P\zeta_y^{-1}\{-30(K_{11}^y)^2(r_{vi}^y)^2 - 60(K_{11}^y)^2 r_{vi}^y r_{vj}^y$$
$$+ 60K_{11}^y(r_{vi}^y)^2 r_{vj}^y - 30(K_{11}^y)^2(r_{vj}^y)^2 + 60K_{11}^y r_{vi}^y(r_{vj}^y)^2 - 54(r_{vi}^y)^2(r_{vj}^y)^2\},$$

$$K_{G2,12} = PL\zeta_y^{-1}\{15(K_{11}^y)^2(r_{vi}^y)^2 + 30(K_{11}^y)^2 r_{vi}^y r_{vj}^y$$
$$- 15K_{11}^y(r_{vi}^y)^2 r_{vj}^y + 15(K_{11}^y)^2(r_{vj}^y)^2 - 15K_{11}^y r_{vi}^y(r_{vj}^y)^2 + 6(r_{vi}^y)^2(r_{vj}^y)^2\},$$

$$K_{G3,3} = P\zeta_z^{-1}\{30(K_{11}^z)^2(r_{vi}^z)^2 + 60(K_{11}^z)^2 r_{vi}^z r_{vj}^z$$
$$- 60K_{11}^z(r_{vi}^z)^2 r_{vj}^z + 30(K_{11}^z)^2(r_{vj}^z)^2 - 60K_{11}^z r_{vi}^z(r_{vj}^z)^2 + 54(r_{vi}^z)^2(r_{vj}^z)^2\},$$

$$K_{G3,5} = PL\zeta_z^{-1}\{15(K_{11}^z)^2(r_{vi}^z)^2 + 30(K_{11}^z)^2 r_{vi}^z r_{vj}^z$$
$$- 15K_{11}^z(r_{vi}^z)^2 r_{vj}^z + 15(K_{11}^z)^2(r_{vj}^z)^2 - 15K_{11}^z r_{vi}^z(r_{vj}^z)^2 + 6(r_{vi}^z)^2(r_{vj}^z)^2\},$$

$$K_{G3,9} = P\zeta_z^{-1}\{-30(K_{11}^z)^2(r_{vi}^z)^2 - 60(K_{11}^z)^2 r_{vi}^z r_{vj}^z$$
$$+ 60K_{11}^z(r_{vi}^z)^2 r_{vj}^z - 30(K_{11}^z)^2(r_{vj}^z)^2 + 60K_{11}^z r_{vi}^z(r_{vj}^z)^2 - 54(r_{vi}^z)^2(r_{vj}^z)^2\},$$

$$K_{G3,11} = PL\zeta_z^{-1}\{15(K_{11}^z)^2(r_{vi}^z)^2 + 30(K_{11}^z)^2 r_{vi}^z r_{vj}^z$$
$$- 15K_{11}^z(r_{vi}^z)^2 r_{vj}^z + 15(K_{11}^z)^2(r_{vj}^z)^2 - 15K_{11}^z r_{vi}^z(r_{vj}^z)^2 + 6(r_{vi}^z)^2(r_{vj}^z)^2\},$$

$$K_{G4,4} = Pr^2/L, \quad K_{G4,10} = -Pr^2/L,$$

$$K_{G5,5} = PL^2\zeta_z^{-1}\{10(K_{11}^z)^2(r_{vi}^z)^2 + 20(K_{11}^z)^2 r_{vi}^z r_{vj}^z$$
$$+ 5K_{11}^z(r_{vi}^z)^2 r_{vj}^z + 10(K_{11}^z)^2(r_{vj}^z)^2 + 5K_{11}^z r_{vi}^z(r_{vj}^z)^2 + 4(r_{vi}^z)^2(r_{vj}^z)^2\},$$

$$K_{G5,9} = PL\zeta_z^{-1}\{-15(K_{11}^z)^2(r_{vi}^z)^2 - 30(K_{11}^z)^2 r_{vi}^z r_{vj}^z$$
$$+ 15K_{11}^z(r_{vi}^z)^2 r_{vj}^z - 15(K_{11}^z)^2(r_{vj}^z)^2 + 15K_{11}^z r_{vi}^z(r_{vj}^z)^2 - 6(r_{vi}^z)^2(r_{vj}^z)^2\},$$

$$K_{G5,11} = PL^2\zeta_z^{-1}\{5(K_{11}^z)^2(r_{vi}^z)^2 + 10(K_{11}^z)^2 r_{vi}^z r_{vj}^z$$
$$- 5K_{11}^z(r_{vi}^z)^2 r_{vj}^z + 5(K_{11}^z)^2(r_{vj}^z)^2 - 5K_{11}^z r_{vi}^z(r_{vj}^z)^2 - (r_{vi}^z)^2(r_{vj}^z)^2\},$$

$$K_{G6,6} = PL^2\zeta_y^{-1}\{10(K_{11}^y)^2(r_{vi}^y)^2 + 20(K_{11}^y)^2 r_{vi}^y r_{vj}^y$$
$$+ 5K_{11}^y(r_{vi}^y)^2 r_{vj}^y + 10(K_{11}^y)^2(r_{vj}^y)^2 + 5K_{11}^y r_{vi}^y(r_{vj}^y)^2 + 4(r_{vi}^y)^2(r_{vj}^y)^2\},$$

$$K_{G6,8} = PL\zeta_y^{-1}\{-15(K_{11}^y)^2(r_{vi}^y)^2 - 30(K_{11}^y)^2 r_{vi}^y r_{vj}^y$$
$$+ 15K_{11}^y(r_{vi}^y)^2 r_{vj}^y - 15(K_{11}^y)^2(r_{vj}^y)^2 + 15K_{11}^y r_{vi}^y(r_{vj}^y)^2 - 6(r_{vi}^y)^2(r_{vj}^y)^2\},$$

$$K_{G6,12} = PL^2\zeta_y^{-1}\{5(K_{11}^y)^2(r_{vi}^y)^2 + 10(K_{11}^y)^2 r_{vi}^y r_{vj}^y$$

$$- 5K_{11}^y(r_{vi}^y)^2 r_{vj}^y + 5(K_{11}^y)^2(r_{vj}^y)^2 - 5K_{11}^y r_{vi}^y(r_{vj}^y)^2 - (r_{vi}^y)^2(r_{vj}^y)^2\},$$

$$K_{G7,7} = 0,$$

$$K_{G8,8} = P\zeta_y^{-1}\{30(K_{11}^y)^2(r_{vi}^y)^2 + 60(K_{11}^y)^2 r_{vi}^y r_{vj}^y$$

$$- 60K_{11}^y(r_{vi}^y)^2 r_{vj}^y + 30(K_{11}^y)^2(r_{vj}^y)^2 - 60K_{11}^y r_{vi}^y(r_{vj}^y)^2 + 54(r_{vi}^y)^2(r_{vj}^y)^2\},$$

$$K_{G8,12} = PL\zeta_y^{-1}\{-15(K_{11}^y)^2(r_{vi}^y)^2 - 30(K_{11}^y)^2 r_{vi}^y r_{vj}^y$$

$$+ 15K_{11}^y(r_{vi}^y)^2 r_{vj}^y - 15(K_{11}^y)^2(r_{vj}^y)^2 + 15K_{11}^y r_{vi}^y(r_{vj}^y)^2 - 6(r_{vi}^y)^2(r_{vj}^y)^2\},$$

$$K_{G9,9} = P\zeta_z^{-1}\{30(K_{11}^z)^2(r_{vi}^z)^2 + 60(K_{11}^z)^2 r_{vi}^z r_{vj}^z$$

$$- 60K_{11}^z(r_{vi}^z)^2 r_{vj}^z + 30(K_{11}^z)^2(r_{vj}^z)^2 - 60K_{11}^z r_{vi}^z(r_{vj}^z)^2 + 54(r_{vi}^z)^2(r_{vj}^z)^2\},$$

$$K_{G9,11} = PL\zeta_z^{-1}\{-15(K_{11}^z)^2(r_{vi}^z)^2 - 30(K_{11}^z)^2 r_{vi}^z r_{vj}^z$$

$$+ 15K_{11}^z(r_{vi}^z)^2 r_{vj}^z - 15(K_{11}^z)^2(r_{vj}^z)^2 + 15K_{11}^z r_{vi}^z(r_{vj}^z)^2 - 6(r_{vi}^z)^2(r_{vj}^z)^2\},$$

$$K_{G10,10} = Pr^2/L,$$

$$K_{G11,11} = PL^2\zeta_z^{-1}\{10(K_{11}^z)^2(r_{vi}^z)^2 + 20(K_{11}^z)^2 r_{vi}^z r_{vj}^z$$

$$+ 5K_{11}^z(r_{vi}^z)^2 r_{vj}^z + 10(K_{11}^z)^2(r_{vj}^z)^2 + 5K_{11}^z r_{vi}^z(r_{vj}^z)^2 + 4(r_{vi}^z)^2(r_{vj}^z)^2\},$$

$$K_{G12,12} = PL^2\zeta_y^{-1}\{10(K_{11}^y)^2(r_{vi}^y)^2 + 20(K_{11}^y)^2 r_{vi}^y r_{vj}^y$$

$$+ 5K_{11}^y(r_{vi}^y)^2 r_{vj}^y + 10(K_{11}^y)^2(r_{vj}^y)^2 + 5K_{11}^y r_{vi}^y(r_{vj}^y)^2 + 4(r_{vi}^y)^2(r_{vj}^y)^2\}$$

$$(3.18)$$

where

$$K_{11}^z = 12EI_y/L^3, \qquad\qquad K_{11}^y = 12EI_z/L^3,$$

$$\zeta_z = 60L(K_{11}^z r_{vi}^z + K_{11}^z r_{vj}^z + r_{vi}^z r_{vj}^z)^2, \qquad \zeta_y = 60L(K_{11}^y r_{vi}^y + K_{11}^y r_{vj}^y + r_{vi}^y r_{vj}^y)^2.$$

Therefore, using the updated Lagrangian approach, the tangent stiffness matrix of the beam-column element accounting for the connection spring stiffness can be evaluated directly from

$$[K_T] = [K_L] + [K_G] \tag{3.19}$$

The consistent mass and damping matrices of the hybrid element can be found in Law et al., (2003).

Application to the Millennium Bridge

Dynamic forces generated by moving crowds of people while walking produce unexpected lateral movements with the London Millennium Footbridge on its opening day

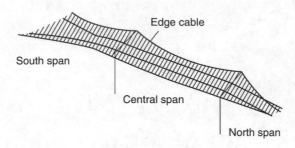

Figure 3.6   Elevation view of the bridge

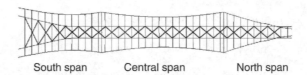

Figure 3.7   Deck with 38 hybrid members

(Dallard et al., 2001). The movements took place mainly on the south and central spans at frequencies corresponding to their first and second lateral modes, which are very low and between 0.5 and 1.0 Hz. This section presents a vibration mitigation study on the footbridge using a bracing member which includes two connecting rods and a frictional damper, i.e. a slotted, bolted connection element (SBCE).

The bridge structure has three suspended spans. The finite element model is constructed from the published data of the Millennium Footbridge with a central span of 144 m, a south span of 100 m and a north span of 81 m (Dallard et al., 2001). The 4 m wide deck is in segments that are each 16 m long and the steel box sections of the deck are linked together via a tongue and groove movement joint. The segments span over 4 m between the edge steel tubes on each side of the bridge. Transverse arms extending from the deck sections clamp onto a 200 mm diameter steel cable at each side of the bridge at 8 m intervals. The cables provide the bridge with both vertical and lateral stiffnesses with a horizontal dead load tension force of approximately 22.5 MN. There is only a small sag of cable of 2.3 m over the central span in the tension cable. The cables are in turn supported by two V-shape steel piers found on a reinforced concrete pier foundation. The arrangement of the structure is shown in Figure 3.6.

The hybrid element
The finite element model of the structure is referred to by Law et al. (2004b). The dynamic response of the structure is mitigated by a friction damper installed inside a hybrid-bracing member between corresponding nodal points underneath the bridge deck as shown in Figure 3.7. The SBCE is modelled as a virtual shear spring between two rod members, and its physical size is assumed infinitely small. Nodes I and J are external nodes, and nodes i and j are internal nodes. The SBCE is between nodes i and j. Rods 1 and 2 are between nodes I and i, and between nodes J and j, respectively as shown in Figure 3.8.

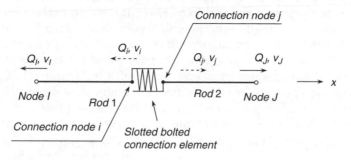

Figure 3.8 Hybrid element with SBCE

Static condensation

The Guyan static condensation technique (Guyan, 1965) is employed to eliminate the internal DOFs i and j. The three basic governing conditions, i.e. the compatibility, the equilibrium and the constitutive relations for a finite element are considered. The SBCE has two end displacements, u_i and u_j, at each end with corresponding nodal forces, Q_i and Q_j. The stiffness matrix of the SBCE is $\begin{bmatrix} r_v & -r_v \\ -r_v & r_v \end{bmatrix}$ from Equation (3.10).

Denoting the member end nodal axial displacements by v_I and v_J, gives

$$\begin{Bmatrix} Q_I \\ Q_i \\ Q_j \\ Q_J \end{Bmatrix} = \begin{bmatrix} r_1 & -r_1 & 0 & 0 \\ -r_1 & r_1 + r_v & -r_v & 0 \\ 0 & -r_v & r_2 + r_v & -r_2 \\ 0 & 0 & -r_2 & r_2 \end{bmatrix} \begin{Bmatrix} v_I \\ v_i \\ v_j \\ v_J \end{Bmatrix} \tag{3.20}$$

with $r_1 = A_1 E_1 / L_1$ and $r_2 = A_2 E_2 / L_2$. A_1 and A_2 are the cross-sectional areas of the two rods; L_1 and L_2 are their lengths; and Q_I and Q_J are the nodal forces at the ends of these members. The loads are assumed to be applied only at the global nodes I and J, and hence the shear forces Q_i and Q_j are equal to zero. Taking the conditions $Q_i = 0$ and $Q_j = 0$, Equation (3.20) gives

$$\begin{Bmatrix} v_i \\ v_j \end{Bmatrix} = \begin{bmatrix} r_1 + r_v & -r_v \\ -r_v & r_2 + r_v \end{bmatrix}^{-1} \begin{bmatrix} r_1 & 0 \\ 0 & r_2 \end{bmatrix} \begin{Bmatrix} v_I \\ v_J \end{Bmatrix} = [T_1] \begin{Bmatrix} v_I \\ v_J \end{Bmatrix} \tag{3.21}$$

in which

$$[T_1] = \begin{bmatrix} r_1 + r_v & -r_v \\ -r_v & r_2 + r_v \end{bmatrix}^{-1} \begin{bmatrix} r_1 & 0 \\ 0 & r_2 \end{bmatrix} \tag{3.22}$$

Matrix transformation

If the lateral displacements at the end nodes I and J of the element are expressed as v_{yI}, v_{zI} and v_{yJ}, v_{zJ}, respectively, and letting $\{\hat{Q}\} = \{Q_i \; Q_j \; Q_I \; Q_J \; Q_{YI} \; Q_{ZI} \; Q_{YJ} \; Q_{ZJ}\}^T$

be the vector and $\{\hat{V}\} = \{v_i \quad v_j \quad v_I \quad v_J \quad v_{YI} \quad v_{ZI} \quad v_{YJ} \quad v_{ZJ}\}^T$ be the displacement vector of the element, respectively, then there exists

$$\{\hat{Q}\} = [K]\{\hat{V}\} \tag{3.23}$$

and

$$[K] = \begin{bmatrix} [K_1] & [0]_{4\times4} \\ [0]_{4\times4} & [0]_{4\times4} \end{bmatrix} \tag{3.24}$$

in which $[K]$ is the 8×8 elemental stiffness matrix before condensation and $[K_1]$ is a 4×4 matrix defined as

$$[K_1] = \begin{bmatrix} r_1 + r_v & -r_v & -r_1 & 0 \\ -r_v & r_2 + r_v & 0 & -r_2 \\ -r_1 & 0 & r_1 & 0 \\ 0 & -r_2 & 0 & r_2 \end{bmatrix} \tag{3.25}$$

The 8×8 displacement vector, $\{\hat{V}\}$, is further transformed to the 6×6 displacement vector, $\{V\} = \{v_I \quad v_{YI} \quad v_{ZI} \quad v_J \quad v_{YJ} \quad v_{ZJ}\}^T$, of the hybrid element as

$$\{\hat{V}\} = [T_2][B]\{V\} \tag{3.26}$$

with

$$[T_2]_{8\times6} = \begin{bmatrix} [T_1]_{2\times2} & [0]_{2\times4} \\ & [I]_{6\times6} \end{bmatrix}, \quad [B]_{6\times6} = \begin{bmatrix} 1 & 0 & 0 & 0 & 0 & 0 \\ 0 & 0 & 0 & 1 & 0 & 0 \\ 0 & 1 & 0 & 0 & 0 & 0 \\ 0 & 0 & 1 & 0 & 0 & 0 \\ 0 & 0 & 0 & 0 & 1 & 0 \\ 0 & 0 & 0 & 0 & 0 & 1 \end{bmatrix} \tag{3.27}$$

The 8×8 elemental stiffness matrix, $[K]$, is condensed to the 6×6 stiffness matrix of the hybrid element, $[\overline{K}]$, using the same transformation matrices as

$$[\overline{K}] = ([T_2][B])^T[K]([T_2][B]) \tag{3.28}$$

Similarly, the 8×8 elemental lump mass matrix corresponding to the displacement vector $\{\hat{U}\}$ is given by

$$[M] = dia\left[\frac{1}{2}\overline{m}L_1 + M_d \quad \frac{1}{2}\overline{m}L_2 + M_d \quad \frac{1}{2}\overline{m}L_1 \quad \frac{1}{2}\overline{m}L_2 \quad \frac{1}{2}\overline{m}L \quad \frac{1}{2}\overline{m}L \quad \frac{1}{2}\overline{m}L \quad \frac{1}{2}\overline{m}L\right] \tag{3.29}$$

in which $L = L_1 + L_2$; $\overline{m}$ is the mass per unit length of the rod; and M_d is the lumped mass of the SBCE at the internal nodes. The 6×6 transformed element mass matrix of the hybrid element is then given by

$$[\overline{M}] = ([T_2][B])^T[M]([T_2][B]) \tag{3.30}$$

Since matrices $[\overline{K}]$ and $[\overline{M}]$ are in the local coordinates system, they are further transformed to the global coordinates system by

$$[\widetilde{K}] = [T_3]^T[\overline{K}][T_3] \quad [\widetilde{M}] = [T_3]^T[\overline{M}][T_3] \tag{3.31}$$

in which $[T_3]$ is the coordinate transformation matrix. Matrices $[\widetilde{K}]$ and $[\widetilde{M}]$ are then assembled to form the stiffness and mass matrices of the system.

Responses with dampers in structure

The loading case with 150 people distributed evenly along the south and central spans of the bridge deck is studied. Details of the pedestrian loading model and the parameters of the dampers are referred to in Law et al. (2004b). The responses at the mid-span of the central span for two cases of (a) having 38 dampers in all spans and (b) only 28 dampers in the south and central spans, are shown in Figure 3.9. The time responses show a significant reduction in the vibration amplitude to meet the requirement of comfort serviceability for the pedestrian with an acceleration below 20 milli-g and a lateral displacement at the mid-span of the central span of under 10 mm (Dallard et al., 2001). There is also a large reserve damping capacity with a clear reducing trend in

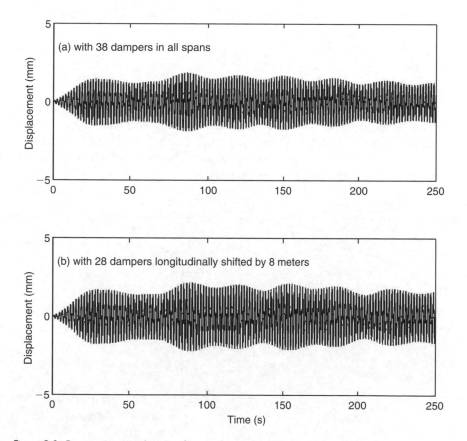

Figure 3.9 Response at mid-span of central span for different damper layouts

the vibration amplitude with time. Both time histories also maintain a close to balance state in the latter part of the time history indicating that significant damping to the vibration motion can be provided by the dampers.

3.2.1.2 Hybrid beam with both shear and flexural flexibilities

This section studies the dynamic properties of a bolted joint with flexibility in both tangential and rotational directions. The slotted bolted connection shown in Figure 3.1 is referenced.

The Richard–Abbott model

There are many existing models to describe the rotational flexibility in a semi-rigid joint. The Richard–Abbott model (Richard and Abbott, 1975) exhibits the virgin loading path of the moment-rotation behaviour as

$$
M = \frac{(k_0 - k_p)|\phi|}{\left[1 + \left|\frac{(k_0 - k_p)|\phi|}{M_0}\right|^n\right]^{1/n}} + k_p|\phi| \tag{3.32a}
$$

with the corresponding tangent connection stiffness S_c as

$$
S_c = \frac{\mathrm{d}M}{\mathrm{d}\phi} = \frac{(k_0 - k_p)}{\left[1 + \left|\frac{(k_0 - k_p)|\phi|}{M_0}\right|^n\right]^{(n+1)/n}} + k_p \tag{3.32b}
$$

in which k_0 is the initial rotational stiffness; k_p is the strain-hardening stiffness; M_0 is a reference moment; and n is a parameter defining the curvature of the curve. The Richard–Abbott model is shown in Figure 3.10(a) for different values of the parameter, n. The slope of the curve decreases when the moment is close to the ultimate moment for increasing values of the parameter, n.

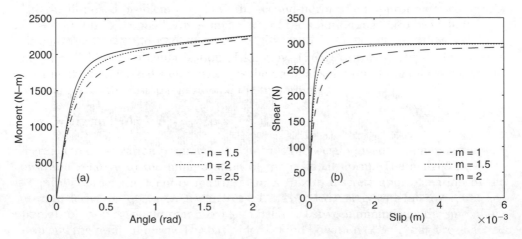

Figure 3.10 (a) Moment-rotation model (b) Shear-slip model

The unloading and reloading part of the $M - \phi$ curve can be written similarly from Equation (3.32a) as

$$\left| \frac{M - M^*}{2} \right| = \frac{(k_0 - k_p) \left| \frac{\phi^* - \phi}{2} \right|}{\left[1 + \left| \frac{(k_0 - k_p)|\phi^* - \phi|}{2M_0} \right|^n \right]^{1/n}} + k_p \left| \frac{\phi^* - \phi}{2} \right| \tag{3.33a}$$

and the corresponding tangent rotational stiffness of the joint is

$$S_c = \frac{dM}{d\phi} = \frac{(k_0 - k_p)}{\left[1 + \left| \frac{(k_0 - k_p)(\phi^* - \phi)}{2M_0} \right|^n \right]^{(n+1)/n}} + k_p \tag{3.33b}$$

in which (ϕ^*, M^*) is the reversal point the loading path has just gone through, similar to that for the shear-slip model. If point B is the next reversal point, the reloading stiffness is obtained by replacing (ϕ^*, M^*) with (ϕ_b^*, M_b^*) in Equation (3.33).

The Richard–Abbott model requires only four parameters in its definition, and it can represent a nonlinear and smooth $M - \phi$ curve accurately. It has been used widely for the description of the rotational stiffness of a semi-rigid joint.

The shear-slip model

The same shear-slip model as described in Section 3.2.1.1 is adopted, and the shear–slip relationship of such model is shown in Figure 3.10(b) for different values of the parameter, m. A hybrid element with end springs representing both types of flexibility is presented in the next section and its stiffness matrix is given.

Connection spring element

This semi-rigid joint incorporating both in-plane shear and moment flexibilities modelled with a virtual connection spring element attached to the end of a beam-column element to form the hybrid element is shown in Figure 3.11. These connection spring elements are located at the member ends, and they are assumed infinitely small. The complete hybrid element includes a uniform member and the corresponding connections. Nodes I and J are the external nodes, and nodes i and j are the internal nodes. The beam-column element is between nodes i and j. The connections I and J are between nodes I and i and between nodes J and j, respectively.

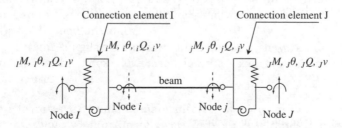

Figure 3.11 Hybrid beam-column element with end connection spring elements

In the derivation of the spring stiffness matrix, the three basic governing conditions, i.e. the compatibility, the equilibrium and the constitutive relations for an element, are considered. Considering a spring with stiffnesses S_c and r_v; two end displacements, $_Iv$ and $_iv$; and two rotations, $_I\theta$ and $_i\theta$, on either side of the spring, the corresponding lateral shear forces, $_IQ$ and $_iQ$, and corresponding moments $_IM$ and $_iM$, the equilibrium conditions require

$$_IQ + {}_iQ = 0, \quad {}_IM + {}_iM = 0 \tag{3.34}$$

Assuming that v is the relative shear displacement between the joint interfaces, and θ is the connection rotation angle, the compatibility conditions require

$$v = {}_Iv - {}_iv, \quad \theta = {}_I\theta - {}_i\theta \tag{3.35}$$

The constitutive relationships of the shearing and rotational actions are

$$r_v = \frac{{}_IQ}{v} = \frac{{}_iQ}{v}, \quad S_c = \frac{{}_IM}{\theta} = \frac{{}_iM}{\theta} \tag{3.36}$$

Substituting Equation (3.34) into Equation (3.36), gives

$$\begin{Bmatrix} {}_IQ \\ {}_iQ \end{Bmatrix} = \begin{bmatrix} r_v & -r_v \\ -r_v & r_v \end{bmatrix} \begin{Bmatrix} {}_Iv \\ {}_iv \end{Bmatrix}, \quad \begin{Bmatrix} {}_IM \\ {}_iM \end{Bmatrix} = \begin{bmatrix} S_c & -S_c \\ -S_c & S_c \end{bmatrix} \begin{Bmatrix} {}_I\theta \\ {}_i\theta \end{Bmatrix} \tag{3.37}$$

Therefore, the in-plane stiffness matrix of the connection element is given by:

$$\begin{bmatrix} r_v & 0 & -r_v & 0 \\ 0 & S_c & 0 & -S_c \\ -r_v & 0 & -r_v & 0 \\ 0 & S_c & 0 & S_c \end{bmatrix} \tag{3.38}$$

The shape function

By using the cubic Hermitian function, the lateral deflection, v, in the y-direction at a location x along the centre-line of a straight element can be written as

$$v = a_0 + a_1 x + a_2 x^2 + a_3 x^3 \tag{3.39}$$

in which $(a_j, j = 0, 1, 2, 3)$ are coefficients. Denoting the beam end rotations by θ_i and θ_j, and the beam end lateral deflections by v_i and v_j at the end nodes i and j of the element respectively, gives the following boundary conditions

$$v = v_i, \quad \frac{dv}{dx} = \theta_i, \quad \text{at } x = 0,$$

$$v = v_j, \quad \frac{dv}{dx} = \theta_j, \quad \text{at } x = L, \tag{3.40}$$

in which L is the length of the element.

The four unknowns coefficients $(a_j, j = 0, 1, 2, 3)$ in Equation (3.39) can be solved from these four boundary conditions as

$$
\begin{bmatrix} a_0 \\ a_1 \\ a_2 \\ a_3 \end{bmatrix} = \begin{bmatrix} 1 & 0 & 0 & 0 \\ 0 & 1 & 0 & 0 \\ -3/L^2 & -2/L & 3/L^2 & -1/L \\ 2/L^3 & 1/L^2 & -2/L^3 & 1/L^2 \end{bmatrix} \begin{bmatrix} v_i \\ \theta_i \\ v_j \\ \theta_j \end{bmatrix} \tag{3.41}
$$

Substituting the coefficients into Equation (3.39), the lateral deflection function, v, becomes

$$
v = [(3 - 2\rho_1)\rho_1^2 \quad (3 - 2\rho_2)\rho_2^2] \begin{bmatrix} v_i \\ v_j \end{bmatrix} + [\rho_1^2\rho_2 L \quad -\rho_1\rho_2^2 L] \begin{bmatrix} \theta_i \\ \theta_j \end{bmatrix} \tag{3.42}
$$

in which

$$
\rho_1 = 1 - \frac{x}{L}, \quad \rho_2 = \frac{x}{L}. \tag{3.43}
$$

The lateral deflection, v, in Equation (3.42) has not yet accounted for the effect of connection flexibility at the ends of the beam-column element.

Considering the inner beam-column between the connection springs at the two ends of the hybrid element, the elemental stiffness matrix links the forces and deformations as

$$
\begin{Bmatrix} Q_i \\ M_i \\ Q_j \\ M_j \end{Bmatrix} = \begin{bmatrix} K_{11} & K_{12} & K_{13} & K_{14} \\ K_{21} & K_{22} & K_{23} & K_{24} \\ K_{31} & K_{32} & K_{33} & K_{34} \\ K_{41} & K_{42} & K_{43} & K_{44} \end{bmatrix} \begin{Bmatrix} v_i \\ \theta_i \\ v_j \\ \theta_j \end{Bmatrix} \tag{3.44}
$$

in which K_{ij} is the elemental stiffness of the beam-column element given by

$$
K_{11} = K_{33} = -K_{13} = -K_{31} = \frac{12EI}{L^3}
$$

$$
K_{12} = K_{21} = K_{14} = K_{41} = -K_{23} = -K_{32} = -K_{34} = -K_{43} = \frac{6EI}{L^2} \tag{3.45}
$$

$$
K_{22} = K_{44} = 2K_{24} = 2K_{42} = \frac{4EI}{L}
$$

The elemental stiffness matrix for the complete hybrid element, with the connection springs at two ends of the beam-column element, can be written as

$$
\begin{Bmatrix} Q_i \\ M_i \\ Q_j \\ M_j \\ Q_I \\ M_I \\ Q_J \\ M_J \end{Bmatrix} =
\begin{bmatrix}
r_{vi}+K_{11} & K_{12} & K_{13} & K_{14} & -r_{vi} & 0 & 0 & 0 \\
K_{21} & S_{ci}+K_{22} & K_{23} & K_{24} & 0 & -S_{ci} & 0 & 0 \\
K_{31} & K_{32} & r_{vj}+K_{33} & K_{34} & 0 & 0 & -r_{vj} & 0 \\
K_{41} & K_{42} & K_{43} & S_{cj}+K_{44} & 0 & 0 & 0 & -S_{cj} \\
-r_{vi} & 0 & 0 & 0 & r_{vi} & 0 & 0 & 0 \\
0 & -S_{ci} & 0 & 0 & 0 & S_{ci} & 0 & 0 \\
0 & 0 & -r_{vj} & 0 & 0 & 0 & r_{vj} & 0 \\
0 & 0 & 0 & -S_{cj} & 0 & 0 & 0 & S_{cj}
\end{bmatrix}
\begin{Bmatrix} v_i \\ \theta_i \\ v_j \\ \theta_j \\ v_I \\ \theta_I \\ v_J \\ \theta_J \end{Bmatrix}
$$

(3.46)

The loads are assumed to be applied only at the global nodes I and J, and hence the shear forces, Q_i and Q_j, and moments, M_i and M_j, are equal to zero. Thus,

$$
\begin{Bmatrix} v_i \\ \theta_i \\ v_j \\ \theta_j \end{Bmatrix} =
\begin{bmatrix}
r_{vi}+K_{11} & K_{12} & K_{13} & K_{14} \\
K_{21} & S_{ci}+K_{22} & K_{23} & K_{24} \\
K_{31} & K_{32} & r_{vj}+K_{33} & K_{34} \\
K_{41} & K_{42} & K_{43} & S_{cj}+K_{44}
\end{bmatrix}^{-1}
\begin{bmatrix}
r_{vi} & 0 & 0 & 0 \\
0 & S_{ci} & 0 & 0 \\
0 & 0 & r_{vj} & 0 \\
0 & 0 & 0 & S_{cj}
\end{bmatrix}
\begin{Bmatrix} v_I \\ \theta_I \\ v_J \\ \theta_J \end{Bmatrix}
$$

(3.47)

The internal DOFs are eliminated by substituting Equation (3.47) into Equation (3.46), thus the relationship between the external nodal force and nodal displacements/rotations of the element can be obtained.

In the assembling of the elemental stiffness matrix of a structure, the above matrix is rewritten in the global coordinate system for geometrical compatibility. Substituting Equation (3.47) into Equation (3.42) and transforming the external nodal deformations, v_I, θ_I and v_J, θ_J, about the local axis at the ends of the element to the global nodal deformations $\bar{v}_i$, $\bar{\theta}_i$ and $\bar{v}_j$, $\bar{\theta}_j$, the displacement function v can finally be written as

$$
v = \begin{bmatrix} (3-2\rho_1)\rho_1^2 & \rho_1^2\rho_2 L & (3-2\rho_2)\rho_2^2 & -\rho_1\rho_2^2 L \end{bmatrix}
$$

$$
\times
\begin{bmatrix}
r_{vi}+K_{11} & K_{12} & K_{13} & K_{14} \\
K_{21} & S_{ci}+K_{22} & K_{23} & K_{24} \\
K_{31} & K_{32} & r_{vj}+K_{33} & K_{34} \\
K_{41} & K_{42} & K_{43} & S_{cj}+K_{44}
\end{bmatrix}^{-1}
\begin{bmatrix}
r_{vi} & 0 & 0 & 0 \\
0 & S_{ci} & 0 & 0 \\
0 & 0 & r_{vj} & 0 \\
0 & 0 & 0 & S_{cj}
\end{bmatrix}
\times [T]
\begin{Bmatrix} \bar{v}_i \\ \bar{\theta}_i \\ \bar{v}_j \\ \bar{\theta}_j \end{Bmatrix}
$$

(3.48)

in which $[T]$ is the coordinate transformation matrix.

3.2.2 Decomposition of system matrices

A number of new finite element model updating methodologies have been proposed and applied to structural engineering for solving the inverse problem of model updating or damage assessment (Mottershead and Friswell, 1993). There are two important issues in the use of the model updating methods: (a) the selection strategy for the updating parameters; and (b) the choice of dynamic characteristics of the structure from measurement. While the dynamic characteristics that are sensitive to the parameters should be chosen to correlate the finite element model predictions and the test results with a small parameter adjustment, further reviews show that there are two types of problems with the first issue:

- Type A. The initial analytical model includes all the potential physical effects. The updating problem is merely the choosing of relevant physical parameters from the analytical model and finding out the correct numerical values of them in the updating process.
- Type B. The initial analytical model neglects some significant physical effects, and the relevant effects must be re-introduced into the updated model.

If researchers effectively limit themselves to the type A problem, they will adopt a *physical parameter* selection strategy for model updating such as Chen and Garba (1980) and Hoff and Natke (1989). Li et al. (2006) have taken out the corresponding terms for axial, transverse, rotational and transverse-rotational stiffness of a planar beam element into sparse matrices in an attempt to identify the different load-carrying stiffnesses with a damage index method. These methods always have clear physical meanings, in that the updated models keep the same connective topology between the structural members defined by the initial analytical model. In practice, however, these methods are too restrictive by keeping the assumed element shape functions unchanged. Given a complex structure, if the initial analytical model neglects some significant effects that exist in the real structure, it would be very difficult to find the 'real' model by merely correcting parameters from the inadequate analytical model.

3.2.2.1 The generic element

Generic element theory (Gladwell and Ahmadian, 1995) may be a good solution for the type B problem. This method displays a good balance between the direct matrix method and the physical parameter method, such that the updated model correlates the experimental data by automatically introducing relevant effects, while keeping the correct connectivity topology specified by the finite element modelling.

The generic element method studies how an element stiffness matrix and mass matrix can be updated by adjusting its eigenvalues and eigenvectors. There are three generic parameters defining the elastic axial stiffness, symmetric bending stiffness and the anti-symmetric bending stiffness of a planar frame element. Without loss of generality, the model error is assumed to be related to the stiffness properties of the structure, and therefore only the generic stiffness matrix of an element is studied.

The finite element model of the structure is constructed with the global stiffness matrix of the structure from n global DOFs as $\{u_G\}$ and it is modelled using an assemblage of n_e finite elements. Each of the n_e elements itself connects the elemental DOF, e.g. $\{u_\alpha^E\}$ for the αth element which has size $(n_\alpha \times 1)$. The corresponding elemental matrices in this coordinate system are $[k_\alpha^E]$ and $[m_\alpha^E]$ of $(n_\alpha \times n_\alpha)$ size. The free vibration of an undamped element is itself governed by the equation

$$([k_\alpha^E] - \omega_j^2 [m_\alpha^E])\{\phi\}_j = \{0\} \quad j = 1, 2, \ldots, n_\alpha \tag{3.49}$$

if the element has $r(\leq 6)$ rigid body modes, the $(n_\alpha \times n_\alpha)$ mode shape matrix can be written as $[\Phi_\alpha] = [\phi_1, \ldots, \phi_r | \phi_{r+1}, \ldots, \phi_{n_\alpha}] = [\Phi_R, \Phi_S]$, where the rigid body modes are specified in Φ_R, and the other strain modes in Φ_S. With the mode shapes normalized with respect to the mass and stiffness matrices,

$$[\Phi_\alpha]^T [m_\alpha^E][\Phi_\alpha] = I, \quad [\Phi_\alpha]^T [k_\alpha^E][\Phi_\alpha] = [\Lambda_\alpha] \tag{3.50}$$

where $[\Lambda_\alpha] = diag(0, \ldots, 0, \omega_1^2, \ldots, \omega_{n_\alpha - r}^2)$, and $[k_\alpha^E][\Phi_R] = 0$. One way of constructing a generic family of the element matrices can be found as

$$[m_\alpha^E] = [\Phi_\alpha]^{-T}[\Phi_\alpha]^{-1}, \quad [k_\alpha^E] = [\Phi_\alpha]^{-T}[\Lambda_\alpha][\Phi_\alpha]^{-1} \tag{3.51}$$

Let $[m_\alpha^E]_0$ and $[k_\alpha^E]_0$, which are extracted from the system matrices of the initial analytical model, be the initial pair of the mass and stiffness matrices for the element with corresponding mode shape matrix $[\Phi_\alpha]_0$. Let $[m_\alpha^E]$ and $[k_\alpha^E]$ be any other member of the generic family and its corresponding modes that form the columns of $[\Phi_\alpha]$. Since the columns of $[\Phi_\alpha]_0$ are independent of each other and they form the initial vectors for the n_α-dimension space, there is a unique matrix, $[S_\alpha]$, relating $[\Phi_\alpha]$ and $[\Phi_\alpha]_0$ by the equation

$$[\Phi_\alpha] = [\Phi_\alpha]_0 [S_\alpha]^{-1} \quad \text{or} \quad [\Phi_\alpha]_0 = [\Phi_\alpha][S_\alpha] \tag{3.52}$$

Inserting Equation (3.52) into Equation (3.51), and noticing that $[\Phi_\alpha]_0^{-T} = [m_\alpha^E]_0[\Phi_\alpha]_0$, the formulation of the generic element family is given as

$$[m_\alpha^E] = [m_\alpha^E]_0[\Phi_\alpha]_0[U_\alpha][\Phi_\alpha]_0^T[m_\alpha^E]_0, \quad [k_\alpha^E] = [m_\alpha^E]_0[\Phi_\alpha]_0[V_\alpha][\Phi_\alpha]_0^T[m_\alpha^E]_0 \tag{3.53}$$

where

$$[U_\alpha] = [S_\alpha]^T[S_\alpha], \quad [V_\alpha] = [S_\alpha]^T[\Lambda_\alpha][S_\alpha] \tag{3.54}$$

Matrices $[U_\alpha]$ and $[V_\alpha]$ consist of the generic parameters of the generic family of elements $[m_\alpha^E]$ and $[k_\alpha^E]$. When the unknowns comprising $[S_\alpha]$ and $[\Lambda_\alpha]$ are correlated to the measured global modal data, the 'real' member can be identified from the generic family.

Before going further, the following two points in the generic theory have to be noted: (a) the modes making up $[\Phi_\alpha]$ are from the free vibration of an element and they have nothing to do with the measured global modes; (b) the transform matrix, $[S_\alpha]$, has $(n_\alpha \times n_\alpha)$ unknowns. For the system with n_e elements, there would be a huge number of unknowns to be solved. The first barrier can be resolved by assembling the generic elements into a generic structure, while the second obstacle can be solved by restricting the matrix, $[S_\alpha]$, on physical grounds.

Let us follow the assembling procedure of standard finite element modelling. Since the elemental DOFs are related by a transformation matrix to the global DOF as

$$\{u_G\} = [T_\alpha]\{u_\alpha^E\} \tag{3.55}$$

in which $[T_\alpha]$ represents the transformation matrix for the αth element, then the global system matrices can be formed by assembling all the elemental matrices according to

$$[K] = \sum_{\alpha=1}^{n_e} [T_\alpha][k_\alpha^E][T_\alpha]^T, \quad [M] = \sum_{\alpha=1}^{n_e} [T_\alpha][m_\alpha^E][T_\alpha]^T \qquad (3.56)$$

The $(n \times n_\alpha)$ elemental-to-global DOF transformation matrices, $[T_\alpha]$, include the index lookup corresponding between the elemental and global DOFs, the coordinate transformation from the elemental frames to the global frame and the effect of applied constraints.

The generic form of the element matrices in Equation (3.53) can then be substituted into Equation (3.56) to get the generic form of the structure

$$[K] = \sum_{\alpha=1}^{n_e} [T_\alpha][m_\alpha^E]_0[\Phi_\alpha]_0[V_\alpha][\Phi_\alpha]_0^T[m_\alpha^E]_0^T[T_\alpha]^T, \qquad (3.57a)$$

$$[M] = \sum_{\alpha=1}^{n_e} [T_\alpha][m_\alpha^E]_0[\Phi_\alpha]_0[U_\alpha][\Phi_\alpha]_0^T[m_\alpha^E]_0^T[T_\alpha]^T \qquad (3.57b)$$

This expression can be further simplified into

$$[K] = [A][V][A]^T, \quad [M] = [A][U][A]^T \qquad (3.58)$$

where the sparse topology matrix, $[A]$, is defined by

$$[A] = [([T]_1[m_1^E]_0[\Phi_1]_0) \quad ([T]_2[m_2^E]_0[\Phi_2]_0) \quad \cdots \quad ([T]_{n_e}[m_{n_e}^E]_0[\Phi_{n_e}]_0)] \qquad (3.59)$$

and $[U]$ and $[V]$ are diagonal block matrices consisting of all the assembled elemental generic parameters, where

$$[U] = diag([U_1], \ [U_2], \ \ldots, \ [U_{n_e}]), \quad [V] = diag([V_1], \ [V_2], \ \ldots, \ [V_{n_e}]) \qquad (3.60)$$

Assuming the structural connectivity of the analytical model is correct, and the model errors only relate to the stiffness of the structural members, the topology matrix, $[A]$, can be regard as independent of the elemental generic parameters. The derivatives of global stiffness matrix with respect to the elemental generic parameters, e.g. $\{P\}_{1 \times n_p}$, can then be directly computed as

$$\frac{\partial[K]}{\partial p_i} = [A]\frac{\partial[V]}{\partial p_i}[A]^T, \quad \frac{\partial[M]}{\partial p_i} = [A]\frac{\partial[U]}{\partial p_i}[A]^T \quad i = 1, 2, \ldots, n_p \qquad (3.61)$$

with n_p unknown generic parameters.

Case study

The European Space Agency Structure shown in Figure 3.12 is used to illustrate the applicability and effectiveness of the generic elements. The structure is modelled by 48 frame elements with three DOFs at each node for the translation and rotational deformations. Each frame element is constructed by integrating an Euler–Bernoulli

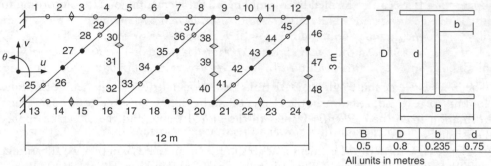

All units in metres

'Measured' DOFs for Cases 2 and 3: ◇x-direction; ◊ y-direction; ● both x-y direction

Figure 3.12 Finite element model of the European Space Agency Structure

beam element with a rod element. The elastic modulus of material is assumed to be $E = 7.5 \times 10^{10}$ N/m^2 and the density is assumed to be $\rho = 2800$ kg/m^3. The total number of DOFs specified in the analytical finite element model is 126. Another finite element model is constructed to provide the 'experimental' modal data. The model has the same finite element topology as the analytical one. However, the Euler–Bernoulli beam elements are replaced by Timoshenko beam elements that incorporate the transverse shear deformation. Local manufacturing faults are also introduced by reducing the elastic modulus of the 2nd, 19th, 36th and 47th elements by 20%. Only stiffness modelling errors are considered in this case study, although the method is equally applicable to the case with mass modelling errors.

Modelling with the element

The strain energy of the Timoshenko beam element can be expressed as

$$U_\varepsilon = \int_l \frac{P^2(x)}{2EA} dx + \int_l \frac{M^2(x)}{2EI} dx + \int_l \frac{k_s Q^2(x)}{2GA} dx \tag{3.62}$$

where P, M and Q are internal axial force, bending moment and shear force, respectively; G represents the elastic shear modulus of material; and k_s is the transverse shear coefficient determined from the shape of the beam cross-section, which, in this case, equals 2.11 for an I/wide flange section. The first and second terms on the right-hand-side of Equation (3.62) represent the axial and bending strain energy of the beam, while the third term, which is the shear strain energy, does not exist with the Euler–Bernoulli beam element. For a long-span beam with a cross-section with a small k_s value, the shear strain energy could be small and negligible. However, the frame structure as shown in Figure 3.12 is composed of deep and thick frame elements and the transverse shear effect could be significant and therefore should not be neglected. From the above discussions, it can be concluded that:

(a)   the modelling error between the initial analytical model and the 'experimental' data originates from two sources: systemic error due to the neglecting of the transverse shear effects and local stiffness error; and

(b) the error due to the shear effect cannot be corrected by a conventional physical parameter updating method since the analytical model does not include the relevant variables, i.e. k_s and G, at all.

The following paragraphs give detailed considerations on how to construct a generic family for a plane frame element. Starting with an Euler–Bernoulli element in the initial analytical model with its initial lump mass and stiffness matrices as

$$[m^E]_0 = A\rho l \begin{bmatrix} \frac{1}{2} & & & & & \\ & \frac{1}{2} & & & 0 & \\ & & \frac{1}{24} & & & \\ & & & \frac{1}{2} & & \\ & 0 & & & \frac{1}{2} & \\ & & & & & \frac{1}{24} \end{bmatrix}, \quad [k^E]_0 = E \begin{bmatrix} \frac{A}{l} & 0 & 0 & \frac{-A}{l} & 0 & 0 \\ & \frac{12I}{l^3} & \frac{6I}{l^2} & 0 & \frac{-12I}{l^3} & \frac{6I}{l^2} \\ & & \frac{4I}{l} & 0 & \frac{-6I}{l^2} & \frac{2I}{l} \\ & & & \frac{A}{l} & 0 & 0 \\ & sym & & & \frac{12I}{l^3} & \frac{-6I}{l^2} \\ & & & & & \frac{4I}{l} \end{bmatrix}$$

(3.63)

Its eigenvalues and mass-normalized mode shapes are

$$[\Lambda]_0 = \begin{bmatrix} 0 & & & & & \\ & 0 & & 0 & & \\ & & 0 & & & \\ & & & 4 & & \\ & 0 & & & 48 & \\ & & & & & 192 \end{bmatrix}, \quad [\Phi]_0 = \begin{bmatrix} k_1 & 0 & 0 & -k_1 & 0 & 0 \\ 0 & k_2 & -(\sqrt{3}/2)k_2 & 0 & 0 & 0.5k_2 \\ 0 & 0 & \sqrt{3}k_3 & 0 & 2\sqrt{3}k_3 & 3k_3 \\ k_1 & 0 & 0 & k_1 & 0 & 0 \\ 0 & k_2 & (\sqrt{3}/2)k_2 & 0 & 0 & -0.5k_2 \\ 0 & 0 & \sqrt{3}k_3 & 0 & -2\sqrt{3}k_3 & 3k_3 \end{bmatrix}$$

(3.64)

where $k_1 = \sqrt{EA/l}$, $k_2 = \sqrt{EI/l}$ and $k_3 = \sqrt{EI/l^3}$. The first three modes are rigid body modes, and the last three are strain modes, as shown in Equation (3.64). It is important to note that the second, fourth and fifth modes are symmetric about the centre of the element while the others are of anti-symmetry. If two assumptions on the family of the generic element are included, namely: (a) the number of rigid modes remains the same and the new rigid modes are linear combinations of the original ones; and (b) the modal symmetry of the element is preserved, then the transformation matrix, $[s]$, can be partitioned into the following

$$[s] = \begin{bmatrix} s_R & s_{RS} \\ 0 & s_S \end{bmatrix} = \begin{bmatrix} s_{11} & & & s_{14} & & \\ & s_{22} & & & s_{25} & \\ & & s_{33} & & & s_{36} \\ & & & s_{44} & & \\ & 0 & & & s_{55} & \\ & & & & & s_{66} \end{bmatrix}$$

(3.65)

This then leads to a diagonal matrix, $[V]$, in Equation (3.53), such that the stiffness matrix of the generic element family can be expressed as (Law et al., 2001a)

$$[k^E] = [m^E]_0 \begin{bmatrix} -k_1 & 0 & 0 \\ 0 & 0 & 0.5k_2 \\ 0 & 2\sqrt{3}k_3 & 3k_3 \\ k_1 & 0 & 0 \\ 0 & 0 & -0.5k_2 \\ 0 & -2\sqrt{3}k_3 & 3k_3 \end{bmatrix} \begin{bmatrix} v_1 & & 0 \\ & v_2 & \\ 0 & & v_3 \end{bmatrix}$$

$$\times \begin{bmatrix} -k_1 & 0 & 0 & k_1 & 0 & 0 \\ 0 & 0 & 2\sqrt{3}k_3 & 0 & 0 & -2\sqrt{3}k_3 \\ 0 & 0.5k_2 & 3k_3 & 0 & -0.5k_2 & 3k_3 \end{bmatrix} [m^E]_0 \qquad (3.66)$$

where $v_1 = \omega_1^2 s_{44}^2$, $v_2 = \omega_2^2 s_{55}^2$ and $v_3 = \omega_3^2 s_{66}^2$; and ω_i is the ith elastic circular frequency. There are only three generic parameters to define the stiffness matrix for each element.

Since the transfer matrix, $[s]_0$, corresponding to the initial finite element model is a unity matrix, the generic parameters take the initial values of $v_1 = 4$, $v_2 = 48$ and $v_3 = 192$. It is noted that v_1, v_2 and v_3 correspond to the elastic axial mode, symmetric bending mode and the anti-symmetric bending mode, respectively. When these parameters take different values, the stiffness matrix represents elements with different stiffness properties. For example, when only one generic parameter, s_{66}, takes the form as $s_{66} = \sqrt{1/(1+b)}$,

$$v_3 = 192/(1+b) \qquad (3.67)$$

in which $b = 12EIk_s/GAl^2$. Substituting v_3 into Equation (3.66) gives the familiar stiffness matrix of a Timoshenko beam element

$$[k^E] = \frac{E}{1+b} \begin{bmatrix} \dfrac{A(1+b)}{l} & 0 & 0 & \dfrac{-A(1+b)}{l} & 0 & 0 \\ & \dfrac{12I}{l^3} & \dfrac{6I}{l^2} & 0 & \dfrac{-12I}{l^3} & \dfrac{6I}{l^2} \\ & & \dfrac{(4+b)I}{l} & 0 & \dfrac{-6I}{l^2} & \dfrac{(2-b)I}{l} \\ & & & \dfrac{A(1+b)}{l} & 0 & 0 \\ & & & & \dfrac{12I}{l^3} & \dfrac{-6I}{l^2} \\ & sym & & & & \dfrac{(4+b)I}{l} \end{bmatrix} \qquad (3.68)$$

It is noted that taking all the three generic parameters as variables gives us a very large family of generic elements that can include many types of modelling errors and unknown effects.

3.2.2.2 The eigen-decomposition

A uniform stiffness (or mass) reduction is widely adopted by researchers to model local damage sites. In practice, however, the damage may be of different patterns, which affect the system dynamic properties differently and it cannot be modelled as a uniform change in the stiffness or mass matrix without much error. To describe the damage of all possible patterns with one set of indicators, a more generalized parameter set, called the elemental eigen-parameters, is formulated.

The elemental matrices of a frame element are decomposed into their eigenvalue and eigenvector matrices. The eigenvalues represent physically the stiffnesses of the element corresponding to its different deformed shapes. The different eigenvalues give a detailed description of the different load-carrying resistances of the structure.

The global stiffness matrix of a structure, consisting of n global DOFs, $\{u_G\}$, can be modelled using an assembly of n_e finite elements. The elemental local DOFs are related to the global DOFs by

$$\{u_G\} = \sum_{\alpha=1}^{ne} [T]_\alpha \{u_E\}_\alpha \tag{3.69}$$

in which $[T]_\alpha$ is the transformation matrix for the αth element. The global system matrices are assembled as

$$[K] = \sum_{\alpha=1}^{n_e} [T]_\alpha [k_E]_\alpha [T]_\alpha^T, \quad [M] = \sum_{\alpha=1}^{n_e} [T]_\alpha [m_E]_\alpha [T]_\alpha^T \tag{3.70}$$

It is important to note that Equation (3.70) is not a minimum-rank definition of the structural disassembly problem, because the elemental matrices $[k_E]_\alpha$ and $[m_E]_\alpha$ are always rank deficient. Although $[k_E]_\alpha$ is an $(n_\alpha \times n_\alpha)$ symmetric matrix with $(n_\alpha \times (n_\alpha + 1)/2)$ unknown entries, only a few of them are actually independent. Consider as an example, a linear spring element connecting two nodes, each of which includes three displacement $\{u, \quad v, \quad w\}$ DOFs. There are potentially 21 unknown entries in this (6×6) element stiffness matrix. However, the rank of the elemental stiffness matrix is one, as there is only one axial stiffness of the spring in the matrix.

The rank deficient elemental matrices can be decomposed into their static eigenvalues and eigenvectors (Doebling, 1995; Doebling et al., 1998a) such that

$$[k_E]_\alpha = [\eta]_\alpha [p^k]_\alpha [\eta]_\alpha^T, \quad [m_E]_\alpha = [\mu]_\alpha [p^m]_\alpha [\mu]_\alpha^T \tag{3.71}$$

where $[p^k]_\alpha$ and $[p^m]_\alpha$ are the corresponding diagonal matrices of nonzero static eigenvalues. If the ranks of the stiffness and mass matrices for the αth element are r_α and τ_α, respectively, then $[\eta]_\alpha$ is an $(n_\alpha \times r_\alpha)$ eigenvector matrix of the elemental stiffness matrix and $[\mu]_\alpha$ is an $(n_\alpha \times \tau_\alpha)$ eigenvector matrix of the mass matrix. Physically, the columns of matrix $[\eta]_\alpha$ are the distinct and statically equilibrated deformation shapes of the element which have nonzero strain energy, while those of matrix $[\mu]_\alpha$ are rigid displacements related to the kinetic energy. Both these matrices can be normalized as

$$[\eta]_\alpha^T [\eta]_\alpha = [I]_{r_\alpha}, \quad [\mu]_\alpha^T [\mu]_\alpha = [I]_{\tau_\alpha} \tag{3.72}$$

Substituting the static decomposition of Equation (3.71) into Equation (3.70) gives

$$[K] = \sum_{\alpha=1}^{n_e} [T]_\alpha [\eta]_\alpha [p^k]_\alpha [\eta]_\alpha^T [T]_\alpha^T, \quad [M] = \sum_{\alpha=1}^{n_e} [T]_\alpha [\mu]_\alpha [p^m]_\alpha [\mu]_\alpha^T [T]_\alpha^T \qquad (3.73)$$

or

$$[K] = [A][P_k][A]^T, \quad [M] = [B][P_m][B]^T \qquad (3.74)$$

where the sparse matrices, $[A]$ and $[B]$, are called the 'stiffness topology matrix' and the 'mass topology matrix', respectively. They are defined by

$$[A] = [([T]_1[\eta]_1) \quad ([T]_2[\eta]_2) \quad \cdots \quad ([T]_{n_e}[\eta]_{n_e})]$$

$$[B] = [([T]_1[\mu]_1) \quad ([T]_2[\mu]_2) \quad \cdots \quad ([T]_{n_e}[\mu]_{n_e})] \qquad (3.75)$$

and $[P_k]$ and $[P_m]$ are diagonal matrices of size $(n_{pk} \times n_{pk})$ and $(n_{pm} \times n_{pm})$, respectively, of the assembled elemental stiffness and mass eigenvalues, where

$$[P_k] = diag([p^k]_1, [p^k]_2, \ldots, [p^k]_{n_e}), [P_m] = diag([p^m]_1, [p^m]_2, \ldots, [p^m]_{n_e}) \quad (3.76)$$

It should be noticed that Equation (3.74) does not imply that $[P_k]$ contains the eigenvalues of the global stiffness matrix, $[K]$, because the columns of $[A]$ do not in general form an orthogonal basis mathematically. The columns of matrix $[A]$ physically embody the stiffness contribution to the global stiffness matrix in terms of the eigen-parameters, p_i^k. The same argument applies to matrix $[P_m]$. It is noted that methods have also been developed to calculate the eigenvectors for a general type of finite element by Doebling (1996) and Robertson et al. (1996).

Frame element

Consider a two-dimensional, six DOFs frame element with only three displacement modes, the eigen-decomposition on $[k_E]$ yields the following stiffness eigenvectors and eigenvalues as

$$[\eta] = \begin{bmatrix} \frac{1}{\sqrt{2}} & 0 & 0 \\ 0 & 0 & \frac{\sqrt{2}}{\sqrt{L^2+4}} \\ 0 & \frac{-1}{\sqrt{2}} & \frac{L}{\sqrt{2}\sqrt{L^2+4}} \\ \frac{-1}{\sqrt{2}} & 0 & 0 \\ 0 & 0 & \frac{-\sqrt{2}}{\sqrt{L^2+4}} \\ 0 & \frac{1}{\sqrt{2}} & \frac{L}{\sqrt{2}\sqrt{L^2+4}} \end{bmatrix}, \quad [p^k] = \begin{bmatrix} \frac{2EA}{L} & & 0 \\ & \frac{2EI}{L} & \\ 0 & & \frac{6EI(L^2+4)}{L^3} \end{bmatrix} \qquad (3.77)$$

The diagonal elements of the eigenvalues in Equation (3.77) consist of the sectional properties of the element. They are renamed as an eigen-parameter set $\{p^k\}_\alpha = \left\{ \frac{2EA}{L}, \frac{2EI}{L}, \frac{6EI(L^2+4)}{L^3} \right\}_\alpha$ and are indicators of the state of the local stiffness of the element. These parameters exceed the conventional single design parameter, e.g. the elastic modulus of the material, E_α, which lacks a correlation with the damage pattern. The type of damage is represented by the pattern in the set of eigen-parameter changes. Damage, and hence load resistance quantification, can then be assessed from these changes.

For the European Space Agency structure in Figure 3.12, the stiffness matrix of each member can be decomposed as Equation (3.71) with

$$
[p^k]_\alpha = diag \left[\frac{EA}{L} \quad \frac{EI_z}{L} \quad \frac{EI_y}{L} \quad \frac{GJ}{L} \quad \frac{3EI_z(L^2+4)}{L^3} \quad \frac{3EI_y(L^2+4)}{L^3} \right]
$$
$$
= diag[G_1 \quad G_2 \quad G_3 \quad G_4 \quad G_5 \quad G_6]
$$

where $[p^k]_\alpha$ is the diagonal matrix containing the nonzero eigenvalues. The six eigenvalues are related to the elemental stiffness parameters such as EA, EI and GJ defining the corresponding axial, torsional, the two cross-sectional bending and shear stiffnesses, respectively. They represent the stiffness moduli of the corresponding deformed shapes, where G_2 and G_5 represent the moduli corresponding to the symmetric and anti-symmetric bending shapes in the x–y plane, and G_3 and G_6 correspond to the symmetric and anti-symmetric bending shapes in the x–z plane. If taking the elemental length as constant, the proportional relations of $G_5 = \sigma G_2$ and $G_6 = \sigma G_3$ can be found with $\sigma = 3(L^2+4)/L^2$, where L denotes the length of the member.

Ten-bay cantilever frame structure

The ten-bay three-dimensional frame structure shown in Figure 3.13 was assembled using the Meroform M12 construction system. The structure consists of 80, 22 mm-diameter alloy steel tubes jointed together with 40 standard Meroform ball nodes. Each tube is fitted with a screwed end connector which, when tightened into the node, also clamps the tube by means of an internal compression fitting. The length of all the horizontal and vertical tube members between the centres of two adjacent balls is exactly 0.5 m after assembly. The structure orients horizontally and is fixed into a rigid concrete support at four nodes at one end.

A Link Dynamic System model 450 shaker is used to apply a continuous random force in the range of 0~100 Hz through the concrete support to the fixed end of the frame. The appropriate amplitude of the force input is selected to have a satisfactory measured response spectrum. The fundamental frequency of the frame is 3.78 Hz.

The artificial damage is a perforated slot cut in the central length of the beam. The slot is 137 mm long, and the remaining depth of the tube in the cut cross-section is 14.375 mm. The slot in element 17 in the third bay opens horizontally, and that in element 43 in the sixth bay opens vertically.

Figure 3.13 Ten-bay cantilever frame structure

The four independent eigenvalues, G_1, G_2, G_3 and G_4, of an element are selected as updating parameters for each element to correct for the local stiffness errors, and the proportional constant σ is maintained. There are $4 \times 80 = 320$ unknown eigen-parameters in the identification.

A regularization technique with a truncated SVD is applied for the identification (Ren et al., 2000). Figure 3.14 shows the updated correction results of the stiffness eigen-parameters, in which G_1 represents the stiffness modulus for the axial elastic deformation, G_4 indicates the modulus of elastic twisting of the member, and G_2 and G_3 are proportional to the bending stiffness moduli in the horizontal and vertical planes, respectively. It is seen that the twisting stiffness, G_4, dramatically degrades almost equally in the damaged members because of the open cut in the originally closed circular cross-section of the member. Another interesting observation is that the direction of the open slot in the damage member can be seen through the different patterns in G_2 and G_3. The damaged member, element 17, has the slot open horizontally, and therefore the bending stiffness in the horizontal plane reduces much more than that in the vertical plane, as indicated by the patterns of G_2 and G_3. A similar observation can be made for element 43, where the slot opens vertically.

Shell element
The in-plane, shear and bending deformations in different directions of a shell element are coupled. The formulations for such an element are more complex compara-tively because of the higher-order partial differential equations in the finite element formulation.

The Mindlin–Reissner plate including the effect of transverse shear deformation is adopted for this study with the shear deformable plate element and the membrane element combined. The transverse displacement, w, and slope, θ, are independent, as are the shape functions required for the interpolation. As a result, the shear deformable

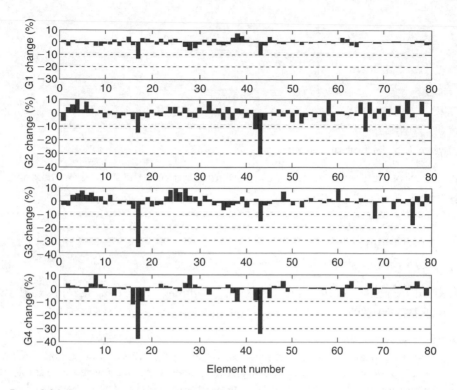

Figure 3.14 Percentage correction of local faults

plate element requires C^0 compatibility and isoparametric shape functions are used for the finite element formulation. For an isotropic homogeneous plate with constant thickness, t, the elemental stiffness matrix for the plate in bending can be expressed as

$$[K_b^e] = \frac{t^3}{12} \int_{\Omega^e} [B_b]^T [D_b][B_b] d\Omega + \kappa t \int_{\Omega^e} [B_s]^T [D_s][B_s] d\Omega \qquad (3.78)$$

in which, Ω^e is the two-dimensional element domain in the $x - y$ plane as shown in Figure 3.15; D_b is the elasticity matrix for plate bending; D_s is the elasticity matrix for shear deformation; κ is the shear correction factor; and B_b and B_s are the strain-displacement transformation matrices for bending and shear, respectively.

Applying the small-deformation theory, the membrane stretching and bending effects are decoupled in a shell element. A plate-bending element has a transverse deflection and two rotations at a node while a plane-stress element has two in-plane displacements at a node. All together, there are five DOFs for a node in a shell element, i.e. three displacements and two rotations. Now, consider the assembled structure with the shell elements at different orientations, a drilling DOF about the local z-axis is included at a node leading to a total of six DOFs at a node. The shell element formulation is based

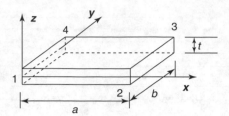

Figure 3.15 Rectangular element

on a superposition of the membrane and bending actions, the stiffness matrix of which is given as

$$[K^e] = \begin{bmatrix} [K_b^e] & [0] & [0] \\ [0] & [K_m^e] & [0] \\ [0] & [0] & [0] \end{bmatrix} \tag{3.79}$$

where subscripts b and m denote the bending and membrane deformations of the shell element, respectively.

The rectangular element

A four-node bilinear isoparametric plate-shell element is shown in Figure 3.15. Considering the shear-locking phenomenon with thin plates, the selective reduced integration (SRI) technique is adopted to generate the stiffness matrix of the element. The 2×2 Gauss–Legendre quadrature is used for the bending term while 1-point integration is used for the shear term. The eigenvalues and eigenvectors of the stiffness matrix after decomposition are

$$[p]_b = \begin{Bmatrix} \dfrac{\kappa G A_x}{4a}(4+a^2) \\[2mm] \dfrac{\kappa G A_y}{4b}(4+b^2) \\[2mm] \dfrac{1}{3}\dfrac{E}{(1-v^2)}\left(\dfrac{I_y}{a}+\dfrac{1}{2}\dfrac{I_x}{b}(1-v)\right) \\[2mm] \dfrac{1}{3}\dfrac{E}{(1-v^2)}\left(\dfrac{I_x}{b}+\dfrac{1}{2}\dfrac{I_y}{a}(1-v)\right) \\[2mm] \dfrac{1}{2}\dfrac{E}{(1-v^2)}\left(\left(\dfrac{I_y}{a}+\dfrac{I_x}{b}\right)-C_1t^3\right) \\[2mm] \dfrac{1}{2}\dfrac{E}{(1-v^2)}\left(\left(\dfrac{I_y}{a}+\dfrac{I_x}{b}\right)+C_1t^3\right) \\[2mm] \dfrac{1}{2}\dfrac{E}{(1-v^2)}\left(\dfrac{I_y}{a}+\dfrac{I_x}{b}\right)(1-v) \end{Bmatrix}, \quad [p]_m = \begin{Bmatrix} \dfrac{G}{3}\left(2\dfrac{A_x}{a(1-v)}+\dfrac{A_y}{b}\right) \\[2mm] \dfrac{G}{3}\left(2\dfrac{A_y}{b(1-v)}+\dfrac{A_x}{a}\right) \\[2mm] \dfrac{G}{(1-v)}\left(\dfrac{A_x}{a}+\dfrac{A_y}{b}-\dfrac{C_1}{2}\left(\dfrac{A_x}{b}+\dfrac{A_y}{a}\right)\right) \\[2mm] \dfrac{G}{(1-v)}\left(\dfrac{A_x}{a}+\dfrac{A_y}{b}+\dfrac{C_1}{2}\left(\dfrac{A_x}{b}+\dfrac{A_y}{a}\right)\right) \\[2mm] G\left(\dfrac{A_y}{b}-\dfrac{A_x}{a}\right) \end{Bmatrix}$$

$$\tag{3.80}$$

in which, G is the shear elastic modulus; v is the Poisson ratio; a and b are the length and width of the element; t is the thickness; $C_1 = \left\|\sqrt{(a/b-b/a)^2+4v^2}\right\|$ is a constant;

$A_x = b \times t$ and $A_y = a \times t$ are defined as the cross-sectional areas normal to the x- and y-axes; and $I_x = a \times t^3/12$ and $I_y = b \times t^3/12$ are the second moment of inertia of the cross-section about the x- and y-axes, respectively.

Each term in the resulting stiffness matrix through multiplication of the eigenvectors and eigenvalues is compared with the corresponding term obtained from numerical integration. The error is very small and is believed to be due to the numerical integration.

Each term in vectors $[p]_b$ and $[p]_m$ denotes the macro-stiffness of the corresponding deformation mode shown later in this section. An inspection of the vector of eigenvalues shows that the first two eigenvalues in $[p]_b$ and all those in $[p]_m$ are related to the parameters A_x and A_y. The remaining eigenvalues of $[p]_b$ are related to the parameters I_x and I_y. This observation shows that some of the deformation modes are contributing to the in-plane stiffness while others are contributing to the flexural stiffness of the element. This uncoupling of the stiffness resistance contribution may be useful for the study of the load path within a structure under external load and for the assessment of the contribution of the load resistance from the different macro-stiffnesses.

The eigenvectors defining the corresponding deformation modes of the four-node shell element are given as follows and they are plotted in Figures 3.16 and 3.17 for further illustration with

$$[\eta]_b = \begin{bmatrix} 0 & 1 & 0 & 1 & -C_2 & -C_3 & -1 \\ -a/2 & 0 & 1 & 0 & 1 & 1 & a/b \\ 0 & 1 & 0 & -1 & -C_2 & -C_3 & 1 \\ -a/2 & 0 & -1 & 0 & -1 & -1 & a/b \\ 0 & 1 & 0 & 1 & C_2 & C_3 & 1 \\ -a/2 & 0 & 1 & 0 & -1 & -1 & -a/b \\ 0 & 1 & 0 & -1 & C_2 & C_3 & -1 \\ -a/2 & 0 & -1 & 0 & 1 & 1 & -a/b \\ 1 & 2/b & 0 & 0 & 0 & 0 & 0 \\ -1 & 2/b & 0 & 0 & 0 & 0 & 0 \\ -1 & -2/b & 0 & 0 & 0 & 0 & 0 \\ 1 & -2/b & 0 & 0 & 0 & 0 & 0 \end{bmatrix} = [\Phi_1 \quad \Phi_2 \quad \cdots \quad \Phi_7] \quad (3.81)$$

$$[\eta]_b = [\bar{\eta}]_b / diag[\|\Phi_1\| \quad \|\Phi_2\| \quad \cdots \quad \|\Phi_7\|]$$

$$[\bar{\mu}]_m = \begin{bmatrix} 1 & 0 & 1 & 1 & a/b \\ 0 & 1 & C_2 & C_3 & 1 \\ -1 & 0 & -1 & -1 & a/b \\ 0 & -1 & C_2 & C_3 & -1 \\ 1 & 0 & -1 & -1 & -a/b \\ 0 & 1 & -C_2 & -C_3 & -1 \\ -1 & 0 & 1 & 1 & -a/b \\ 0 & -1 & -C_2 & -C_3 & 1 \end{bmatrix} = [\Psi_1 \quad \Psi_2 \quad \cdots \quad \Psi_5] \quad (3.82)$$

$$[\mu]_m = [\bar{\mu}]_m / diag[\|\Psi_1\| \quad \|\Psi_2\| \quad \cdots \quad \|\Psi_5\|]$$

where $\|\Phi_i\|, \|\Psi_j\| (i = 1, \ldots 7; j = 1, \ldots 5)$ are the norms of the column vectors in $[\bar{\eta}]_b$ and $[\bar{\mu}]_m$, respectively; $C_2 = (a/b - b/a - C_1)/(2v)$ and $C_3 = (a/b - b/a + C_1)/(2v)$.

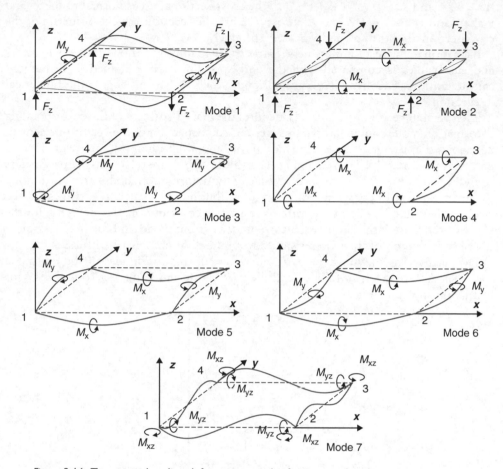

Figure 3.16 Transverse bending deformation modes for rectangular elements

It should be noted that C_2 is negative and C_3 is positive for all values of a and b. The elements in a column of $[\eta]_b$ correspond to the DOFs $(\theta_{x1}, \theta_{y1}, \theta_{x2}, \theta_{y2}, \theta_{x3}, \theta_{y3}, \theta_{x4}, \theta_{y4}, w_1, w_2, w_3, w_4)$, while those in a column of $[\eta]_m$ correspond to the DOFs $(u_1, v_1, u_2, v_2, u_3, v_3, u_4, v_4)$. Each column in $[\eta]_b$ and $[\mu]_m$ describes a distinct deformation mode with nonzero strain energy. It is noted that each deformation mode corresponds to a set of nodal forces that cause such deformations, and Figures 3.16 and 3.17 include these sets of nodal forces for illustration.

The deformed modes

The first two deformation modes in $[\eta]_b$ result from the transverse shear deformation. Since their eigenvalues consist of physical parameters, κ, G, A_x and A_y, and the corresponding deformation mode bends about the y-axis and the x-axis, respectively. It is noted that the associated two eigenvalues in $[p]_b$ are the same as the shear stiffness terms in a Timoshenko beam arising from the transverse deformation and rotation at a node. These eigenvalues are called the *shear stiffness* in the x- and y-directions, respectively.

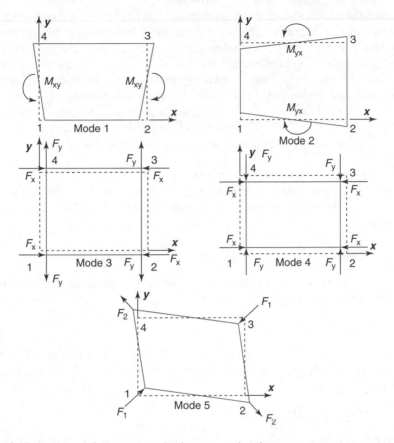

Figure 3.17 In-plane deformation modes for rectangular elements

The last five deformation modes in $[\eta]_b$ correspond to pure transverse bending deformations of the element. This also agrees with the observation that their corresponding eigenvalues are related to parameters I_x and I_y. The third deformation mode in $[\eta]_b$ consists of pure bending deformation in the x-direction. However, this pure bending mode varies linearly along the y-direction. The same observation is found in the fourth deformation mode, but with references to the x- and y-directions interchanged. The corresponding eigenvalues for these two modes are the two *symmetric twisting stiffnesses* in the y- and x-directions, respectively.

The fifth mode has a sagging bending moment in the x-direction and a hogging bending moment in the y-direction with a saddle-shape deformation mode. The sixth mode has sagging moments in both the x- and y-directions. The corresponding eigenvalues are called the *saddle* and *symmetric bending stiffnesses* of the element about the two centroidal axes.

The seventh mode consists of twisting deformation in all the four edges of the element, while the deformation mode shape is symmetric about the catercorner lines. The corresponding eigenvalue is called the *anti-symmetric torsional stiffness* of the element.

The five deformation modes in $[\mu]_m$ correspond to the pure in-plane stretching deformation since their corresponding eigenvalues consist of the parameters A_x and A_y only. The first two modes define the deformation of the element under the in-plane bending moment M_{xy} and M_{yx}, respectively. The corresponding eigenvalues are called the *in-plane bending stiffnesses* in the x- and y-directions, respectively. The deformation of the third mode is under in-plane x-direction compression and y-direction tension, while the fourth mode is under both in-plane x- and y-direction compression. The corresponding eigenvalues are called the *in-plane tension–compression and compression–compression stiffnesses*, respectively. The fifth mode is a set of in-plane shear deformation, and the corresponding eigenvalue is called the *in-plane shear distortional stiffness*.

The system matrix and elemental forces

Assuming a plate structure is under the action of external force $\{F\}$ subject to a set of constraints at the boundary DOFs, the static equilibrium equation of the system is expressed as

$$[K]\{d^G\} = \{F\} \tag{3.83}$$

where $\{d^G\}$ is the displacement vector at all the DOFs. Substituting Equation (3.74) into Equation (3.83), gives

$$[A][P_k][A]^T\{d^G\} = \{F\} \tag{3.84}$$

For the αth element in the system, the displacement vector and the equilibrated applied force vector at the local DOFs arising from the applied external force $\{F\}$ are expressed as,

$$\{d^e\}_\alpha = [B]_\alpha^d\{d\}, \quad \{F^e\}_\alpha = [B]_\alpha^F\{F\} \tag{3.85}$$

where $[B]_\alpha^d$ and $[B]_\alpha^F$ are the transformation matrices from the global displacement and the equilibrated force vectors to the local matrices of the αth element, respectively. Here, $\{d^e\}_\alpha$ is the set of combined displacement with nonzero strain energy and the rigid displacement of the element with $\{d^e\}_\alpha = \{d_f^e\}_\alpha + \{d_r^e\}_\alpha$, where the subscripts f and r denote the strain and rigid displacements, respectively.

Being similar to the equilibrium equation of the system, the force–deformation equation for the αth element is given as

$$[K^e]_\alpha\{d_f^e\}_\alpha = \{F^e\}_\alpha \tag{3.86}$$

Substituting Equation (3.71) into Equation (3.86) and omitting the subscript α for convenience, gives

$$[\eta_f^e][p^e][\eta_f^e]^T\{d_f^e\} = \{F^e\} \tag{3.87}$$

and multiplying $[U_f^e]^T$ to both sides of the equation, gives

$$[p^e][\eta_f^e]^T\{d_f^e\} = [\eta_f^e]^T\{F^e\}$$

$$[p^e]\{\bar{d}_f\} = \{\bar{F}\} \tag{3.88}$$

in which

$$\{\bar{d}_f\} = [\eta_f^e]^T\{d_f^e\}, \quad \{\bar{F}\} = [\eta_f^e]^T\{F^e\} \tag{3.89}$$

where, $\{\bar{d}_f\}$ is a $(\gamma + \tau) \times 1$ vector, and is defined as the vector of weights on the corresponding deformation modes. Any one of these weights is defined as a multiplier on each deformation mode that contributes to the elemental nodal deformation, $\{d_f^e\}$. $\{\bar{F}\}$ is the set of weights associated with the distribution of the elemental nodal forces in association with the different deformation modes, with each contributing a fraction of the elemental nodal forces $\{F^e\}$.

Similarly, the rigid displacement component of an element, $\{d_r^e\}$, can also be expressed in terms of the six rigid modes as

$$\{\bar{d}_r\} = [\eta_r^e]^T\{d_r^e\} \tag{3.90}$$

Combining Equations (3.89) and (3.90) forms,

$$\{d^e\} = [\eta]\{Y\} \tag{3.91}$$

where $[\eta] = [[\eta_f^e] \quad [\eta_r^e]]$ is an $n_{dof} \times (\gamma + \tau + 6)$ matrix, including all the nonzero strain deformation modes and the rigid modes of the element. $\{Y\} = \{\{\bar{d}_f\} \quad \{\bar{d}_r\}\}^T$ with each term denoting the contribution of the corresponding deformation mode to $\{d^e\}$.

The Scordelis–Lo roof

The geometry of the Scordelis–Lo roof (Andelfinger and Ramm, 1992), a deep cylindrical shell, under self-weight is given in Figure 3.18. The structure, which is supported through rigid end diaphragms, is 50 ft long with a thickness of 3in. The roof has a radius of 25 ft with a subtended angle of 80 degrees. The structure has elastic modulus of 30×10^6 psi, and a weight of 90 lb/ft^2. A quarter of the structure is modelled due to the symmetry. The original Poisson ratio is 0.0 and it has been changed to 0.15 for the purpose of this study. The analytical solution of the vertical deflection at the centre of

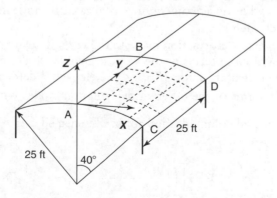

Figure 3.18 The Scordelis–Lo roof and mesh

the free edge is 3.7033 in. When 12×14 elements are used, the vertical deflection is 3.4598 in, which is satisfactory. The roof is allowed to slide in the longitudinal direction under load while it is restrained in three directions at the four lower corners. The quarter plate is divided into 14×14 elements for the following analysis.

Weights for the deformed modes

The weights on the eigen-modes in Equation (3.89) are shown in Figures 3.19 and 3.20. The elemental x- and y-axes are in the direction of lines AC and AB in Figure 3.18.

The X-direction shear-bending mode in Figure 3.19 (mode 1) shows a larger deformation in a wide conical zone at the middle portion of the structure around line BD, while the Y-direction shear-bending mode (mode 2) shows some large values in a zone close to the mid-span of the free edge at D. These two plots show the path of the shear transference onto the support at C. The Y-direction edge-twisting mode (mode 3) has some large weights in a zone along the free edge CD and along the catercorner line AD. The weights are zero along line AC at the top of the roof. The X-direction edge-twisting mode (mode 4) shows similar observations to mode 3, with the sign of the weights reversed. The unsymmetrical bending mode 5 has some large weights in a wide conical zone at the middle of the structure. The symmetrical bending mode 6 has close to zero weights at the mid-span along line BD with diminishing weights towards point A at the top of the structure at the support. The anti-symmetrical edge-twisting mode 7 has mostly zero weights except in a zone at the mid-span close to the free edge at point D.

All the in-plane modes show large deformations in the elements at and close to the mid-span at point D. This matches the engineering prediction to have large in-plane stresses along the bottom edge of the roof near point D when under self-weight. The weights for mode 2 are larger than those for mode 1 with in-plane edge moments M_{yx} and much larger than M_{xy} along the two pairs of edges of an element. Mode 3 shows some large weights in a zone at the mid-span along BD. The weights change sign from a positive value at the top of the roof to a negative value close to the mid-span at point D. This mode, with tension–compression in the element, dominates the in-plane deformation. The compression–compression deformed mode 4 exhibits small weights, except at the mid-span of the free edge at point D. The in-plane torsional mode 5 also has small weights, except in some elements close to point D and along the catercorner line AD. Both modes 2 and 3 dominate the in-plane deformation.

Work done in each element

The contribution of each deformation mode to the work done by the external applied force, $\{F\}$, can be written in terms of the work done in each deformation mode. The work done by the elemental forces, $\{F^e\}$, on the elemental deformation vector, $\{d_f^e\}$, can be expressed in terms of the different deformation modes with an associated set of nodal forces as

$$
\begin{aligned}
W &= \frac{1}{2} \sum \{F^e\}^T \{d_f^e\} \\
&= \frac{1}{2} \sum \{\bar{d}_f\}^T [P^e] \{\bar{d}_f\} = \frac{1}{2} \sum \{\bar{F}\}^T \{\bar{d}_f\}
\end{aligned}
\tag{3.92}
$$

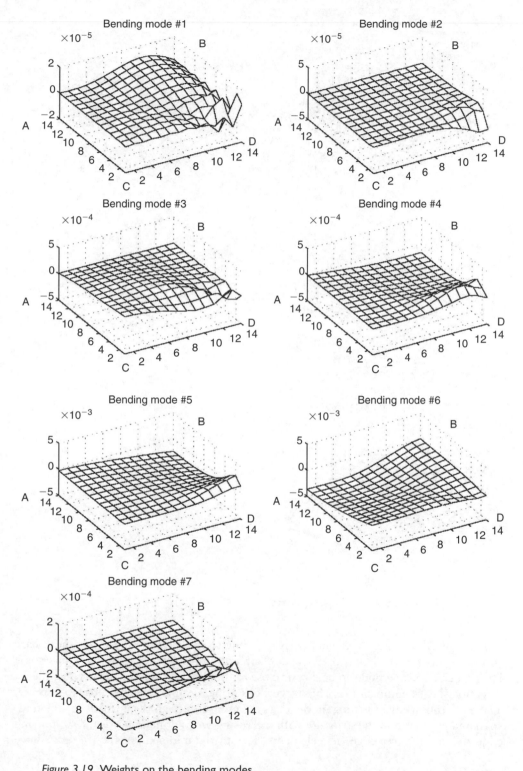

Figure 3.19 Weights on the bending modes

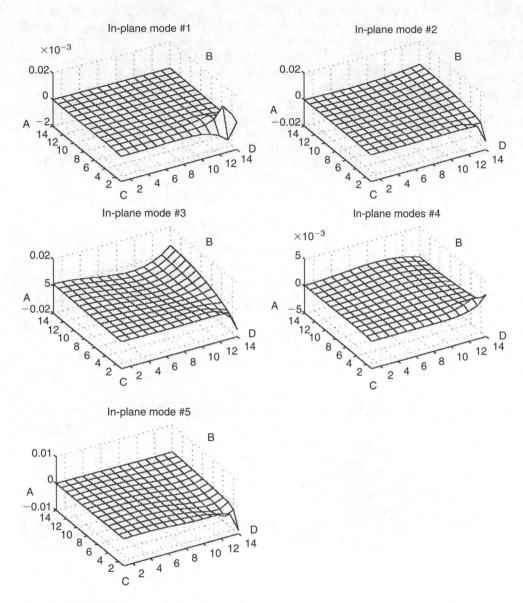

Figure 3.20 Weights on the in-plane modes

The work done in each element corresponding to the different deformed modes are shown in Figures 3.21 and 3.22. The X-direction shear-bending mode 1 has some large work done in a wide conical zone close to mid-span of the structure while the Y-direction shear-bending mode 2 has large work done in some elements close to point D. The Y-direction edge-twisting mode 3 has zero work done in the elements at the mid-span and at the top of the structure with increasing work done towards the supporting edge AC and the free edge. The elements close to the mid-span of the free edge have

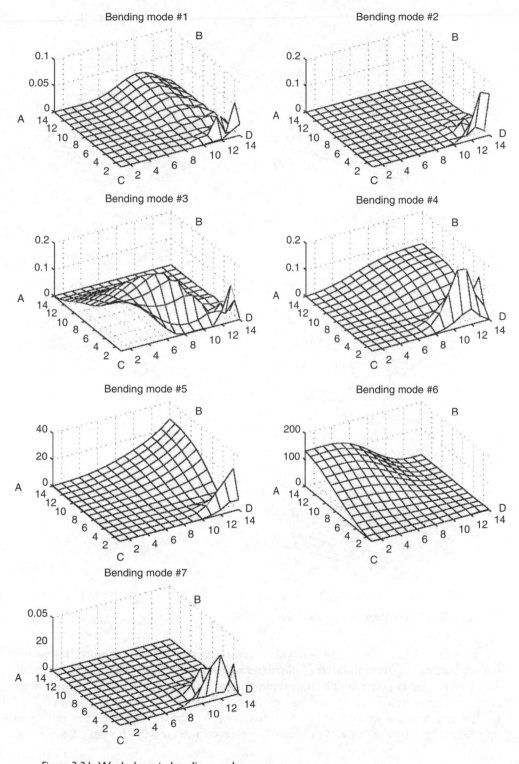

Figure 3.21 Work done in bending modes

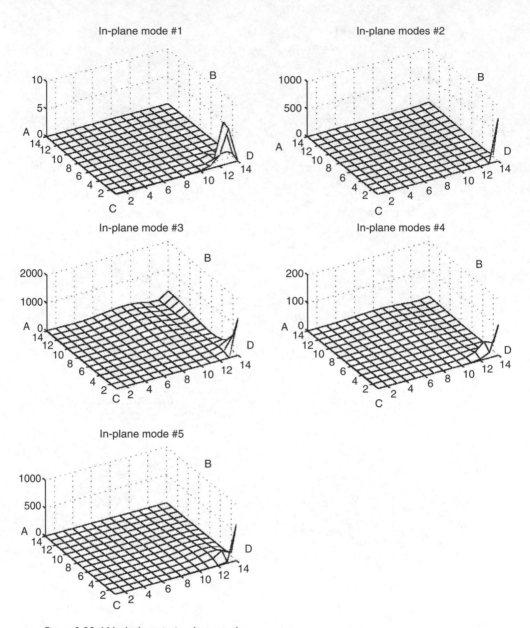

Figure 3.22 Work done in in-plane modes

some large weights. The*X*-direction edge-twisting mode 4 has zero work done along the supporting edge *AC* and part of the free edge *CD*, with large values at the mid-span of the free edge at point *D*. The anti-symmetric bending mode 5 has zero work done along the supporting edge *AC* and part of the free edge, with the largest value at corner *B*. The symmetric bending mode 6 has close to zero work done along the mid-span line *BD* and the free edge *CD*, with increasing large values towards corner *A* at the top

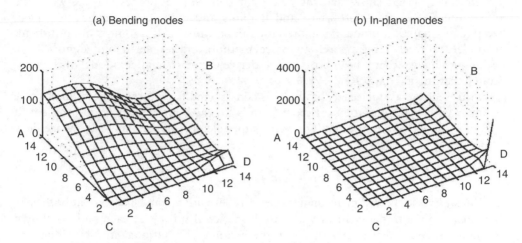

Figure 3.23 Total work done in each type of deformed mode

of the roof. The effect of the anti-symmetric edge-twisting mode 7 is very small and it is confined to elements close to the mid-span at point D.

All the in-plane modes show a distinct large peak in the elements close to the mid-span at point D. The compression–compression mode 3 also shows large work done in some of the elements at the mid-span along line BD. Both modes 2 and 3 dominate the contribution of the in-plane modes in the load resistance of the roof.

The contribution of each deformed mode to the total work done is 0.006%, 0.004%, 0.05%, 0.03%, 4.21%, 34.54% and 0.001% for the bending deformed modes and 0.09%, 6.07%, 40.98%, 3.51% and 10.52% for the five in-plane deformed modes.

The distributions of the work done in each element corresponding to the two types of deformed modes are shown in Figure 3.23 for further illustration. The work done from the bending deformed modes in elements close to point A is larger with decreasing values towards the other three corners of the quarter plate. The in-plane work done is in general small in all the elements except in the element at the mid-span of the free edge at point D.

3.2.3 Super-element

A super-element representing a segment of a large-scale structure is presented in this section. Each individual structural component is represented by a sub-element in the model. The large number of DOFs in the analytical model is reduced, while the modal sensitivity relationship of the structural model to small physical changes can be retained at the sub-element level. These properties are significant to structural damage assessment.

All civil engineering structures consist of connections or *joints* between the structural components. The joints are fixed with different types of fasteners, depending on the materials used to build the structure. However, in practical analysis and design practice, only three types of framing joints are considered: I) fully rigid joints; II) perfectly pinned joints; III) semi-rigid joints. Experimental investigations (Jones et al., 1982;

Lewitt et al., 1969) show that the true joint behaviour is somewhere between the two simplified extremes of Type I and II joints and they are somewhat nonlinear. Despite an analytical model being presented in Section 3.2.1, a joint is often difficult to model accurately using a purely analytical method, and a range of joint identification techniques have recently been proposed to derive the mathematical model of the joint from experimental data (Ren, 1992).

In this section, a super-element model is constructed for a three-dimensional frame member with semi-rigid joints at the two ends for the joint identification problem. It is later extended to the formulation for the super-element of the Tsing Ma Bridge deck structure.

3.2.3.1 Beam element with semi-rigid joints

The linear hybrid three-dimensional frame member with a semi-rigid joint at both ends is modelled by a frame element with three rotational springs at each end, as shown in Figure 3.24(a). The springs represent the bending or rotational stiffness about the three local axes.

Considering the set of springs belonging to a massless connection element between the beam and the joint as shown in Figure 3.24(b), the following equilibrium relations can be obtained for moments at the member ends

$$
\begin{bmatrix} K_1 & 0 & 0 \\ 0 & K_2 & 0 \\ 0 & 0 & K_3 \end{bmatrix} \begin{Bmatrix} \theta_1 \\ \theta_2 \\ \theta_3 \end{Bmatrix} = \begin{Bmatrix} M_1 \\ M_2 \\ M_3 \end{Bmatrix}
\tag{3.93}
$$

in which

$$
K_i = \begin{bmatrix} r_m^i & -r_m^i & 0 & 0 \\ -r_m^i & r_m^i + k_{11}^i & k_{12}^i & \\ & k_{21}^i & r_n^i + k_{22}^i & -r_n^i \\ & & -r_n^i & r_n^i \end{bmatrix}, \theta_i = \begin{Bmatrix} {}_e\theta_m^i \\ {}_i\theta_m^i \\ {}_i\theta_n^i \\ {}_e\theta_n^i \end{Bmatrix}, M_i = \begin{Bmatrix} {}_em_m^i \\ {}_im_m^i \\ {}_im_n^i \\ {}_em_n^i \end{Bmatrix}
\tag{3.94}
$$

where $i = 1, 2, 3$ denote the terms about the local x-, y- and z-axes, respectively; r_m and r_n are the joint stiffnesses at the two ends, written as $r^i = \frac{p}{1-p} \left(\frac{4EI}{L} \right)$ for $i = 2, 3$

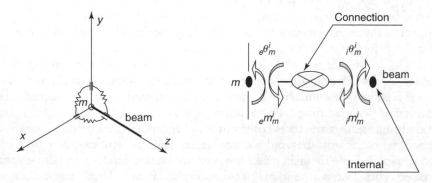

Figure 3.24 (a) 3D Semi-rigid joint model (b) Nomenclature at connection

and $r^1 = \frac{p}{1-p}\left(\frac{GJ}{L}\right)$, in which p is the fixity factor, which is zero for perfectly pinned joints and one for perfectly rigid joints; the second moment of inertia of the member cross-section about the x- and y-axes are assumed to be the same; $_e\theta_m^i$, $_e\theta_n^i$, $_i\theta_m^i$ and $_i\theta_n^i$ are the external and internal rotation about the ith axis, respectively, as shown in Figure 3.24(b); and k_{pq}^i are the corresponding components of the stiffness matrix of a conventional beam element if the bowing effect is ignored, given by

$$k_{pq}^1 = \frac{GJ}{L}\begin{bmatrix} 1 & -1 \\ -1 & 1 \end{bmatrix}, \quad k_{pq}^i = \frac{EI}{L}\begin{bmatrix} 4 & 2 \\ 2 & 4 \end{bmatrix}, \quad (i = 2,3) \tag{3.95}$$

Since the external forces and moments are applied at the external nodes of the jointed member only, the applied moments about each axis at the internal nodes connecting the beam element and the spring can be obtained by considering equilibrium of the connection shown in Figure 3.24(b):

$$\begin{bmatrix} r_m^i + k_{11}^i & k_{12}^i \\ k_{21}^i & r_n^i + k_{22}^i \end{bmatrix}\begin{Bmatrix} _i\theta_m^i \\ _i\theta_n^i \end{Bmatrix} - \begin{bmatrix} r_m^i & 0 \\ 0 & r_n^i \end{bmatrix}\begin{Bmatrix} _e\theta_m^i \\ _e\theta_n^i \end{Bmatrix} = \begin{Bmatrix} 0 \\ 0 \end{Bmatrix}, \quad (i = 1,2,3) \tag{3.96}$$

or

$$\begin{Bmatrix} _i\theta_m^i \\ _i\theta_n^i \end{Bmatrix} = \begin{bmatrix} r_m^i + k_{11}^i & k_{12}^i \\ k_{21}^i & r_n^i + k_{22}^i \end{bmatrix}^{-1}\begin{bmatrix} r_m^i & 0 \\ 0 & r_n^i \end{bmatrix}\begin{Bmatrix} _e\theta_m^i \\ _e\theta_n^i \end{Bmatrix}, \quad (i = 1,2,3) \tag{3.97}$$

Substituting Equation (3.97) into Equation (3.93), the stiffness terms related to the internal nodes are eliminated, and the condensed stiffness matrix relating the moments and rotations about the external nodes can be written as:

$$\begin{Bmatrix} _em_m^1 \\ _em_m^2 \\ _em_m^3 \\ _em_n^1 \\ _em_n^2 \\ _em_n^3 \end{Bmatrix} = \begin{bmatrix} K_{mn} & K_{21} \\ K_{12} & K_{nm} \end{bmatrix}\begin{Bmatrix} _e\theta_m^1 \\ _e\theta_m^2 \\ _e\theta_m^3 \\ _e\theta_n^1 \\ _e\theta_n^2 \\ _e\theta_n^3 \end{Bmatrix} \tag{3.98}$$

where $K_{ij} = \begin{bmatrix} r_i^1 - (r_i^1)^2(r_j^1 + k_{22}^1)/\rho_1 & 0 & 0 \\ 0 & r_i^2 - (r_i^2)^2(r_j^2 + k_{22}^2)/\rho_2 & 0 \\ 0 & 0 & r_i^3 - (r_i^3)^2(r_j^3 + k_{22}^3)/\rho_3 \end{bmatrix}$

$$\text{for } \begin{Bmatrix} i = m,n \\ j = m,n \\ i \neq j \end{Bmatrix},$$

and $K_{12} = K_{21} = \begin{bmatrix} k_{12}^1 r_m^1 r_n^1/\rho_1 & 0 & 0 \\ 0 & k_{12}^2 r_m^2 r_n^2/\rho_2 & 0 \\ 0 & 0 & k_{12}^3 r_m^3 r_n^3/\rho_3 \end{bmatrix},$

where

$$\rho_i = \det \begin{bmatrix} k_{11}^i + r_m^i & k_{12}^i \\ k_{21}^i & k_{22}^i + r_n^i \end{bmatrix} \tag{3.99}$$

Adding the terms for the axial force to Equation (3.98), and transforming the resulting 8×8 matrix to the nodal DOFs of the element, the stiffness matrix of the semi-rigid jointed beam element can be written as:

$$K_e = T \cdot \begin{bmatrix} EA/L & 0 & 0 & 0 \\ 0 & K_{mn} & 0 & K_{21} \\ 0 & 0 & EA/L & 0 \\ 0 & K_{12} & 0 & K_{nm} \end{bmatrix}_{8 \times 8} \cdot T^T \tag{3.100}$$

in which T is the 12×8 transformation matrix mapping the six external moments and two axial forces to the nodal forces and moments in the local coordinates, respectively, given by

$$T = \begin{bmatrix} 1 & 0 & 0 & 0 & 0 & 0 & 0 & 0 & 0 & 0 & 0 & 0 \\ 0 & 0 & 0 & 1 & 0 & 0 & 0 & 0 & 0 & 0 & 0 & 0 \\ 0 & 1/L & 0 & 0 & 1 & 0 & 0 & -1/L & 0 & 0 & 0 & 0 \\ 0 & 0 & 1/L & 0 & 0 & 1 & 0 & 0 & -1/L & 0 & 0 & 0 \\ 0 & 0 & 0 & 0 & 0 & 0 & 1 & 0 & 0 & 0 & 0 & 0 \\ 0 & 0 & 0 & 0 & 0 & 0 & 0 & 0 & 0 & 1 & 0 & 0 \\ 0 & 1/L & 0 & 0 & 0 & 0 & 0 & -1/L & 0 & 0 & 1 & 0 \\ 0 & 0 & 1/L & 0 & 0 & 0 & 0 & 0 & -1/L & 0 & 0 & 1 \end{bmatrix} \tag{3.101}$$

It can be easily verified that the semi-rigid jointed beam element expressed in Equation (3.100) is equal to a pinned joint beam element if the spring stiffness is zero and to a rigid joint beam element if the spring stiffness is infinite.

Modelling of a ten-bay cantilever frame structure

The ten-bay cantilever frame structure shown in Figure 3.13 is divided into five super-elements in the longitudinal direction, each consists of four longitudinal-beam type sub-elements and twelve cross-beam type sub-elements connected by semi-rigid joints. The super-element model of the beam segment is shown in Figure 3.25. There are four master nodes at the corners and one slave node at the centre of the end-section. Each master node has three translational DOFs parallel to the local coordinate axes. Each slave node has three rotational DOFs about the three global axes of the end-section. There are 15 DOFs in each end-section, and there are 10 nodes (8 master and 2 slave) and 30 DOFs for each super-element. Altogether, there are 30 nodes and 75 DOFs for the whole structure. This model is much smaller than the conventional finite element model which has 240 DOFs.

The super-element model has been used to achieve good updated results for the lower frequency modes but not for the higher frequency modes (Law et al., 2001c). A perfectly refined model that represents the real structure is not achieved, and the updated results have errors which increase as the modal order increases. The first type of error is due to the incorrect assumptions made while condensing the finite element DOFs to those of

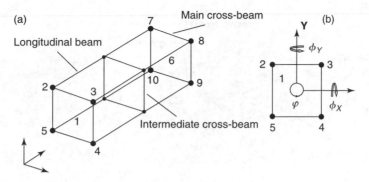

Figure 3.25 Super-element model of truss segment

the super-element. The second type of error is from the difference in the macro updating parameter in different members resulting from different ways of assembling the test structure. But the more significant error comes from the incomplete measurement, particularly from the rotational DOFs which cannot be measured in this study.

Half of the DOFs in a conventional finite element model of the truss are rotational DOFs. The super-element model has the advantage that the rotational DOFs are condensed and it has a smaller proportion of rotational DOFs than the conventional finite element model. There are only 15 rotational DOFs within a total of 75 DOFs for the whole test structure. It is therefore considered that the use of all the lower mode shapes in the identification reduces the first type of error, while the other types of errors cannot be avoided.

3.2.3.2   The Tsing Ma bridge deck

The formulation of the bridge deck model (Lantau, 1998) of the Tsing Ma suspension bridge in Hong Kong is presented for further illustration of the super-element modelling. The bridge serves as the link between the new airport of Hong Kong and the commercial centres. The bridge deck is a two-level enclosed structure, which carries a dual three-lane highway at the upper level and two railway tracks and two traffic lanes at the lower level. The whole deck structure can be divided into segments between adjacent sets of suspenders 18m apart, the arrangement of which is shown in Figure 3.26(a). Each segment consists of 66 structural components of longitudinal beams, cross-beams, bracings and stiffened plates.

All the nodes of the super-element are allocated on the two outermost sections along the longitudinal axis. Each end section consists of a number of nodes and several sub-elements as shown in Figure 3.26(b), depending on the specific deck configuration. In the type of bridge deck under consideration, there are ten master nodes and one slave node in the section. The primary longitudinal beams and auxiliary longitudinal beams are connected to nodes 2, 5, 7 and 10 and 3, 4, 8 and 9, respectively. Nodes 6 and 11 are at the intersection of cross-beams at the edges, and there are 14 cross-beams in the section, which are represented by line segments in the figure (the thicker lines indicate that there is also a stiffened plate between two adjacent longitudinal beams). Bracing members and stiffened plates are not shown in this figure, as they are not in the same

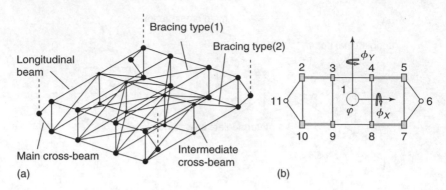

Figure 3.26 The super-element model for deck segment of Tsing Ma Bridge

section. The 66 structural members are: 8 longitudinal beams, 38 cross-beams, 16 bracing members and 4 stiffened plates modelled as four groups of sub-elements in the super-element. Nodes 3, 4, 8 and 9 each has three translational DOFs parallel to the local coordinate axes. To take account of the rigidity of the triangular part enclosed by nodes 5, 6 and 7, it is assumed that these three nodes have the same translational DOFs in the X-Y plane, and each has one different translational DOF in the longitudinal direction. A similar assumption is made for nodes 2, 10 and 11. In addition to the master nodal DOFs, slave node 1 has three DOFs around the three global coordinate axes of the cross-section. Consequently, the model has 25 DOFs in one cross-section and 50 DOFs in the super-element. Whereas there would be more than 300 DOFs if a standard three-dimension finite element model is adopted. It is evident that the number of DOFs is significantly reduced by the use of this super-element.

The following paragraphs give the formulation of the contribution of one type of sub-element, i.e. the longitudinal beams, to the stiffness matrix of the super-element from the variational principle of minimum potential energy.

Figure 3.27 shows that the origin C of the longitudinal sub-element is seated at the centre of gravity of the cross-section of the longitudinal beam. Let the three independent translational displacements of C be $\{\bar{u}, \bar{v}, \bar{w}\}^T$, and the global rotations of the super-element section around the X-, Y- and Z-axes be $\{\phi_X, \phi_Y, \varphi\}^T$, the deflections of an arbitrary point (x, y) in the beam section can be expressed as

$$\begin{cases} u(x,y) = \bar{u} - \varphi(Y_C + y) \\ v(x,y) = \bar{v} + \varphi(X_C + x) \\ w(x,y) = \bar{w} + \phi_X(Y_C + y) - \phi_Y(X_C + x) \end{cases} \tag{3.102}$$

The strain in the z-direction can be obtained as

$$\begin{cases} \varepsilon_z = \dfrac{\partial w(x,y)}{\partial z} = \dfrac{d\bar{w}}{dz} - (X_C + x)\dfrac{d\phi_Y}{dz} + (Y_C + y)\dfrac{d\phi_X}{dz} \\[2mm] \gamma_{zx} = \dfrac{\partial w(x,y)}{\partial x} + \dfrac{\partial u(x,y)}{\partial z} = -\phi_Y + \dfrac{d\bar{u}}{dz} - (Y_C + y)\dfrac{d\varphi}{dz} \\[2mm] \gamma_{zy} = \dfrac{\partial w(x,y)}{\partial y} + \dfrac{\partial v(x,y)}{\partial z} = \phi_X + \dfrac{d\bar{v}}{dz} + (X_C + x)\dfrac{d\varphi}{dz} \end{cases} \tag{3.103}$$

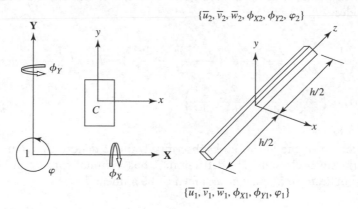

Figure 3.27 Longitudinal sub-element

Considering the longitudinal beam as a Timoshenko beam with two-way bending, its strain energy can be written as

$$H_{lb} = \frac{1}{2} \iiint (E\varepsilon_z^2 + G\gamma_{zx}^2 + G\gamma_{zy}^2)\, dz\, dx\, dy \qquad (3.104)$$

Substituting Equation (3.103) into Equation (3.104), and with some mathematical simplification,

$$
\begin{aligned}
H_{lb} = \frac{1}{2} \int \Bigg\{ & EA\Bigg[\left(\frac{d\overline{w}}{dz}\right)^2 + X_C^2 \left(\frac{d\phi_Y}{dz}\right)^2 + Y_C^2 \left(\frac{d\phi_X}{dz}\right)^2 \\
& - 2X_C \frac{d\overline{w}}{dz}\frac{d\phi_Y}{dz} + 2Y_C \frac{d\overline{w}}{dz}\frac{d\phi_X}{dz} - 2X_c Y_C \frac{d\phi_X}{dz}\frac{d\phi_Y}{dz}\Bigg] \\
& + EI_y \left(\frac{d\phi_Y}{dz}\right)^2 + EI_x \left(\frac{d\phi_X}{dz}\right)^2 - 2EI_{xy}\frac{d\phi_Y}{dz}\frac{d\phi_X}{dz} \\
& + GA\Bigg[\phi_Y^2 + \phi_X^2 + \left(\frac{d\overline{u}}{dz}\right)^2 + \left(\frac{d\overline{v}}{dz}\right)^2 - 2\phi_Y\frac{d\overline{u}}{dz} + 2\phi_X\frac{d\overline{v}}{dz} \\
& + 2Y_C\phi_Y\frac{d\varphi}{dz} + 2X_C\phi_X\frac{d\varphi}{dz} - 2Y_C\frac{d\overline{u}}{dz}\frac{d\varphi}{dz} + 2X_C\frac{d\overline{v}}{dz}\frac{d\varphi}{dz}\Bigg] \\
& + G[A(X_C^2 + Y_c^2) + I_x + I_y]\left(\frac{d\varphi}{dz}\right)^2 \Bigg\}\, dz \qquad (3.105)
\end{aligned}
$$

where A is the cross-sectional area; I_x and I_y are the sectional moments of inertia of the beam cross-section with respect to the x- and y-axes, respectively; and $I_{xy} = \int xy\, dA$. The deflections, i.e. $\overline{u}$ and $\overline{v}$, of an arbitrary section along the longitudinal beam can be assumed to be the form of a Hermite polynomial to include the contribution of bending modes:

$$\bar{u} = \bar{u}_1 \left(\frac{1}{2} - \frac{3z}{2h} + \frac{2z^3}{h^3} \right) + \bar{u}_2 \left(\frac{1}{2} + \frac{3z}{2h} - \frac{2z^3}{h^3} \right) + \phi_{Y1} h \left(\frac{1}{8} - \frac{z}{4h} - \frac{z^2}{2h^2} + \frac{z^3}{h^3} \right)$$

$$+ \phi_{Y2} h \left(-\frac{1}{8} - \frac{z}{4h} + \frac{z^2}{2h^2} + \frac{z^3}{h^3} \right) \tag{3.106}$$

where h is length of the beam, and subscripts 1 and 2 denote the values in the end sections of the super-element. The deflection, $\bar{v}$, has a similar form to $\bar{u}$. However, the longitudinal deflection, $\bar{w}$, may be assumed to be a linear form:

$$\bar{w} = \bar{w}_1 \left(\frac{1}{2} - \frac{z}{h} \right) + \bar{w}_2 \left(\frac{1}{2} + \frac{z}{h} \right) \tag{3.107}$$

The global rotations of an arbitrary section $\{\phi_X, \phi_Y, \varphi\}^T$ also take the same linear form as $\bar{w}$. Substituting these six shape functions in the forms of Equations (3.106) and (3.107) into Equation (3.105), the strain energy of the beam can be expressed in terms of the nodal displacements. After applying the second partial derivation of the strain energy with respect to the nodal displacements, the equilibrium equations are given as:

$$\{F\}_{lb} = [K]_{lb}\{U\}_{lb}, \ \{U\}_{lb} = \{\bar{u}_1, \bar{v}_1, \bar{w}_1, \varphi_1, \phi_{X1}, \phi_{Y1}, \bar{u}_2, \bar{v}_2, \bar{w}_2, \varphi_2, \phi_{X2}, \phi_{Y2}\}^T \tag{3.108}$$

in which $\{U\}_{lb}$ is the nodal displacement vector; $\{F\}_{lb}$ is the nodal force vector; and the 12×12 matrix, $[K]_{lb}$, is the stiffness matrix contribution of the longitudinal beam sub-element to the super-element.

Super-element with semi-rigid joints

To take into account the effects of the flexible joints at the two ends of the longitudinal beam, the formulation in Equations (3.93) to (3.100) can be repeated at the sub-element level. However, the components, k^i_{pq}, in the conventional stiffness matrix of a beam element should be replaced by the terms of the longitudinal beam sub-element, i.e.

$$k^1_{pq} = G(AX_C^2 + AY_C^2 + I_x + I_y)/h \begin{bmatrix} 1 & -1 \\ -1 & 1 \end{bmatrix}, \tag{3.109}$$

$$k^2_{pq} = \begin{bmatrix} E(AY_C^2 + I_x)/h + GAh/3 & GAh/6 - E(AY_C^2 + I_x)/h \\ GAh/6 - E(AY_C^2 + I_x)/h & E(AY_C^2 + I_x)/h + GAh/3 \end{bmatrix}, \tag{3.110}$$

$$k^3_{pq} = \begin{bmatrix} E(AX_C^2 + I_y)/h + GAh/3 & GAh/6 - E(AX_C^2 + I_y)/h \\ GAh/6 - E(AX_C^2 + I_y)/h & E(AX_C^2 + I_y)/h + GAh/3 \end{bmatrix} \tag{3.111}$$

In addition, the mapping matrix T in Equation (3.100) should be reconstructed as

$$T^* = T \cdot \begin{bmatrix} 1 & 0 & 0 & 0 & Y_c & -X_c & 0 & 0 & 0 & 0 & 0 & 0 \\ 0 & 1 & 0 & -Y_c & 0 & 0 & 0 & 0 & 0 & 0 & 0 & 0 \\ 0 & 0 & 1 & X_c & 0 & 0 & 0 & 0 & 0 & 0 & 0 & 0 \\ 0 & 0 & 0 & 1 & 0 & 0 & 0 & 0 & 0 & 0 & 0 & 0 \\ 0 & 0 & 0 & 0 & 1 & 0 & 0 & 0 & 0 & 0 & 0 & 0 \\ 0 & 0 & 0 & 0 & 0 & 1 & 0 & 0 & 0 & 0 & 0 & 0 \\ 0 & 0 & 0 & 0 & 0 & 0 & 1 & 0 & 0 & 0 & Y_c & -X_C \\ 0 & 0 & 0 & 0 & 0 & 0 & 0 & 1 & 0 & -Y_c & 0 & 0 \\ 0 & 0 & 0 & 0 & 0 & 0 & 0 & 0 & 1 & X_c & 0 & 0 \\ 0 & 0 & 0 & 0 & 0 & 0 & 0 & 0 & 0 & 1 & 0 & 0 \\ 0 & 0 & 0 & 0 & 0 & 0 & 0 & 0 & 0 & 0 & 1 & 0 \\ 0 & 0 & 0 & 0 & 0 & 0 & 0 & 0 & 0 & 0 & 0 & 1 \end{bmatrix} \quad (3.112)$$

Mass contribution of longitudinal beam

The displacement at an arbitrary point within the longitudinal beam has been expressed in the form of field functions in Equation (3.102). Since the displacement vector is also a function of time, the accelerations from the displacements can be obtained as

$$\begin{cases} \ddot{u}(x,y,z,t) = \ddot{\bar{u}} - \ddot{\varphi}(Y_C + y) \\ \ddot{v}(x,y,z,t) = \ddot{\bar{v}} + \ddot{\varphi}(X_C + x) \\ \ddot{w}(x,y,z,t) = \ddot{\bar{w}} + \ddot{\phi}_X(Y_C + y) - \ddot{\phi}_Y(X_C + x) \end{cases} \quad (3.113)$$

Substituting Equations (3.106) and (3.107), the accelerations at an arbitrary section of the longitudinal beam can be expressed in terms of the nodal accelerations:

$$\begin{Bmatrix} \ddot{\bar{u}} \\ \ddot{\bar{v}} \\ \ddot{\bar{w}} \\ \ddot{\varphi} \\ \ddot{\phi}_X \\ \ddot{\phi}_Y \end{Bmatrix} = \begin{bmatrix} a_1 & & & & a_2 & a_3 & & & & a_4 & & \\ & a_1 & & & & a_2 & & & a_3 & & a_4 & \\ & & a_5 & & & & & a_6 & & & & \\ & & & a_5 & & & & & a_6 & & & \\ & & & & a_5 & & & & & & & a_5 & a_6 \\ & & & & & a_5 & & & & & & a_5 \end{bmatrix} \begin{Bmatrix} \ddot{\bar{u}}_1 \\ \ddot{\bar{v}}_1 \\ \vdots \\ \ddot{\phi}_{Y1} \\ \ddot{\bar{u}}_2 \\ \ddot{\bar{v}}_2 \\ \vdots \\ \ddot{\phi}_{Y2} \end{Bmatrix} = [H]\{\ddot{U}\}_{lb} \quad (3.114)$$

where $a_i(i = 1, 2, 3, 4)$ are the coefficients of the Hermite polynomial.

$$a_1 = \frac{1}{2} - \frac{3z}{2h} + \frac{2z^3}{h^3}, \quad a_2 = \frac{1}{8} - \frac{z}{4h} - \frac{z^2}{2h^2} + \frac{z^3}{h^3}, \quad a_3 = \frac{1}{2} + \frac{3z}{2h} - \frac{2z^3}{h^3},$$

$$a_4 = -\frac{1}{8} - \frac{z}{4h} + \frac{z^2}{2h^2} + \frac{z^3}{h^3}, \quad a_5 = \frac{1}{2} - \frac{z}{h}, \quad a_6 = \frac{1}{2} + \frac{z}{h}.$$

The inertia force vector of an arbitrary point (with unit volume) within the longitudinal beam can be written as:

$$\{f(t)\} = -\rho \cdot \{\ddot{u} \quad \ddot{v} \quad \ddot{w}\}^T \qquad (3.115)$$

in which ρ is the mass density of the material. Substituting Equation (3.113) into Equation (3.115), and rewriting it in matrix form, gives

$$\{f(t)\} = -\rho \begin{bmatrix} 1 & & & -(Y_C + y) \\ & 1 & & (X_C + x) \\ & & 1 & & (Y_C + y) & -(X_C + x) \end{bmatrix} \begin{Bmatrix} \ddot{u} \\ \ddot{v} \\ \ddot{w} \\ \ddot{\varphi} \\ \ddot{\phi}_X \\ \ddot{\phi}_Y \end{Bmatrix} = [N]\{\ddot{q}\} \qquad (3.116)$$

Upon substitution of Equation (3.114), Equation (3.116) becomes

$$\{f(t)\} = -\rho[N][H]\{\ddot{U}\}_{lb} \qquad (3.117)$$

Integrating Equation (3.117) over the volume, the inertia force of the longitudinal beam is then obtained as

$$\{Q(t)\} = -\int [H]^T[N]^T \rho[N][H]dV \cdot \{\ddot{U}\}_{lb}$$
$$= -[M]_{lb}\{\ddot{u}\}_{lb} \qquad (3.118)$$

where $[M]_{lb}$ is the contribution of the longitudinal beam to the mass matrix of the super-element.

Structural members other than the longitudinal beams also make a significant contribution to the super-element matrices. Their contribution matrices to the stiffness and mass matrices of the super-element can be derived similarly to the longitudinal beams but with different dimensions in the matrices.

3.2.4 Concrete beam with flexural crack and debonding at the steel and concrete interface

This section is devoted to damage in reinforced concrete in the form of flexural cracks and debonding between the steel reinforcement and concrete interface. This is the most common type of damage with structural concrete, which leads to a reduction in the local stiffness and overall damping of the structure. Soh et al. (1999) have presented a damage model, which included the normal and tangential damage factors, to describe the concrete–steel interface mechanism. A reinforced concrete element is

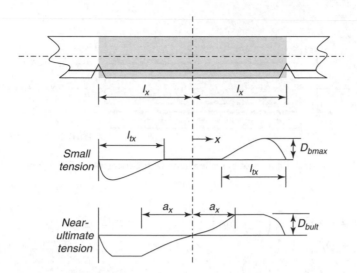

Figure 3.28 Cracked reinforced concrete beam

developed based on this damage model to simulate the bond deterioration in reinforced concrete structures (Soh et al., 2003). Spacone and Limkatanyu (2000) showed the importance of including the bond slip in the response of reinforced concrete members by displacement-based formulation. Later Limkatanyu and Spacone (2002) presented the general theoretical framework of the displacement-based, force-based and mixed formulations of reinforced concrete frame elements with bond slip in the reinforcing bars.

The following section gives the formulation of a two-node reinforced concrete frame element with damage parameters representing the flexural and debonding damages at the steel concrete interface for parameter identification in the condition assessment.

Bond stress distribution function

To evaluate the bond forces in the steel bar, the bond stress is assumed to vary parabolically along the steel bar as shown in Figure 3.28. The magnitude of the peak bond stress is determined by relating its value to the slip at that point satisfying the equilibrium of forces acting on the steel bar. Figure 3.28 shows the bond stress distribution and slip of the reinforced concrete between the two pairs of adjacent flexural cracks in the reinforced concrete beam. The embedment length of the reinforcing bar in the free body is $2l_x$. The transfer length, $l_{tx}(l_{tx} \leq l_x)$, is defined as the embedment length from the crack to the first point when the strains of the reinforcing bar and concrete are equal to each other. The bond stress distribution function can be defined as (Chan et al., 1992; Chan et al., 1993)

$$D_b(x) = 2.502 D_{b\max} \left(\frac{x}{l_{tx}}\right)^2 \sin\left(\frac{\pi x}{l_{tx}}\right) \quad \text{when } l_{tx} < l_x$$

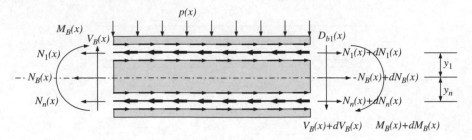

Figure 3.29 Reinforced concrete beam element with concrete-steel interface

$$
\begin{cases}
D_b(x) = 2.502 D_{bult} \left(\dfrac{x}{l_x}\right)^2 \sin\left(\dfrac{\pi x}{l_x}\right), & 0.729 l_x \le x \le l_x \\[2mm]
D_b(x) = D_{bult}, & a_x \le x \le 0.729 l_x \quad \text{when } l_{tx} = l_x \\[2mm]
D_b(x) = 1.328 D_{bult} \left(\dfrac{x}{a_x}\right)^2 \sin\left(\dfrac{\pi x}{1.373 a_x}\right), & 0 \le x \le a_x
\end{cases}
$$

$$(3.119)$$

where $D_{b\,max}$ and D_{bult} are the peak and ultimate bond stress; l_{tx} is the transfer length; and a_x is the distance from the starting point of the ultimate bond stress plateau to the mid-point between two cracks.

Equilibrium and compatibility

A reinforced concrete beam element with n bars and bonding interfaces is shown in Figure 3.29. Only bond stresses tangential to the bars are considered. $u_B(x) = \{u_B(x), v_B(x)\}^T$ and $u_S(x) = \{u_1(x), \ldots, u_n(x)\}^T$ are the section displacements, where $u_B(x), v_B(x), u_i(x)$ are the concrete beam axial and transverse displacements and the axial displacement of the ith bar. The section deformations are grouped in vectors $d_B(x) = \{\varepsilon_B(x), \kappa_B(x)\}^T$ and $d_S(x) = \{\varepsilon_1(x), \ldots \varepsilon_n(x)\}^T$, where $\varepsilon_B(x) = du_B(x)/dx$ is the concrete beam axial strain, $\kappa_B(x) = d^2 v_B(x)/dx^2$ is the concrete beam curvature and $\varepsilon_i(x) = du_i(x)/dx$ is the axial strain of the ith bar. They can be written in matrix form as

$$
\begin{cases}
\mathbf{d}_B(x) = \partial_B \mathbf{u}_B(x) \\
\mathbf{d}_S(x) = \partial_S \mathbf{u}_S(x)
\end{cases}
$$

$$(3.120)$$

$$
\text{where } \partial_B = \begin{bmatrix} d/dx & 0 \\ 0 & d^2/dx^2 \end{bmatrix}, \quad
\partial_S = \begin{bmatrix} d/dx & \cdots & 0 \\ \vdots & \cdots & \vdots \\ 0 & \cdots & d/dx \end{bmatrix}.
$$

The section forces corresponding to the section deformations $d_B(x)$ and $d_S(x)$ are $\mathbf{D}_B(x) = \{N_B(x), M_B(x)\}^T$ and $\mathbf{D}_S(x) = \{N_1(x), \ldots N_n(x)\}^T$, where $N_B(x), M_B(x), N_i(x)$ are the beam-sectional axial force and bending moment and the axial force of the *i*th bar, respectively. Based on the small-deformation assumption, the equilibrium conditions can be obtained as

$$
\begin{cases}
\dfrac{dN_B(x)}{dx} + \sum\limits_{i=1}^{n} D_{bi}(x) = 0 \\[2mm]
\dfrac{dN_i(x)}{dx} - D_{bi}(x) = 0, \quad (i = 1, 2, \ldots, n) \\[2mm]
\dfrac{d^2 M_B(x)}{dx^2} - p(x) - \sum\limits_{i=1}^{n} y_i D_{bi}(x) = 0
\end{cases}
\tag{3.121}
$$

where $p(x)$ is the transverse distribution load. Equation (3.121) represents the governing equilibrium equations of the reinforced concrete beam element with bond slip and it can be written in the following matrix form:

$$
\{\partial_B^T \mathbf{D}_B(x), \partial_S^T \mathbf{D}_S(x)\}^T - \partial_b^T \mathbf{D}_b(x) - \mathbf{p}(x) = 0
\tag{3.122}
$$

where $\partial_b = \begin{bmatrix} -1 & y_1\dfrac{d}{dx} & 1 & \cdots & 0 \\ \vdots & \vdots & \vdots & \vdots & \vdots \\ -1 & y_n\dfrac{d}{dx} & 0 & \cdots & 1 \end{bmatrix}$, $\quad \mathbf{p}(x) = \{0, p_y(x), 0, \ldots, 0\}^T$.

The bond slip is determined by the following compatibility relation between the displacement of the beam and steel bar as

$$
u_{bi}(x) = u_i(x) - u_B(x) + \frac{dv_B(x)}{dx} y_i, \quad i = 1, 2, \ldots, n
\tag{3.123}
$$

where $u_{bi}(x)$ is the bond slip between the beam and the *i*th bar, and y_i is the distance of the *i*th bar from the centroidal axis of the beam. Equation (3.123) can be rewritten in matrix form as

$$
\mathbf{d}_b(x) = \partial_b \mathbf{u}(x)
\tag{3.124}
$$

with $\mathbf{d}_b(x) = \{u_{bi}(x), i = 1, 2, \cdots, n\}^T$ and $\mathbf{u}(x) = \{\mathbf{u}_B(x), \mathbf{u}_S(x)\}^T$.

Force-deformation relations

With the assumption of linear elasticity, the force–deformation relationship is written as

$$
\begin{cases}
\mathbf{D}_B(x) = \mathbf{k}_B \mathbf{d}_B(x) \\
\mathbf{D}_S(x) = \mathbf{k}_S \mathbf{d}_S(x) \\
\mathbf{D}_b(x) = \mathbf{k}_b \mathbf{d}_b(x)
\end{cases}
\tag{3.125}
$$

where $\mathbf{D}_b(x) = \{D_{bi}, i = 1, 2, \ldots, n\}^T$ and $\mathbf{d}_b(x) = \{u_{bi}(x), i = 1, 2, \ldots, n\}^T \cdot \mathbf{k}_B, \mathbf{k}_S, \mathbf{k}_b$ are the concrete beam section stiffness, steel bar section stiffness and bond-interface stiffness matrices, respectively, which can be expressed as:

$$
k_B = \begin{bmatrix} E_B A_B & 0 & \vdots & 0 & \cdots & 0 \\ 0 & E_B I_B & \vdots & 0 & \cdots & 0 \\ \cdots & \cdots & \cdots & \cdots & \cdots & \cdots \\ 0 & 0 & \vdots & E_{s1} A_{s1} & \cdots & 0 \\ \vdots & \vdots & \vdots & \vdots & \cdots & \vdots \\ 0 & 0 & \vdots & 0 & \cdots & E_{sn} A_{sn} \end{bmatrix}, \quad k_S = \begin{bmatrix} E_{s1} A_{s1} & \cdots & 0 \\ \vdots & \cdots & \vdots \\ 0 & \cdots & E_{sn} A_{sn} \end{bmatrix},
$$

$$
k_b = \begin{bmatrix} E_{b1} P_{b1} & \cdots & 0 \\ \vdots & \cdots & \vdots \\ 0 & \cdots & E_{bn} P_{bn} \end{bmatrix} \tag{3.126}
$$

where E_B and A_B are the elastic modulus and the area of the concrete beam section; E_{si} and A_{si} are the elastic modulus and area of the ith reinforced bar; and E_{bi} and P_{bi} are the equivalent bond stiffness and the perimeter of the ith bar, respectively.

Elemental damage indicators

The bond-interface stress distributions in an element are denoted by $\{D_{bi}(x), i = 1, 2, \ldots, n\}$, defined above, and the bond force for one reinforced bar in the element can be written as

$$
\Delta N_i = \int_0^l D_{bi}(x) dx, \quad (i = 1, 2, \ldots, n) \tag{3.127}
$$

where l is the length of the element and ΔN_i is the bond force of the ith bar in the element. The relationship between the bond force and slip of the concrete–steel bar interface can be expressed as (Soh et al., 2003)

$$
\Delta N_i = K_{bi} \Delta u_{bi} = E_{bi} P_{bi} l \, \Delta u_{bi}, \quad (i = 1, 2, \ldots, n) \tag{3.128}
$$

where K_{bi} is the equivalent bonding stiffness of the ith bar and Δu_{bi} is the slip along the ith bar in the element. The equivalent bonding stiffness can be obtained from Equations (3.127) and (3.128) as

$$
K_{bi} = E_{bi} P_{bi} l = \frac{\int_0^l D_{bi}(x) \, dx}{\Delta u_{bi}}, \quad (i = 1, 2, \ldots, n) \tag{3.129}
$$

The bond damage index, α_{bei}, of the ith bar of the element, which includes the effect of the bond degradation, can be defined as

$$
\alpha_{bei} = 1 - \frac{K_{bi}}{\overline{K}_{bi}} = 1 - \frac{E_{bi}}{\overline{E}_{bi}}, \quad (i = 1, 2, \ldots, n) \tag{3.130}
$$

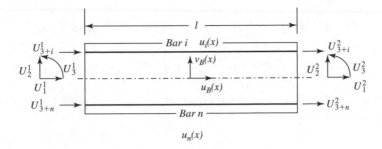

Figure 3.30 Two-node displacement-based reinforced concrete beam element

where $\overline{K}_{bi}$ is the equivalent bonding stiffness of the ith bar without damage or slippage and $\overline{E}_{bi}$ is the equivalent elastic modulus of the bond-interface between the ith bar and concrete without damage or slippage. When $\alpha_{bei} = 0$, there is no damage in this element.

Similarly, the damage index of the concrete beam element α_{Be} can be defined as

$$\alpha_{Be} = 1 - \frac{E_B}{\overline{E}_B} \tag{3.131}$$

where $\overline{E}_B$ is the equivalent elastic modulus of the concrete beam element without damage. When $\alpha_{Be} = 0$, there is no damage in the concrete beam.

Damaged reinforced concrete element

The two-node reinforced concrete beam element model is shown in Figure 3.30. The reinforced concrete beam element consists of the following components: a two-node concrete beam and n two-node bars that can have slip with respect to the concrete beam. $U = \{U_1^1, U_2^1, U_3^1, \ldots, U_{3+n}^1, U_1^2, U_2^2, U_3^2, \ldots, U_{3+n}^2\}^T$ is the vector of the elemental nodal displacements. The element displacements are written as functions of the nodal displacements, U, through the displacement shape functions, $\phi(x)$

$$u(x) = \begin{Bmatrix} u_B(x) \\ u_S(x) \end{Bmatrix} = \phi(x)U = \begin{Bmatrix} \phi_B(x) \\ \phi_u(x) \end{Bmatrix} U \tag{3.132}$$

where

$$\phi_B(x)$$

$$= \begin{bmatrix} 1 - \dfrac{x}{l} & 0 & 0 & 0 \cdots 0 & \dfrac{x}{l} & 0 & 0 & 0 \cdots 0 \\ 0 & 1 - \dfrac{3x^2}{l^2} + \dfrac{2x^3}{l^3} & x - \dfrac{2x^2}{l} + \dfrac{x^3}{l^2} & 0 \cdots 0 & 0 & 1 - \dfrac{3x^2}{l^2} + \dfrac{2x^3}{l^3} & x - \dfrac{2x^2}{l} + \dfrac{x^3}{l^2} & 0 \cdots 0 \end{bmatrix}$$

For a linear slip distribution assumption, the shape function of a single bar with bond slip can be written as

$$\phi_{ui}(x) = \{1 \quad -x/l \quad x/l\}, \quad (i = 1, 2, \ldots, n) \tag{3.133}$$

Then for all the bars,

$$\phi_u(x) = \begin{bmatrix} 0 & 0 & 0 & 1-x/l & \cdots & 0 & 0 & 0 & 0 & x/l & \cdots & 0 \\ \vdots & \cdots & \vdots & \vdots & \cdots & \vdots & \vdots & \cdots & \vdots & \vdots & \cdots & \vdots \\ 0 & 0 & 0 & 0 & \cdots & 1-x/l & 0 & 0 & 0 & 0 & \cdots & x/l \end{bmatrix}.$$

The displacement-based finite element formulation can be obtained by integrating over the length of the element as

$$\int_l \delta \mathbf{u}^T(x)(\{\partial_B^T \mathbf{D}_B(x), \partial_S^T \mathbf{D}_S(x)\}^T - \partial_b^T \mathbf{D}_b(x) - \mathbf{p}(x)) = 0 \tag{3.134}$$

Substituting Equations (3.125) and (3.132) into (1.134), the element stiffness matrix can be obtained as

$$K_e = K_{Be} + K_{Se} + K_{be} \tag{3.135}$$

where

$$K_{Be} = \int_0^l \partial_B \phi_B(x)^T k_B \partial_B \phi_B(x) dx, \quad K_{Se} = \int_0^l \partial_S \phi_u(x)^T k_S \partial_S \phi_u(x) dx$$

$$K_{be} = \int_0^l \partial_b \phi(x)^T k_b \partial_b \phi(x) dx.$$

$\partial_B \phi_B(x)$, $\partial_b \phi(x)$ are calculated from $\phi(x)$ by the differential operator (Limkatanyu and Spacone, 2002). Matrices K_{Be} and K_{be} can be written as

$$K_{Be} = K_{Ce} + K_{Se}, \quad K_{Ce} = \begin{bmatrix} K_{Be1} & 0 & K_{Be2} & 0 \\ 0 & 0 & 0 & 0 \\ K_{Be5} & 0 & K_{Be6} & 0 \\ 0 & 0 & 0 & 0 \end{bmatrix}, \quad K_{Se} = \begin{bmatrix} 0 & 0 & 0 & 0 \\ 0 & K_{Be3} & 0 & -K_{Be3} \\ 0 & 0 & 0 & 0 \\ 0 & -K_{Be3} & 0 & K_{Be3} \end{bmatrix}$$

$$K_{Be1} = \begin{bmatrix} \dfrac{E_B A_B}{l} & 0 & 0 \\ 0 & \dfrac{12 E_B I_B}{l^3} & \dfrac{6 E_B I_B}{l^2} \\ 0 & \dfrac{6 E_B I_B}{l^2} & \dfrac{4 E_B I_B}{l} \end{bmatrix}, \quad K_{Be2} = K_{Be5}^T = \begin{bmatrix} -\dfrac{E_B A_B}{l} & 0 & 0 \\ 0 & -\dfrac{12 E_B I_B}{l^3} & \dfrac{6 E_B I_B}{l^2} \\ 0 & -\dfrac{6 E_B I_B}{l^2} & \dfrac{2 E_B I_B}{l} \end{bmatrix},$$

$$K_{Be6} = \begin{bmatrix} \dfrac{E_B A_B}{l} & 0 & 0 \\ 0 & \dfrac{12 E_B I_B}{l^3} & -\dfrac{6 E_B I_B}{l^2} \\ 0 & -\dfrac{6 E_B I_B}{l^2} & \dfrac{4 E_B I_B}{l} \end{bmatrix}, \quad K_{Be3} = diag\left\{\dfrac{E_{si} A_{si}}{l}, i = 1, 2, \ldots, n\right\}$$

$$K_{be} = \begin{bmatrix} K_{be1} & K_{be2} & K_{be3} & K_{be4} \\ K_{be2}^T & K_{be5} & K_{be6} & K_{be7} \\ K_{be3}^T & K_{be6}^T & K_{be8} & K_{be9} \\ K_{be4}^T & K_{be7}^T & K_{be9}^T & K_{be10} \end{bmatrix},$$

$$K_{be1} = \begin{bmatrix} \dfrac{l}{3}\sum_{i=1}^{n}E_{bi}P_{bi} & \dfrac{1}{2}\sum_{i=1}^{n}E_{bi}P_{bi}y_i & -\dfrac{l}{12}\sum_{i=1}^{n}E_{bi}P_{bi}y_i \\ \dfrac{1}{2}\sum_{i=1}^{n}E_{bi}P_{bi}y_i & \dfrac{6}{5l}\sum_{i=1}^{n}E_{bi}P_{bi}y_i^2 & -\dfrac{2}{5}\sum_{i=1}^{n}E_{bi}P_{bi}y_i^2 \\ -\dfrac{l}{12}\sum_{i=1}^{n}E_{bi}P_{bi}y_i & -\dfrac{2}{5}\sum_{i=1}^{n}E_{bi}P_{bi}y_i^2 & \dfrac{2l}{15}\sum_{i=1}^{n}E_{bi}P_{bi}y_i^2 \end{bmatrix}$$

$$K_{be2} = \begin{bmatrix} -l/3 & \cdots & -l/3 & \cdots & -l/3 \\ -y_1/2 & \cdots & -y_i/2 & \cdots & -y_n/2 \\ ly_1/12 & \cdots & ly_i/12 & \cdots & ly_n/12 \end{bmatrix} k_b,$$

$$K_{be4} = \begin{bmatrix} -l/6 & \cdots & -l/6 & \cdots & -l/6 \\ -y_1/2 & \cdots & -y_i/2 & \cdots & -y_n/2 \\ -ly_1/12 & \cdots & -ly_i/12 & \cdots & -ly_n/12 \end{bmatrix} k_b$$

$$K_{be3} = \begin{bmatrix} \dfrac{l}{6}\sum_{i=1}^{n}E_{bi}P_{bi} & -\dfrac{1}{2}\sum_{i=1}^{n}E_{bi}P_{bi}y_i & \dfrac{l}{12}\sum_{i=1}^{n}E_{bi}P_{bi}y_i \\ \dfrac{1}{2}\sum_{i=1}^{n}E_{bi}P_{bi}y_i & -\dfrac{6}{5l}\sum_{i=1}^{n}E_{bi}P_{bi}y_i^2 & \dfrac{1}{10}\sum_{i=1}^{n}E_{bi}P_{bi}y_i^2 \\ \dfrac{l}{12}\sum_{i=1}^{n}E_{bi}P_{bi}y_i & -\dfrac{1}{10}\sum_{i=1}^{n}E_{bi}P_{bi}y_i^2 & \dfrac{17l}{12}\sum_{i=1}^{n}E_{bi}P_{bi}y_i^2 \end{bmatrix}$$

$$K_{be6}^T = \begin{bmatrix} l/6 & \cdots & l/6 & \cdots & l/6 \\ y_1/2 & \cdots & y_i/2 & \cdots & y_n/2 \\ -ly_1/12 & \cdots & -ly_i/12 & \cdots & -ly_n/12 \end{bmatrix} k_b,$$

$$K_{be9} = \begin{bmatrix} -l/3 & \cdots & -l/3 & \cdots & -l/3 \\ y_1/2 & \cdots & y_i/2 & \cdots & y_n/2 \\ ly_1/12 & \cdots & ly_i/12 & \cdots & ly_n/12 \end{bmatrix} k_b$$

$$K_{be8} = \begin{bmatrix} \dfrac{l}{3}\sum_{i=1}^{n}E_{bi}P_{bi} & -\dfrac{1}{2}\sum_{i=1}^{n}E_{bi}P_{bi}y_i & -\dfrac{l}{12}\sum_{i=1}^{n}E_{bi}P_{bi}y_i \\ -\dfrac{1}{2}\sum_{i=1}^{n}E_{bi}P_{bi}y_i & \dfrac{6}{5l}\sum_{i=1}^{n}E_{bi}P_{bi}y_i^2 & -\dfrac{1}{10}\sum_{i=1}^{n}E_{bi}P_{bi}y_i^2 \\ -\dfrac{l}{12}\sum_{i=1}^{n}E_{bi}P_{bi}y_i & -\dfrac{1}{10}\sum_{i=1}^{n}E_{bi}P_{bi}y_i^2 & \dfrac{2l}{15}\sum_{i=1}^{n}E_{bi}P_{bi}y_i^2 \end{bmatrix}$$

$$K_{be5} = l/3k_b, \quad K_{be7} = l/6k_b, \quad K_{be10} = l/3k_b$$

The steel bars are assumed undamaged but with bonding damage with the surrounding concrete. Substituting Equations (3.130) and (3.131) into (3.135), the stiffness matrix of the damaged beam can be written as

$$K_e = K_{Be} + K_{be} + K_{Se} = (1 - \alpha_{Be})\overline{K}_{Be} + (1 - \alpha_{be})\overline{K}_{be} + \overline{K}_{Se} \tag{3.136}$$

where α_{Be} is the damage index of the concrete beam; α_{be} is the index for the bond failure in the damage reinforced concrete beam element; α_{Be} is taken as a scalar for an equivalent damage in the concrete; α_{be} is taken as a scalar here, but it can be in matrix form if the conditions are different in the steel bars; and $\overline{K}_{Be}, \overline{K}_{be}$ and $\overline{K}_{se}$ are element stiffness matrices for the undamaged concrete beam, concrete–steel interface and the steel bars, respectively. The stiffness matrix of the damaged beam structure is the assemblage of all the elemental stiffness matrices K_{ei} as

$$
K = \sum_{i=1}^{N} T_i^T (K_{Bei} + K_{bei} + K_{Sei}) T_i
$$

$$
= \sum_{i=1}^{N} (1 - \alpha_{Bei}) T_i^T \overline{K}_{Bei} T_i + \sum_{i=1}^{N} (1 - \alpha_{bei}) T_i^T \overline{K}_{bei} T_i + \sum_{i=1}^{N} T_i^T \overline{K}_{Sei} T_i \qquad (3.137)
$$

where T_i is the transformation matrix of element nodal displacement that facilitates automatic assembling of global stiffness matrix from the constituent element stiffness matrix; and α_{bei} and α_{Bei} are the ith elemental damage indices due to bonding loss and damage in the concrete of the beam element, respectively.

3.2.5 Beam with unbonded pre-stress tendon

This section gives the model of the pre-stressing effect in a pre-stressed concrete bridge deck modelled as a beam. The more realistic modelling of a pre-stressed concrete box-section bridge deck is discussed in the next section.

Equation of motion

The bridge deck is modelled as a two-span, simply supported, pre-stressed, uniform Timoshenko beam subjected to an external excitation force, $P(t)$, acting at a distance x_p from the left support, as shown in Figure 3.31. The coupled equation for the total deflection, y, and rotation, ψ, of the cross-section under a compressive axial force T_p can be written as (note that the compression is positive and the tension is negative),

$$
\rho A \frac{\partial^2 y(x, t)}{\partial t^2} + c \frac{\partial y(x, t)}{\partial t} - kAG \left(\frac{\partial^2 y}{\partial x^2} - \frac{\partial \psi}{\partial x} \right) + T_p \frac{\partial^2 y(x, t)}{\partial x^2} = P(t) \delta(x - x_p)
$$

$$
\rho I \frac{\partial^2 \psi}{\partial x^2} - kAG \left(\frac{\partial y}{\partial x} - \psi \right) - EI \frac{\partial^2 \psi}{\partial x^2} = 0 \qquad (3.138)
$$

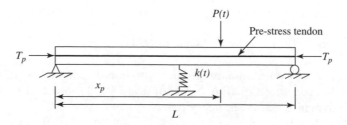

Figure 3.31 The two-span pre-stressed bridge

where y is the total deflection due to bending and shear; ψ is the slope of deflection due to bending; ρ is the mass density of the beam; A is the cross-sectional area; E is the elastic modulus of the material; G is the shear modulus; k is the shear coefficient of the cross-section; c is the viscous damping coefficient of the beam; I is the moment of inertia of the beam cross-section; and $\delta(x)$ is the Dirac delta function.

Modal responses

The kinetic energy, T; the strain energy, U; the work done due to the pre-stress force, W_{Tp}; the work done due to the viscous damping in the beam, W_c; and the work done due to the external force, can be expressed for the beam as

$$T = \frac{1}{2} \int_0^L \left[\rho A \left(\frac{\partial y(x,t)}{\partial t} \right)^2 + \rho I \left(\frac{\partial \psi(x,t)}{\partial t} \right)^2 \right] dx$$

$$U = \frac{1}{2} \int_0^L \left[EI(x) \left(\frac{\partial^2 y(x,t)}{\partial x^2} \right)^2 + kGA \left(\frac{\partial y}{\partial x} - \psi \right)^2 \right] dx$$

$$W_{Tp} = \frac{1}{2} \int_0^L T \frac{\partial^2 y(x,t)}{\partial x^2} dx \qquad (3.139)$$

$$W_c = - \int_0^L y(x,t) c \frac{\partial y(x,t)}{\partial t} dx$$

$$W = \int_0^L P(t) \delta(x - x_p) y(x,t) dx$$

Expressing the vibration responses of the beam, $y(x,t)$, $\psi(x,t)$, in modal coordinates gives

$$y(x,t) = \sum_{i=1}^n Y_i(x) q_i(t) \qquad (3.140)$$

$$\psi(x,t) = \sum_{i=1}^n \phi_i(x) q_i(t) \qquad (3.141)$$

where $Y_i(x)$ and $\phi_i(x)$ are the assumed vibration modes that satisfy the boundary conditions and $q_i(t)$ are the generalized coordinates.

Substituting Equations (3.140) and (3.141) into Equation (3.139), gives

$$T = \frac{1}{2} \int_0^L \left[\rho A \left(\sum_{i=1}^n Y_i(x) \dot{q}_i(t) \sum_{j=1}^n Y_j(x) \dot{q}_j(t) \right) + \rho I \left(\sum_{i=1}^n \phi_i(x) \dot{q}_i(t) \sum_{j=1}^n \phi_j(x) \dot{q}_j(t) \right) \right] dx$$

$$= \frac{1}{2} \sum_{i=1}^n \sum_{j=1}^n \dot{q}_i(t) m_{ij} \dot{q}_j(t)$$

$$U = \frac{1}{2} \int_0^L \left[EI \left(\sum_{i=1}^n q_i(t)\phi_i'(x) \sum_{j=1}^n q_j(t)\phi_j'(x) + kGA \left(\sum_{i=1}^n q_i(t)Y_i'(x) - \sum_{i=1}^n q_i(t)\phi_i(x) \right) \right. \right.$$

$$\left. \left. \times \left(\sum_{j=1}^n q_j(t)Y_j'(x) - \sum_{j=1}^n q_j(t)\phi_j(x) \right) \right) \right] dx$$

$$= \frac{1}{2} \sum_{i=1}^n \sum_{j=1}^n q_i(t)k_{ij}q_i(t) \tag{3.142}$$

$$W_{Tp} = \frac{1}{2} \int_0^L T \sum_{i=1}^n q_i(t)Y_i'(x) \sum_{j=1}^n q_j(t)Y_j'(x)dx = \sum_{i=1}^n \sum_{j=1}^n q_i(t)k_{ij}'q_j(t)$$

$$W_c = - \int_0^L c \sum_{i=1}^n Y_i(x)q_i(t) \sum_{j=1}^n Y_j(x)\dot{q}_j(t)dx = - \sum_{i=1}^n \sum_{j=1}^n q_i(t)c_{ij}\dot{q}_j(t)$$

$$W = \int_0^L P(t)\delta(x - x_p) \sum_{i=1}^n Y_i(x)q_i(t)dx = \sum_{i=1}^n P(t)Y_i(x_p)q_i(t) = \sum_{i=1}^n f_i(t)q_i(t)$$

$$m_{ij} = \int_0^L [\rho A Y_i(x)Y_j(x) + \rho I \phi_i(x)\phi_j(x)]dx$$

where

$$k_{ij} = \int_0^L [EI\phi_i'(x)\phi_j'(x) + kGA(Y_i'(x) - \phi_i(x))(Y_j'(x) - \phi_j(x))dx,$$

$$k_{ij}' = \int_0^L T_p Y_i'(x)Y_j'(x)dx,$$

$$c_{ij} = \int_0^L c Y_i(x)Y_j(x)dx, \quad f_i(t) = P(t)Y_i(x_p).$$

where $\dot{q}_i(t)$ and $\phi_i'(x)$ denote the first derivative of $q_i(t)$ and $\phi_i(x)$ with respect to time t and x, respectively; m_{ij} is the generalized mass; k_{ij} is the generalized stiffness; and $f_i(t)$ is the generalized force. The Lagrange equation can be written as

$$\frac{d}{dt}\left(\frac{\partial T}{\partial \dot{q}}\right) - \frac{\partial T}{\partial q} + \frac{\partial U}{\partial q} - \frac{\partial W_{Tp}}{\partial q} - \frac{\partial W_c}{\partial q} = \frac{\partial W}{\partial q} \tag{3.143}$$

Substituting Equation (3.142) into Equation (3.143), gives

$$\sum_{j=1}^n m_{ij}\ddot{q}_j(t) + \sum_{j=1}^n c_{ij}\dot{q}_j(t) + \sum_{j=1}^n (k_{ij} - k_{ij}')q_j(t) = f_i(t) \quad i = 1, 2, \ldots, n \tag{3.144}$$

Writing Equation (3.144) in matrix form as

$$M\ddot{q}(t) + C\dot{q}(t) + (K - K')q(t) = F(t) \tag{3.145}$$

where

$$M = \{m_{ij}, i = 1, 2, \ldots, n; \ j = 1, 2, \ldots, n\}, \quad C = \{c_{ij}, i = 1, 2, \ldots, n; \ j = 1, 2, \ldots, n\},$$

$$K = \{k_{ij}, i = 1, 2, \ldots n; \ j = 1, 2, \ldots, n\}, \quad K' = \{k'_{ij}, i = 1, 2, \ldots, n; \ j = 1, 2, \ldots, n\},$$

$$q(t) = \{q_1(t), q_2(t), \ldots, q_n(t)\}^T, \qquad\qquad F(t) = \{f_1(t), f_2(t), \ldots, f_n(t)\}^T$$

The assumed mode shapes

The general form of the vibration mode for a uniform Timoshenko beam can be written as

$$Y(x) = A_1 \cos(\alpha x) + A_2 \sin(\alpha x) + A_3 \sinh(\beta x) + A_4 \cosh(\beta x) \tag{3.146}$$

$$\phi(x) = B_1 \cos(\alpha x) + B_2 \sin(\alpha x) + B_3 \sinh(\beta x) + B_4 \cosh(\beta x) \tag{3.147}$$

where A_1 to A_4 and B_1 to B_4 are arbitrary constants, and α and β are frequency parameters.

The vibration modes of a simply supported Timoshenko beam are obtained as (Abramovich, 1992),

$$Y_i(x) = A_i \sin\left(\frac{i\pi}{L}x\right), \quad \phi_i(x) = \cos\left(\frac{i\pi}{L}x\right) \tag{3.148}$$

where

$$A_i = \frac{i\pi L}{[(i\pi)^2 - p^2 b^2]\left(1 - \frac{T_p}{a}\right)}, \quad a = kGA, \quad b^2 = \frac{EI}{aL^2}, \quad p^2 = \frac{(i\pi)^4 - (i\pi)^2 \bar{k}^2}{1 + (i\pi)^2 D},$$

$$\bar{k}^2 = \frac{T_p L^2}{EI\left(1 - \frac{T_p}{a}\right)}, \quad D = R^2\left(1 - \frac{T_p}{a}\right) + b^2, \quad R^2 = \frac{I}{AL^2}$$

Then Equation (3.145) can be expanded into the following form

$$
\begin{bmatrix} m_{11} & 0 & \cdots & 0 \\ 0 & m_{22} & \cdots & 0 \\ & & \ddots & \\ 0 & 0 & \cdots & m_{nn} \end{bmatrix}
\begin{Bmatrix} \ddot{q}_1(t) \\ \ddot{q}_2(t) \\ \vdots \\ \ddot{q}_n(t) \end{Bmatrix}
+
\begin{bmatrix} 2m_1\xi_1\omega_1 & 0 & \cdots & 0 \\ 0 & 2m_2\xi_2\omega_2 & \cdots & 0 \\ & & \ddots & \\ 0 & 0 & \cdots & 2m_n\xi_n\omega_n \end{bmatrix}
\begin{Bmatrix} \dot{q}_1(t) \\ \dot{q}_2(t) \\ \vdots \\ \dot{q}_n(t) \end{Bmatrix}
$$

$$
+
\begin{bmatrix} k_{11} - k'_{11} & 0 & \cdots & 0 \\ 0 & k_{22} - k'_{22} & \cdots & 0 \\ & & \ddots & \\ 0 & 0 & \cdots & k_{nn} - k'_{nn} \end{bmatrix}
\begin{Bmatrix} q_1(t) \\ q_2(t) \\ \vdots \\ q_n(t) \end{Bmatrix}
=
\begin{Bmatrix} f_1(t) \\ f_2(t) \\ \vdots \\ f_n(t) \end{Bmatrix}
\tag{3.149}
$$

The modal response of the beam is computed in the time domain numerically using Newmark's integration method (Newmark, 1959).

Alternate finite element formulation

The pre-stress effect can also be formulated in terms of the general finite element method and the state-space approach. The equation of motion of the pre-stressed beam modelled as an Euler–Bernoulli beam with n DOFs can be written as

$$[M]\{\ddot{x}\} + [C]\{\dot{x}\} + [\overline{K}]\{x\} = [B]\{F\} \tag{3.150}$$

where x is the displacement vector; $\dot{x}$ and $\ddot{x}$ are the first and second derivatives of x with respect to time t; M is the mass matrix; and C is the damping matrix. Rayleigh damping is used, and the C matrix is represented by a linear combination of the system mass and stiffness matrices, i.e.

$$C = a_1[M] + a_2[\overline{K}]$$

where a_1 and a_2 are the two Rayleigh damping coefficients; $\{F\}$ is a vector of the input excitation forces and $[B]$ maps these forces to the associated DOFs of the structure. $\overline{K} = K - K_g$ is the global stiffness matrix of the pre-stressed beam, where K is the global stiffness matrix without pre-stress force and K_g is the global geometrical stiffness matrix expressed as $[K_g] = \sum_{i=1}^{N} [k_g]_e^i$, where N is the total number of elements. The geometrical stiffness matrix of each element can be written as:

$$[k_g]^i = \frac{T^i}{30l} \begin{bmatrix} 30 & 0 & 0 & -30 & 0 & 0 \\ 0 & 36 & 3l & 0 & -36 & 3l \\ 0 & 3l & 4l^2 & 0 & -3l & -l^2 \\ -30 & 0 & 0 & 30 & 0 & 0 \\ 0 & -36 & -3l & 0 & 36 & -3l \\ 0 & 3l & -l^2 & 0 & -3l & 4l^2 \end{bmatrix}, \quad (i = 1, 2, \ldots, N)$$

where T is the axial pre-stress force and l is the length of the element. Writing Equation (3.150) in the state-space formulation,

$$\dot{X} = K^* X + \overline{F} \tag{3.151}$$

$$X = \begin{bmatrix} x \\ \dot{x} \end{bmatrix}_{2n \times 1}, \quad K^* = \begin{bmatrix} 0 & I \\ -M^{-1}\overline{K} & -M^{-1}C \end{bmatrix}_{2n \times 2n}, \quad \overline{F} = \begin{bmatrix} 0 \\ M^{-1}[B]F \end{bmatrix}_{2n \times 1}.$$

where X represents a vector of state variables with a length $2n$ containing the displacements and velocities of the nodes. These differential equations are then converted to discrete equations using exponential matrix representation,

$$X_{k+1} = AX_k + \overline{D}\overline{F}_k \tag{3.152}$$

$$A = e^{K^*h}, \quad \overline{D} = K^{*-1}(A - I)$$

where A is the exponential matrix, $(k + 1)$ denotes the value at the $(k + 1)$th time step of computation. The time step, h, represents the time increment between the variable states X_k and X_{k+1} in the computation. I is a unit matrix. The dynamic response of the system can be obtained from Equation (3.152) with zero initial conditions. Once the displacement and velocity responses are obtained, the acceleration response can be obtained by directly differentiating the velocity response.

Experimental verification

The above model is illustrated with a simply supported, pre-stressed concrete beam in the laboratory. The experimental setup is shown diagrammatically in Figure 3.32. The beam is 4 m long, with a 150 mm × 200 mm uniform cross-section and a 3.8 m clear span. A seven-wire straight strand is placed in a 57 mm diameter duct located at the centroid of the beam cross-section throughout the length of the beam. The duct remains ungrouted. The elastic modulus of the concrete and the steel strand are 31.5×10^9 N/m^2 and 200×10^9 N/m^2, respectively, and the mass density of the concrete and the steel are 2.398×10^3 kg/m^3 and 7.0×10^3 kg/m^3, respectively. The yield strength of the strand is 192 kN. The beam is instrumented with seven equally spaced accelerometers that measured the vertical acceleration responses of the beam.

One load cell is used to measure the true magnitude of the pre-stress force applied at the anchorage of the concrete beam. The pre-stressing strand is tensioned up to 100 kN and the tension force is transferred to the anchorage. A 66.7 kN force is checked to be remaining in the strand at the anchorage after the pre-stress loss. The first three natural frequencies of the intact beam and the pre-stressed beam are listed in Table 3.1, and they are found to increase after pre-stressing. This seems to contradict the prediction from the theoretical formula (Kim et al., 2004)

$$\omega_n^2 = \left(\frac{n\pi}{L}\right)^4 \frac{E_b I_b}{\rho_c A_c} - \left(\frac{n\pi}{L}\right)^2 \frac{T}{\rho_c A_c} \tag{3.153}$$

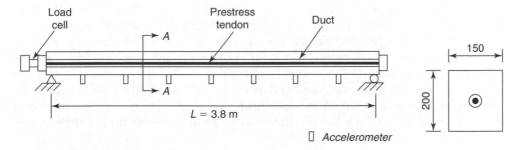

Figure 3.32 Test setup for the pre-stressed beam

Table 3.1 Modal frequencies (Hz) of the non-pre-stressed and pre-stressed beam

	Non-pre-stressed		Pre-stressed	
	FEM	Experimental	FEM	Experimental
Mode 1	23.10	23.21	23.47	23.31
Mode 2	85.62	86.17	86.54	87.98
Mode 3	184.53	183.29	187.12	185.93

where $E_b I_b$ is equivalent to the flexural rigidity of the beam section; $\rho_c A_c$ is the mass of the beam per unit length; and T is the magnitude of the pre-stress force. Equation (3.153) shows that an increase in the axial compressive force reduces the modal frequency and vice versa. But on further checking of the experimental system, the equivalent flexural rigidity of the beam without pre-stress force is found to be $3.13 \times 10^3 \, \text{kN/m}^2$, and it increases to $3.20 \times 10^3 \, \text{kN/m}^2$ after pre-stressing. Also, the equivalent mass per unit length of the beam is increased by 1.49% after pre-stressing. This is due to the presence of the additional equivalent flexural rigidity and the mass of the pre-stressing strand. Thus, the physical presence of the pre-stressing tendon has a dual effect on the natural frequency of the beam. The pre-stressing tendon itself increases the flexural rigidity and hence the natural frequency of the beam, but the self-weight and the compressive axial force it carries reduce the frequency of the beam. However, the stiffening effect from the increase in the equivalent flexural rigidity is greater than the softening effect due to the compressive axial force and the additional inertia effect due to its self-weight. And this results in a net increase in the natural frequency. The other effects, such as an increase in the dynamic modulus of concrete, are considered small and are therefore not discussed.

Identification of pre-stress force

The initial finite element model of the beam before pre-stressing consists of 16 two-dimensional Euler–Bernoulli beam elements with three DOFs at each node. An impulsive force is applied with the impact hammer at $1/4L$ from the left support of the beam. The sampling rate is 2000 Hz. Time histories of both the excitation force and the acceleration are recorded, and data obtained from the third and fourth accelerometers are used in the model updating.

The support stiffnesses are updated using the response sensitivity approach (Lu and Law, 2007a) with one second of measured data from the two accelerometers, and the left and right support stiffnesses are updated to $8.9 \times 10^7 \, \text{N/m}$ and $9.4 \times 10^7 \, \text{N/m}$, respectively. Rayleigh damping is adopted in calculating the structural response.

After the beam is pre-stressed, the pre-stress force is identified using the response sensitivity approach described in Section 7.2 with data from 0.2 second to 1.0 second after the hammer impact. The first 0.2-second data is skipped because of the many high frequency components in the response caused by the impulsive force created by the hammer. An orthogonal polynomial function is used to remove the measurement noise (Zhu and Law, 2001). The pre-stress force is identified using the penalty function method (Mottershead and Friswell, 1993).

The identified pre-stress force and error are shown in Table 3.2, and they are very close to the true force, with a maximum error of 10.9% close to one end of the beam. For convergence of the results, 56 iterations are required, and the corresponding optimal regularization parameter is 1.03×10^{-8}. Figure 3.33 shows the curve of convergence, which indicates a clear converging characteristic, and Figure 3.34 shows the reconstructed acceleration responses and the corresponding measured ones. It is noted that these two sets of time histories match each other very well.

3.2.6 Pre-stressed concrete box-girder with bonded tendon

The bridge deck may lose some of its pre-stress force due to creep resulting from a long period of service, under design or overloaded vehicles. A large reduction of the

Table 3.2 The identified pre-stress force in experiment

Element No.	Pre-stress (kN)/Error (%)	Element No.	Pre-stress (kN)/Error (%)
1	66.6/0.1	9	63.3/5.1
2	69.4/−4.0	10	65.8/1.3
3	63.6/4.7	11	68.9/−3.3
4	65.8/1.3	12	63.0/5.5
5	72.3/−8.4	13	68.0/−2.0
6	61.1/8.4	14	68.6/−2.9
7	67.6/−1.3	15	59.4/10.9
8	70.1/−5.1	16	71.7/−7.5

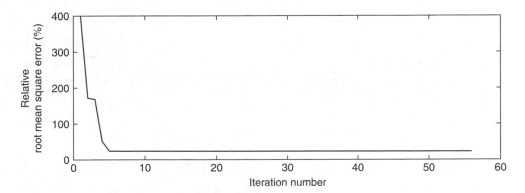

Figure 3.33 Curve of convergence

pre-stress force from the design value could lead to serviceability and safety problems. Therefore, an assessment of the magnitude of the pre-stress force in the bridge deck is important for its load-carrying capacity assessment and for its maintenance programme. However, the existing pre-stress force cannot be estimated directly unless the bridge deck was instrumented when it was constructed. Several researchers (Abraham et al., 1995) tried to predict the loss of pre-stress force based on a damage index derived from the derivatives of mode shapes without success. Others (Miyamoto et al., 2000) studied the behaviour of a beam with unbonded tendons, and a formula was proposed for the prediction of the modal frequency for a given pre-stress force with laboratory and field test verifications. Saiidi et al. (1994) reported a study with modal frequency due to the pre-stress force with laboratory test results. It was shown that the sensitivity of the modal frequency decreases with higher vibration modes, and the pre-stress force affects the first few lower modes more significantly than the higher ones. Consequently, the pre-stress force is difficult to identify from the modal frequencies. Abraham et al. (1995) also reported that the mode shapes remain almost identical with different pre-stress forces in the beam, so it would also be difficult to identify the force from the measured mode shapes.

Stiffness matrix of a shell element

The box-girder bridge deck may be modelled as a combination of different types of shell elements supported on bearings or on a solid foundation. The Mindlin–Reissner plate

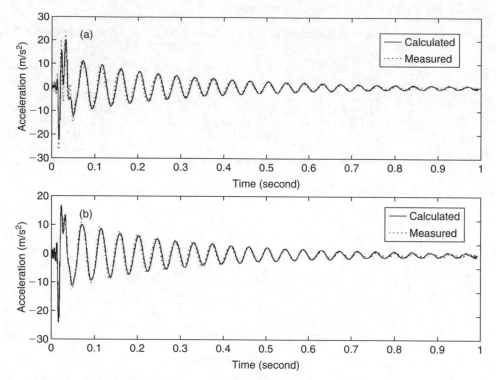

Figure 3.34 Time histories of measured and reconstructed acceleration response (a) from 3rd accelerometer; (b) from 4th accelerometer

discussed in Section 3.2.2.2, including the effect of the transverse shear deformation, is adopted with the shear deformable plate element and the membrane element combined.

Applying the small-deformation theory, the membrane stretching and bending effects are decoupled in a shell element. A plate bending element has a transverse deflection and two rotations at a node while a plane stress element has two in-plane displacements at a node. All together, there are five DOFs for a node in a shell element, i.e. three displacements and two rotations. Further, consider the assembled structure with the shell elements at different orientation, a drilling DOF about the local z-axis is included at a node leading to a total of six DOFs at a node. The shell element formulation is based on a superposition of the membrane and bending actions, the stiffness matrix of which is given as

$$[K^e] = \begin{bmatrix} [K_b^e] & [0] & [0] \\ [0] & [K_m^e] & [0] \\ [0] & [0] & [0] \end{bmatrix} \tag{3.154}$$

where subscripts b and m denote the bending and membrane deformations of the shell element, respectively.

In the case of a shell structure, which is actually flat, the shell stiffness matrix has singular terms associated with the drilling DOFs. To avoid such a problem, a small

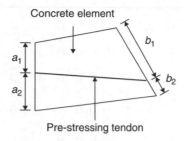

Figure 3.35 Pre-stress tendon in a concrete element

number is added to the diagonal term of the matrix in Equation (3.154) associated with the drilling DOFs.

The pre-stressing tendon

Figure 3.35 shows a segment of the pre-stressing tendon embedded into a flat shell element. Noting that only the axial deformation of the tendon needs to be considered, the three-dimensional truss element is used to model the pre-stressing tendon.

The elemental stiffness of the tendon in terms of local coordinates is expressed as

$$
[k_e] = \frac{E_s A_s}{l}
\begin{bmatrix}
1 & 0 & 0 & -1 & 0 & 0 \\
0 & 0 & 0 & 0 & 0 & 0 \\
0 & 0 & 0 & 0 & 0 & 0 \\
-1 & 0 & 0 & 1 & 0 & 0 \\
0 & 0 & 0 & 0 & 0 & 0 \\
0 & 0 & 0 & 0 & 0 & 0
\end{bmatrix}
\tag{3.155}
$$

The transformation matrix between the local and global coordinates system is given as

$$
T =
\begin{bmatrix}
\xi_1 & \xi_2 & \xi_3 & & & \\
\eta_1 & \eta_2 & \eta_3 & & 0 & \\
\varsigma_1 & \varsigma_2 & \varsigma_3 & & & \\
& & & \xi_1 & \xi_2 & \xi_3 \\
& 0 & & \eta_1 & \eta_2 & \eta_3 \\
& & & \varsigma_1 & \varsigma_2 & \varsigma_3
\end{bmatrix}
\tag{3.156}
$$

where, $\{\xi_1 \quad \eta_1 \quad \varsigma_1\}$ is the direction cosine of local $\bar{x}$-axis with respect to the global coordinate system. Similarly, $\{\xi_2 \quad \eta_2 \quad \varsigma_2\}$ and $\{\xi_3 \quad \eta_3 \quad \varsigma_3\}$ are the direction cosines of the local $\bar{y}$-axis and $\bar{z}$-axis with respect to the global coordinate system.

After transformation into the global coordinate system, the elemental stiffness matrix is expressed as

$$
[\bar{k}_e] = [T]^T [k_e][T]
\tag{3.157}
$$

Since $[\bar{k}_e]$ is a 6×6 matrix and the elemental stiffness matrix of the shell element is of dimension 24×24, a transformation matrix is needed to expand matrix $[\bar{k}_e]$ into a

24×24 dimension matrix. Such a matrix can be written as

$$
TR =
\begin{bmatrix}
\frac{a_2}{l_1} & 0 & 0 & 0 & 0 & 0 & 0 & 0 & 0 & 0 & 0 & 0 & 0 & 0 & 0 & 0 & 0 & 0 & \frac{a_1}{l_1} & 0 & 0 & 0 & 0 & 0 \\
0 & \frac{a_2}{l_1} & 0 & 0 & 0 & 0 & 0 & 0 & 0 & 0 & 0 & 0 & 0 & 0 & 0 & 0 & 0 & 0 & 0 & \frac{a_1}{l_1} & 0 & 0 & 0 & 0 \\
0 & 0 & \frac{a_2}{l_1} & 0 & 0 & 0 & 0 & 0 & 0 & 0 & 0 & 0 & 0 & 0 & 0 & 0 & 0 & 0 & 0 & 0 & \frac{a_1}{l_1} & 0 & 0 & 0 \\
0 & 0 & 0 & 0 & 0 & 0 & \frac{b_2}{l_2} & 0 & 0 & 0 & 0 & 0 & \frac{b_1}{l_2} & 0 & 0 & 0 & 0 & 0 & 0 & 0 & 0 & 0 & 0 & 0 \\
0 & 0 & 0 & 0 & 0 & 0 & 0 & \frac{b_2}{l_2} & 0 & 0 & 0 & 0 & 0 & \frac{b_1}{l_2} & 0 & 0 & 0 & 0 & 0 & 0 & 0 & 0 & 0 & 0 \\
0 & 0 & 0 & 0 & 0 & 0 & 0 & 0 & \frac{b_2}{l_2} & 0 & 0 & 0 & 0 & 0 & \frac{b_1}{l_2} & 0 & 0 & 0 & 0 & 0 & 0 & 0 & 0 & 0
\end{bmatrix}
$$

$$(3.158)$$

$$K_{es} = [TR]^T [\bar{k}_e][TR] \tag{3.159}$$

The consistent mass matrix for the space truss element can also be written as

$$
[\bar{m}_e] = \frac{\rho A l}{6}
\begin{bmatrix}
2\xi_1^2 & 2\xi_1\xi_2 & 2\xi_1\xi_3 & \xi_1^2 & \xi_1\xi_2 & \xi_1\xi_3 \\
2\xi_1\xi_2 & 2\xi_2^2 & 2\xi_2\xi_3 & \xi_1\xi_2 & \xi_2^2 & \xi_2\xi_3 \\
2\xi_1\xi_3 & 2\xi_2\xi_3 & 2\xi_3^2 & \xi_1\xi_3 & \xi_2\xi_3 & \xi_3^2 \\
\xi_1^2 & \xi_1\xi_2 & \xi_1\xi_3 & 2\xi_1^2 & 2\xi_1\xi_2 & 2\xi_1\xi_3 \\
\xi_1\xi_2 & \xi_2^2 & \xi_2\xi_3 & 2\xi_1\xi_2 & 2\xi_2^2 & 2\xi_2\xi_3 \\
\xi_1\xi_3 & \xi_2\xi_3 & \xi_3^2 & 2\xi_1\xi_3 & 2\xi_2\xi_3 & 2\xi_3^2
\end{bmatrix}
\tag{3.160}
$$

Similarly, the elemental mass matrix is expanded into a 24×24 dimension matrix

$$M_{es} = [TR]^T [\bar{m}_e][TR] \tag{3.161}$$

The stiffness matrix of a shell element with pre-stress

The stiffness matrix of a shell element with pre-stress can be given as

$$
K_e = \int_A (B_i^p)^T D^p B_j^p \, dA + \int_A (B_i^s)^T D^s B_j^s \, dA + \int_A (B_i^b)^T D^b B_j^b \, dA
$$
$$
+ \int_A G_i^T \begin{bmatrix} T_x & T_{xy} \\ T_{xy} & T_y \end{bmatrix} G_j \, dA + K_{es} \tag{3.162}
$$

The first term on the right-hand-side of Equation (3.162) is the stiffness corresponding to the planar stress; the second term corresponds to the transverse shear; the third term corresponds to the bending effect; and the fourth term is the geometric stiffness due to the pre-stress force, where T_x and T_y are the two components of the pre-stress force in the x- and y-axes, and T_{xy} is zero. The fifth term is the stiffness due to the pre-stressing tendon.

The elemental mass matrix is expressed as

$$M_e = m_e + M_{es} \tag{3.163}$$

The elemental matrices can then be assembled into the equation of motion for free vibration analysis and subsequent damage assessment.

3.2.7 Models with thin plate

Engineers have made many efforts to model crack-induced flexibility and investigate its effect on the dynamic characteristics of the damaged structure. Dimarogonas (1996) has summarized these works into three categories, namely: continuous model, discrete-continuous model and discrete models, i.e. finite-element models.

Real structures are more complicated than the geometrically simple ones described by the first two types of models. Discrete models are usually used to study the cracked structures – the finite element method (FEM) is the most popular and most commonly used. A literature review of the FEM-based model of a cracked plate shows that there are mainly three groups of methods. The simplest model represents the crack as a reduction in the elasticity modulus of the element at the crack position (Cawley and Adams, 1979) or a reduction in the cross-sectional area of the element (Baschmid et al., 1984). These models have been successfully used in damage localization of plate-like structures (Cornwell et al., 1999; Li et al., 2002). However, as these methods study the approximate crack size and location at the element level, a very fine finite element mesh is required to avoid a large error.

Another method models a crack by separating the nodes of finite elements along the crack line (Zastrau, 1985). To properly model the singular character of the stress and strain fields around the crack tip, a very dense mesh of finite elements or singular-shaped isoparametric elements (Shen and Pierre, 1990) are used to cover the crack tip area. Obviously, this method can model the cracked structure very well, but is not suitable or feasible for damage identification. This is due to the low computation efficiency associated with the large number of finite element. Also, a finite element mesh has to be constructed for each suspected location of damage in the plate at one time.

In the third group of methods, a rectangular plate element with an open and depth-through crack parallel to the plate boundary is modelled (Qian et al., 1991; Krawczuk, 1993; Krawczuk and Ostachowicz, 1994). The stiffness matrix of a cracked element is written as $K = TF^{-1}T^T$, where F is the matrix representing the sum of the flexibility of the non-cracked plate and the additional flexibility due to the crack; and T is a transformation matrix. However, as the derived stiffness matrix cannot be explicitly parameterized in terms of damage variable(s) to indicate the location, orientation and the extend of the crack, it is still difficult to incorporate this model into the inverse problem of structural damage identification.

3.2.7.1 Anisotropic model of elliptical crack with strain energy equivalence

The earliest crack model oriented towards damage identification was probably presented by Lee et al. (1997). Although their work focuses on fracture mechanics and aims to derive a damage evolution equation that is consistent with the continuum damage mechanics. Prior to this effort, there were a number of theories developed on continuum damage mechanics to derive the equivalent constitutive equation of a damaged material and its crack growth law on the basis of the strain equivalence principle introduced by Lemaitre (1985) or the stress equivalence principle introduced by Simo

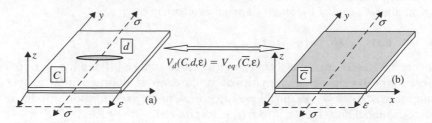

Figure 3.36 The effective stiffness model from the strain-energy equivalence principle

and Ju (1987). By introducing the strain energy equivalence principle as an alternative to the above principles, an effective stiffness model of the damaged material can be obtained in terms of the undamaged material properties and damage variable(s) as illustrated in Figure 3.36, where C denotes the elastic stiffness of the intact host element; $\overline{C}$ is the effective continuum stiffness of the damaged host element; d is the selected damage variable (or tensor); and ε denotes the uniform strain on the boundaries of the intact and damaged host elements.

Restricting the strain-energy principle to a two-dimensional elastic solid under biaxial stress (σ_1 and σ_2) and in-plane shear (τ_{12}) at infinity, Lee et al. (1997) introduced a damage in the form of an elliptical through crack with the major axis (length $2a$) and the minor axis (length $2b$), respectively, aligned with the Cartesian coordinates 1 and 2. For the intact state of the isotropic solid and the effective stiffness model of the damaged solid, the strain energy contained in the circular host element of radius R are expressed, respectively, as

$$
V_0 = \frac{1}{2}\pi R^2 h \begin{Bmatrix} \varepsilon_1 \\ \varepsilon_2 \\ \gamma_{12} \end{Bmatrix}^T \begin{bmatrix} C_{11} & C_{12} & 0 \\ C_{12} & C_{22} & 0 \\ 0 & 0 & C_{66} \end{bmatrix} \begin{Bmatrix} \varepsilon_1 \\ \varepsilon_2 \\ \gamma_{12} \end{Bmatrix},
$$

$$
V_{eq} = \frac{1}{2}\pi R^2 h \begin{Bmatrix} \varepsilon_1 \\ \varepsilon_2 \\ \gamma_{12} \end{Bmatrix}^T \begin{bmatrix} \overline{C}_{11} & \overline{C}_{12} & 0 \\ \overline{C}_{12} & \overline{C}_{22} & 0 \\ 0 & 0 & \overline{C}_{66} \end{bmatrix} \begin{Bmatrix} \varepsilon_1 \\ \varepsilon_2 \\ \gamma_{12} \end{Bmatrix} \tag{3.164}
$$

where h denotes the thickness of the plate-like solid. For the plane stress condition, the intact stiffness coefficients are defined in terms of the usual engineering constants as

$$
C_{11} = C_{22} = \frac{E}{1-v^2}, \quad C_{12} = \frac{vE}{1-v^2}, \quad \text{and } C_{66} = G,
$$

in which E, v and G denote the elastic modulus, the Poisson ratio and the shear modulus, respectively.

The strain energy released during the growth of the elliptical cavity has been derived by Sih and Liebowitz (1968) as

$$V_1 = \frac{1}{2}\pi a^2 h \begin{Bmatrix} \varepsilon_1 \\ \varepsilon_2 \\ \gamma_{12} \end{Bmatrix}^T \begin{bmatrix} C_{11}e_{11} & C_{12}e_{12} & 0 \\ C_{12}e_{12} & C_{22}e_{22} & 0 \\ 0 & 0 & C_{66}e_{66} \end{bmatrix} \begin{Bmatrix} \varepsilon_1 \\ \varepsilon_2 \\ \gamma_{12} \end{Bmatrix} \quad (3.165)$$

where the coefficients, e_{ij}, generally depend on the Poisson ratio and the cavity geometry. For the plane stress condition, $e_{11} = \frac{2v^2}{1-v^2} + \frac{(1-v)s}{1+v} + \frac{2s^2}{1-v^2}$, $e_{22} = \frac{2}{1-v^2} + \frac{(1-v)s}{1+v} + \frac{2v^2s^2}{1-v^2}$, $e_{12} = \frac{2}{1-v^2} - \frac{(1-v)s}{v(1+v)} + \frac{2s^2}{1-v^2}$ and $e_{66} = \frac{(1+s)^2}{1+v}$, with $s = b/a$ denoting the aspect ratio of the elliptical cavity.

According to the strain-energy equivalence principle, the strain energy contained in the effective stiffness continuum model of the damaged circle region can be expressed as

$$V_{eq} = V_d = V_0 - V_1 \quad (3.166)$$

Substituting Equations (3.164) and (3.165) into (3.166), gives the effective stiffness coefficients of the damaged plate cell as

$$\overline{C}_{ij} = C_{ij}(1 - e_{ij}d) \quad (3.167)$$

where the damage variable, $d = (a/R)^2$, represents the ratio of the effective damaged area to the total area of the considered solid host element and the coefficients, e_{ij}, represent the anisotropic behaviour of the host element due to the crack.

By further restricting the aspect ratio of the elliptical hole, $s = 0$, Lee et al. (2003) studied the effective stiffness model of a thin plate with line micro-cracks, and then developed a model updating technique to identify the damage size and orientation by using the frequency-response functions measured from the damaged plate. Numerical examples were simulated for demonstration but there is no experimental evidence to support the validity of the theory.

Lee's theory has been checked with the following limitations:

1) The crack released strain energy calculated from Equation (3.165) is valid only for an infinite plate containing a central crack subjected to uniform stress load. Although the result can be approximately used for a micro-crack away from the plate boundary, corrections must be made for the case of a macro-crack to take into account the finite dimensions of the plate and different load patterns.

2) Equation (3.165) is derived based on the Griffith theory of ideal brittle fracture mechanics. For ductile materials, a plastically deformed area induced by stress concentration around the crack tip will consume a part of the released energy, which is usually not negligible.

3) The model-updating-based damage identification technique requires an initial FEM model to represent the undamaged structure. However, a good quality model for a complex structure is difficult to achieve. Model reduction or simplification could easily result in initial model errors spreading the damage-induced localized changes in the updated model.

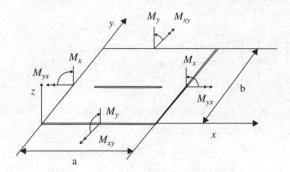

Figure 3.37 Thin plate element hosting a through crack parallel to its edge

A new effective stiffness model for thin plate elements with a central line crack is introduced in next section, in which the damage scalar in Lee et al.'s (1997) model is expanded and replaced by a vector of damage variables. Since the crack released strain energy is difficult to estimate accurately due to the complexity of fracture mechanics, a principle of behaviour equivalence is used to determine the damage variables. The damage model is then verified with experimental results.

3.2.7.2 Thin plates with anisotropic crack from dynamic characteristic equivalence

The problem of a thin plate with a non-propagating, open crack parallel to one side of the plate is studied. It is assumed that the crack is depth-through but narrow so that it does not change the plate mass. With a small rectangular plate element containing the crack, as shown in Figure 3.37, the constitutive relations between the internal moments and rotational displacement derivatives can be written as

$$
M(x, y) = \left\{ \begin{array}{c} M_x \\ M_y \\ M_{xy} \end{array} \right\} = D \begin{bmatrix} \dfrac{\partial}{\partial x} & 0 \\ 0 & \dfrac{\partial}{\partial y} \\ \dfrac{\partial}{\partial y} & \dfrac{\partial}{\partial x} \end{bmatrix} \left\{ \begin{array}{c} \theta_x \\ \theta_y \end{array} \right\} = DL\theta \tag{3.168}
$$

where $M(x, y)$ denotes the internal moments per unit length and $\{\theta_x \quad \theta_y\}^T$ denotes the rotation of the normal planes induced by the moments. The matrix, D, contains the flexural and twisting stiffness components of the plate. For an intact plate with isotropic elasticity and plane stress behaviour, the constitutive matrix, D, is defined as D_0 in the form of

$$
D_0 = \frac{h^3}{12} C \tag{3.169}
$$

where the matrix C represents the stress–strain constitutive stiffness of the isotropic material.

A crack will induce local flexibility in the plate. According to Lee et al.'s (1997) continuum elastic stiffness theory, the cracked plate element will exhibit orthotropic stiffness properties compatible with the orientation of the crack line. Furthermore, based on the observation from the fracture mechanics point of view that the highest stress intensity at the crack edge along the major axis direction will effectively reduce the stiffness in the minor axis direction, a rational supposition can be made that the crack will mainly affect the flexural stiffness normal to the crack line, while contributing relatively little to the plate stiffness parallel to the crack. (In Lee et al.'s model, the stiffness reduction along the major axis is v^2 times that along the minor axis). In order to verify this assumption, the cracked plate element is represented by an effective element of continuum anisotropic material with the major axis of the material parallel to the crack line and the minor axis normal to the crack. Thus, under the plane stress condition, the constitutive matrix, D, of the cracked element can be written as

$$D = \frac{h^3}{12} \overline{C} \tag{3.170}$$

where the stiffness components are

$$\overline{C}_{11} = \frac{E_1}{1 - v_{12}v_{21}}, \quad \overline{C}_{22} = \frac{E_2}{1 - v_{12}v_{21}}, \quad \overline{C}_{12} = \frac{v_{12}E_1}{1 - v_{12}v_{21}} = \frac{v_{21}E_2}{1 - v_{12}v_{21}} \text{ and } \overline{C}_{66} = G.$$

A vector of damage variable, $\{\alpha \quad \beta \quad \chi\}$, is selected to relate the effective continuum stiffness model of the cracked plate element to the isotropic stiffness of the undamaged material as

$$E_1 = E\alpha, \quad v_{21} = v, \quad E_2 = E\beta, \quad G_{12} = G\chi \tag{3.171}$$

where α, β and χ denote the stiffness reduction factors due to the crack. Considering the relationship $v_{12}/E_2 = v_{21}/E_1$, gives

$$v_{12} = v_{21}\frac{E_2}{E_1} = v\frac{\beta}{\alpha} \tag{3.172}$$

Substituting Equations (3.171) and (3.172) into (3.170), gives

$$D = \frac{h^3}{12} \begin{bmatrix} \dfrac{E\alpha^2}{\alpha - \beta v^2} & \dfrac{vE\alpha\beta}{\alpha - \beta v^2} & 0 \\[2ex] \dfrac{vE\alpha\beta}{\alpha - \beta v^2} & \dfrac{E\alpha\beta}{\alpha - \beta v^2} & 0 \\[2ex] 0 & 0 & \dfrac{E\chi}{2(1 + v)} \end{bmatrix} \tag{3.173}$$

According to Kirchhoff's theory for thin plates, the shear deformations are neglected to give

$$\nabla w - \theta = 0 \tag{3.174}$$

Substituting Equation (3.171) into Equation (3.164), gives

$$M(x, y) = D\kappa(x, y) \tag{3.175}$$

where the vector $\kappa(x, y) = \left\{ \frac{\partial^2 w}{\partial x^2} \quad \frac{\partial^2 w}{\partial y^2} \quad 2\frac{\partial^2 w}{\partial x \partial y} \right\} = \{\kappa_x \quad \kappa_y \quad \kappa_{xy}\}$ contains the flexural curvatures and twisting curvature at the middle plane of the plate.

Now the transverse displacement function, $w(x, y)$, can be defined as a biquadratic polynomial

$$w(x, y) = p\alpha \tag{3.176}$$

where the vector $p = \{1 \quad x \quad y \quad x^2 \quad xy \quad y^2 \quad x^3 \quad x^2y \quad xy^2 \quad y^3 \quad x^3y \quad xy^3\}$, and the vector $\alpha = \{\alpha_1 \quad \alpha_2 \quad \cdots \quad \alpha_{12}\}$ denote the unknown coefficients. It is then convenient to express the deflection in terms of displacements at the four nodes of the plate element as

$$w(x, y) = H(x, y)\overline{w} \tag{3.177}$$

where $H(x, y)$ is the vector of isoparametric shape functions of the element, and $\overline{w} = \{w_1 \quad w_2 \quad w_3 \quad w_4 \quad \theta_{x1} \quad \theta_{x2} \quad \theta_{x3} \quad \theta_{x4} \quad \theta_{y1} \quad \theta_{y2} \quad \theta_{y3} \quad \theta_{y4}\}$ is the nodal displacement vector. Substituting Equation (3.177) into the constitutive relation in Equation (3.172) gives

$$M(x, y) = DB(x, y)\overline{w} \tag{3.178}$$

where $B(x, y) = \left[\frac{\partial^2 H}{\partial x^2}; \quad \frac{\partial^2 H}{\partial y^2}; \quad 2\frac{\partial^2 H}{\partial x \partial y} \right]$. Equation (3.178) relates the internal moments at any point of the element with the nodal displacements. The internal moments at the four nodes are found by substituting the corresponding nodal coordinates, $(x_i, y_i) \; i = 1 \ldots 4$, into Equation (3.178). Writing this in matrix form gives

$$\begin{Bmatrix} \overline{M}_B \\ \overline{M}_T \end{Bmatrix} = \begin{bmatrix} K_{BT} & K_{BR} \\ K_{TT} & K_{TR} \end{bmatrix} \begin{Bmatrix} \overline{w}_T \\ \overline{w}_R \end{Bmatrix} = K\overline{w} \tag{3.179}$$

where $\overline{M}_B = \{M_{x1} \quad M_{y1} \quad M_{x2} \quad M_{y2} \quad M_{x3} \quad M_{y3} \quad M_{x4} \quad M_{y4}\}^T$ denotes the bending moments at the four nodes; $\overline{M}_T = \{M_{xy1} \quad M_{xy2} \quad M_{xy3} \quad M_{xy4}\}^T$ denotes the nodal twisting moments; $\overline{w}_T = \{w_1 \quad w_2 \quad w_3 \quad w_4\}$ denotes the translational DOFs at the four nodes; and $\overline{w}_R$ denotes the nodal rotational DOFs. Matrix K represents the general constitutive relations between the nodal loads and nodal displacements. Rewriting Equation (3.179) in partitioned form, the rotational and translational displacements can be expressed as

$$\overline{w}_R = K_{BR}^{-1}\overline{M}_B - K_{BR}^{-1}K_{BT}\overline{w}_T \tag{3.180a}$$

$$\overline{w}_T = K_{TT}^*\overline{M}_T - K_{TT}^*K_{TR}\overline{w}_R \tag{3.180b}$$

where the superscript * denotes the pseudo inverse of the matrix.

To determine the coupled damage variables, $\{\alpha \quad \beta \quad \chi\}$, the above-mentioned principles of strain equivalence and strain energy equivalence do not work. They are extended

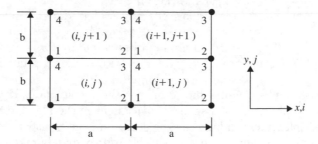

Figure 3.38 Four adjacent rectangular plate elements

to a more general principle that *the effective continuum model of a damaged struc-ture should have identical macro-behaviours with those exhibited by the real damaged structure.* The macro-behaviours here include the static and dynamic characteristics of the structure, such as deformations under loads, natural frequencies, mode shapes and frequency response functions. Based on this principle, the damage variables are related to the nodal deformation as shown below.

Consider a group of four adjacent rectangular plate elements, as shown in Figure 3.38, in which all the elements are suspected of crack damage. For the element, (i, j), the rotation about the y-axis, θ_x, at its second node can be evaluated using Equation (3.180a)

$$\theta_{x2}|_{(i,j)} = \left(\frac{2a}{\alpha E h^3} M_{x1} - \frac{2av}{\alpha E h^3} M_{y1} + \frac{4a}{\alpha E h^3} M_{x2} - \frac{4av}{\alpha E h^3} M_{y2} + \frac{w_2 - w_1}{a} \right)_{(i,j)} \quad (3.181)$$

Similarly, the rotation about the y-axis at the first node of the element, $(i+1, j)$, can be estimated by

$$\theta_{x1}|_{(i+1,j)} = \left(-\frac{4a}{\alpha E h^3} M_{x1} + \frac{4av}{\alpha E h^3} M_{y1} - \frac{2a}{\alpha E h^3} M_{x2} + \frac{2av}{\alpha E h^3} M_{y2} + \frac{w_2 - w_1}{a} \right)_{(i+1,j)}$$

$$(3.182)$$

According to the continuity condition, the slope at node 2 of the element (i, j) equals the slope at node 1 of the element $(i+1, j)$. Thus,

$$\theta_{x2}^{(i,j)} = \theta_{x1}^{(i+1,j)} \quad (3.183)$$

It is important to note that Equations (3.181) to (3.183) are derived for the damaged plate.

The moment–curvature relationship for an intact plate can be obtained from Equations (3.168) and (3.169) as

$$\tilde{M}_x(x, y) = \frac{E h^3}{12(1 - v^2)} [\tilde{\kappa}_x(x, y) + v\tilde{\kappa}_y(x, y)],$$

$$\tilde{M}_y(x,y) = \frac{Eh^3}{12(1-v^2)}[\tilde{\kappa}_y(x,y) + v\tilde{\kappa}_x(x,y)] \tag{3.184}$$

$$\tilde{M}_{xy}(x,y) = \frac{Eh^3}{12(1+v)}\tilde{\kappa}_{xy}(x,y)$$

where the superscript $\sim$ denotes the terms for the intact plate. It is assumed that the applied loads only produce a small displacement in the plate structure, and that the additional displacement induced by the crack damage is also small, such that the internal forces in the damaged structure can be approximately taken as those for the intact structure, which means $M_x \approx \tilde{M}_x$, $M_y \approx \tilde{M}_y$ and $M_{xy} \approx \tilde{M}_{xy}$. Therefore, Equation (3.184) can be directly substituted into Equations (3.181) to (3.183) to give

$$\left(\frac{\tilde{\kappa}_{x1} + 2\tilde{\kappa}_{x2}}{\alpha}\right)_{(i,j)} + \left(\frac{2\tilde{\kappa}_{x1} + \tilde{\kappa}_{x2}}{\alpha}\right)_{(i+1,j)} = 6\kappa_{x2}^{(i,j)} \tag{3.185a}$$

Similarly, considering the slope continuity condition at node 3 of element (i,j) and node 1 of element $(i+1,j)$, gives $\theta_{x3}^{(i,j)} = \theta_{x4}^{(i+1,j)}$, yielding

$$\left(\frac{\tilde{\kappa}_{x4} + 2\tilde{\kappa}_{x3}}{\alpha}\right)_{(i,j)} + \left(\frac{2\tilde{\kappa}_{x4} + \tilde{\kappa}_{x3}}{\alpha}\right)_{(i+1,j)} = 6\kappa_{x3}^{(i,j)} \tag{3.185b}$$

Also, from the slope continuity condition of rotation, θ_y in the y-direction, $\theta_{y4}^{(i,j)} = \theta_{y1}^{(i,j+1)}$ and $\theta_{y3}^{(i,j)} = \theta_{y2}^{(i,j+1)}$, leading to

$$\left(\frac{\tilde{\kappa}_{y1} + 2\tilde{\kappa}_{y4} + v(\tilde{\kappa}_{x1} + 2\tilde{\kappa}_{x4})}{\beta}\right)_{(i,j)} + \left(\frac{\tilde{\kappa}_{y4} + 2\tilde{\kappa}_{y1} + v(\tilde{\kappa}_{x4} + 2\tilde{\kappa}_{x1})}{\beta}\right)_{(i,j+1)}$$
$$= 6(1-v^2)\kappa_{y4}^{(i,j)} + S_{14} \tag{3.186a}$$

and

$$\left(\frac{\tilde{\kappa}_{y2} + 2\tilde{\kappa}_{y3} + v(\tilde{\kappa}_{x2} + 2\tilde{\kappa}_{x3})}{\beta}\right)_{(i,j)} + \left(\frac{\tilde{\kappa}_{y3} + 2\tilde{\kappa}_{y2} + v(\tilde{\kappa}_{x3} + 2\tilde{\kappa}_{x2})}{\beta}\right)_{(i,j+1)}$$
$$= 6(1-v^2)\kappa_{y3}^{(i,j)} + S_{23} \tag{3.186b}$$

respectively, where

$$S_{14} = \left(\frac{v(\tilde{\kappa}_{x1} + 2\tilde{\kappa}_{x4}) + v^2(\tilde{\kappa}_{y1} + 2\tilde{\kappa}_{y4})}{\alpha}\right)_{(i,j)} + \left(\frac{v(\tilde{\kappa}_{x4} + 2\tilde{\kappa}_{x1}) + v^2(\tilde{\kappa}_{y4} + 2\tilde{\kappa}_{y1})}{\alpha}\right)_{(i,j+1)},$$

$$S_{23} = \left(\frac{v(\tilde{\kappa}_{x2} + 2\tilde{\kappa}_{x3}) + v^2(\tilde{\kappa}_{y2} + 2\tilde{\kappa}_{y3})}{\alpha}\right)_{(i,j)} + \left(\frac{v(\tilde{\kappa}_{x3} + 2\tilde{\kappa}_{x2}) + v^2(\tilde{\kappa}_{y3} + 2\tilde{\kappa}_{y2})}{\alpha}\right)_{(i,j+1)}$$

Next, the transverse displacement at the nodes are estimated from Equation (3.180b) and taking into account the continuity condition of deflection w in the x- and y-directions separately, this gives $w_2^{(i,j)} = w_1^{(i+1,j)}$ and $w_4^{(i,j)} = w_1^{(i,j+1)}$, leading to

$$\left(\frac{\tilde{\kappa}_{xy1} + \tilde{\kappa}_{xy2} + \tilde{\kappa}_{xy3} + \tilde{\kappa}_{xy4}}{\chi}\right)_{(i,j)} + \left(\frac{\tilde{\kappa}_{xy1} + \tilde{\kappa}_{xy2} + \tilde{\kappa}_{xy3} + \tilde{\kappa}_{xy4}}{\chi}\right)_{(i+1,j)}$$

$$= 4(\kappa_{xy2} + \kappa_{xy3})_{(i,j)} \tag{3.187ba}$$

$$\left(\frac{\tilde{\kappa}_{xy1} + \tilde{\kappa}_{xy2} + \tilde{\kappa}_{xy3} + \tilde{\kappa}_{xy4}}{\chi}\right)_{(i,j)} + \left(\frac{\tilde{\kappa}_{xy1} + \tilde{\kappa}_{xy2} + \tilde{\kappa}_{xy3} + \tilde{\kappa}_{xy4}}{\chi}\right)_{(i,j+1)}$$

$$= 4(\kappa_{xy3} + \kappa_{xy4})_{(i,j)} \tag{3.187bb}$$

It is noted from Equations (3.185) to (3.187) that

1) The coupled damage indices α, β and χ in the crack model can be decoupled by the curvature expressions of the nodal transverse deflection.
2) The curvature-based formulations are valid for any load condition that satisfies the small-deformation assumption.
3) Provided that the curvatures of the intact and cracked plates, $\tilde{\kappa}$ and κ, respectively, can be measured on a regular mesh of the rectangular plate. Three sets of equations containing the decoupled damage indices can be established separately from Equations (3.185) to (3.187), by taking the reciprocal of the damage indices as unknowns. The equations are linear and determinate.

Experimental verification
An aluminium plate specimen having the dimensions $600\,\text{mm} \times 500\,\text{mm} \times 3\,\text{mm}$ is shown in Figure 3.39, together with the experimental set-up for testing. A rectangular

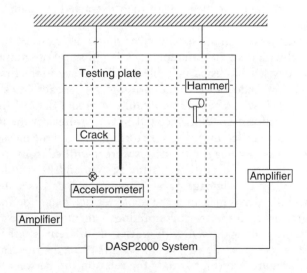

Figure 3.39 Diagram of the experimental system

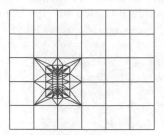

Figure 3.40   (a) Artificial crack in the plate (b) Finite element model

Table 3.3 Scheme of artificial crack and the corresponding natural frequencies

Crack information			Natural frequencies (Hz) in order of modes				
Centre (x, y)		(200, 200)	1	2	3	4	5
Length (mm)	Intact	N/A	31.379	41.578	65.859	78.105	88.263
	State I	80	31.372	41.228	65.571	78.040	88.245
	State II	120	31.372	40.812	65.238	77.991	88.194
	State III	160	31.361	40.133	64.843	77.870	88.028

mesh of 7 × 6 measuring points is outlined on each plate. The intact plate is suspended from a rigid frame by two steel wires of 0.5 mm in diameter and 0.75 m long, to simulate the free boundary condition. An impulsive signal was generated by hitting with a force hammer at each measuring point, and the vibration response of the plate due to the impulse was collected by an accelerometer model B&K 4370 as shown in the figure. Both the force and response signals were amplified and fed into a modal testing and analysis software package. The natural frequencies and corresponding mode shapes of the plate at the rectangular mesh were then extracted through a MISO transfer function analysis.

An artificial crack was cut in the specimen, as shown in Figure 3.40(a). To verify the above crack model with cases of different crack lengths and orientations, a scheme of crack cutting was devised and listed in Table 3.3. The crack in the first state is 0.5 mm wide, and it is 0.25 mm wide for the other two states. After each crack cutting exercise, the above hammer test was repeated to obtain the modal data of the plate for each damage state. The Uniform Load Surface (ULS) Curvatures for the undamaged plate and the cracked plate in different states were estimated from the measured modal data of the first five modes. The damage parameters were identified from an algorithm based on the uniform load surface curvature sensitivity (Wu and Law, 2005b) discussed in Section 6.6.3. The identified damage parameters for each cracks state were plotted in Figure 3.41, where the x-axis measures the relative crack length defined by $2c/a$, with a denoting the dimension of the crack containing element along the x-axis.

Besides testing the plate specimen with cracks of different lengths, a refined finite element model was also constructed by OPENFEM to model the cracked aluminium plate, as shown in Figure 3.40(b), to study the relationship between the values of the damage variables and the crack length. The finite element model initially has a crack of 40 mm long, and then the crack is lengthened in steps of 20 mm each and the finite element mesh was modified for each case. The Uniform Load Surface Curvatures are estimated from the 'measured' modal data of the plate model for each case with a definite crack length, and then they were used as references for crack identification. The resulting damage variables are shown separately in Figure 3.41 by the symbol '*', and they are curve-fitted as shown. The other curves shown in Figure 3.41 by the symbol 'o' are the results from Lee et al. (1997).

The test results are noted consistent with the numerical result using the present model and identification method, with the damage variables exhibiting the same trend of change with the crack propagation. Parameter α decays very slowly and changes little, which means the stiffness reduction in the direction of the crack extension is limited. Parameter β sharply drops with the extension of the crack, indicating a remarkable reduction in the stiffness normal to the crack line. There are two phases for parameter χ, which represents the in-plane twisting stiffness of the cracked plate element. Firstly, the parameter drops uniformly when the relative crack length is less than 0.6 of the element dimension, and then the twisting stiffness degrades abruptly as the crack propagates towards the element edges. However, the results from Lee et al. (1997) are very different. Parameter α remains unchanged with different crack length, which means the line crack never affects the stiffness in its extension direction; while the stiffness normal to the crack line and the twisting stiffness, shown as β and χ, respectively, may take negative values when the crack is close to the element edge. The presence of these illogical results can be explained by the infinite plate assumption in the model of Lee et al. (1997) and that the model is only suitable for micro-cracks.

3.2.8 Model with thick plate

3.2.8.1 Thick plate with anisotropic crack model

The crack model described in last section can only be applied to thin plates in which the shear deformations are neglected and the crack is parallel to the plate edge. If the crack line is not parallel to the edge of the plate, as shown in Figure 3.42, where ξ denotes the angle between the x-axis and the crack line (or the principal direction of the equivalent anisotropic element), the flexural constitutive relation of the plate can be rewritten as

$$\sigma = TC_b T^T \varepsilon \tag{3.188}$$

where $\sigma = \{\sigma_x \quad \sigma_y \quad \tau_{xy}\}^T$ and $\varepsilon = \{\varepsilon_x \quad \varepsilon_y \quad \gamma_{xy}\}^T$ denote the stress and strain in the mid-plane; and $C_b = 12D/h^3$ is the constitutive matrix. The transformation matrix can be defined as

$$T = \begin{bmatrix} \cos^2 \xi & \sin^2 \xi & -2\sin \xi \cos \xi \\ \sin^2 \xi & \cos^2 \xi & 2\sin \xi \cos \xi \\ \sin \xi \cos \xi & -\sin \xi \cos \xi & \cos^2 \xi - \sin^2 \xi \end{bmatrix} \tag{3.189}$$

The constitutive relations between the internal moments and the rotational displacement derivatives of the cracked plate are referred to in Equations (3.168) to (3.173). For a thick plate in which a plane normal to the mid-plane before deformation does not

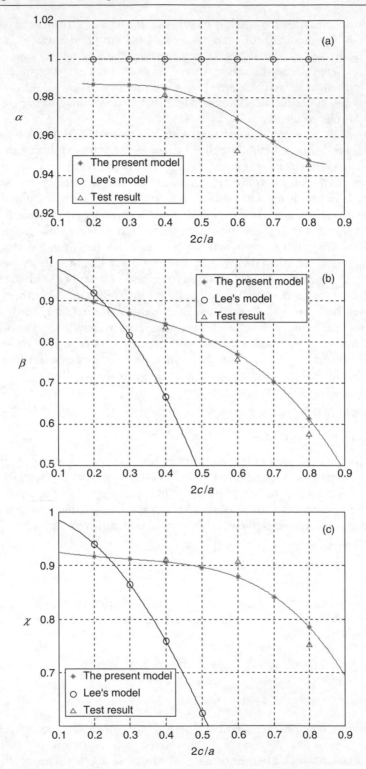

Figure 3.41 Damage variables from experiment and finite element model

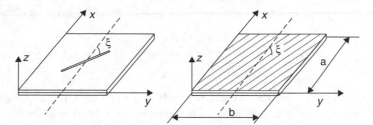

Figure 3.42 Plate element with a through crack and the effective stiffness model of anisotropic material

remain normal to the mid-plane any more after deformation, the effect of the transverse shear deformation should be taken into account. The corresponding constitutive relation of the cracked plate relating to the transverse shear stress is modelled as

$$\begin{Bmatrix} \tau_{13} \\ \tau_{23} \end{Bmatrix} = \begin{bmatrix} G\delta^{13} & 0 \\ 0 & G\delta^{23} \end{bmatrix} \begin{Bmatrix} \gamma_{13} \\ \gamma_{23} \end{Bmatrix} = C_s \begin{Bmatrix} \gamma_{13} \\ \gamma_{23} \end{Bmatrix} \tag{3.190}$$

where it is assumed that the crack affects the shear stiffness of the plate element in 1-3 plane represented by a fraction δ^{13} and the shear stiffness in 2-3 plane with a fraction δ^{23}.

Considering the included angle between the crack line and the plate boundary, the shear constitutive relation can be written as

$$\bar{\tau} = SC_s S^T \bar{\gamma} \tag{3.191}$$

where $\bar{\tau} = \{\tau_{xz} \quad \tau_{yz}\}^T$ and $\bar{\gamma} = \{\gamma_{xz} \quad \gamma_{yz}\}^T$ denote the transverse shear stress and strain, respectively, and the transformation matrix $S = \begin{bmatrix} \cos\xi & -\sin\xi \\ \sin\xi & \cos\xi \end{bmatrix}$.

Sensitivity of effective stiffness to crack damage variables

A vector of damage variables is needed to characterize a thick plate element with an inclined through crack. To formulate the stiffness matrix of the new effective continuum model of the cracked plate element, the internal strains need to be expressed in terms of the nodal variables, which include the transverse displacement, w, and the rotations of the mid-plane about the x- and y-axes, i.e. θ_x and θ_y, respectively. These rotation variables are formed from the bending and transverse shear deformations as

$$\theta_x = \frac{\partial w}{\partial x} - \gamma_{xz}, \quad \theta_y = \frac{\partial w}{\partial y} - \gamma_{yz} \tag{3.192}$$

where γ_{xz} and γ_{yz} are the deformed angles arising from the transverse shear stress. The in-plane displacements are given as

$$u = -z\theta_x(x, y), \quad v = -z\theta_y(x, y) \tag{3.193}$$

Taking the three displacement variables of each node independently, the displacement field of the plate element can be written in terms of the nodal displacements as

$$
\theta_x = \sum_{i=1}^{nn} H_i(x,y)\,(\theta_x)_i\,, \quad \theta_y = \sum_{i=1}^{nn} H_i(x,y)\,(\theta_y)_i\,, \quad w = \sum_{i=1}^{nn} H_i(x,y)w_i \tag{3.194}
$$

where $H_i(x,y)$ is the isoparametric shape function for the ith node and nn denotes the number of nodes of the plate element. Considering the relations in Equation (3.193), the in-plane strains can be expressed in terms of the nodal variables as

$$
\left\{ \begin{array}{c} \varepsilon_x \\ \varepsilon_y \\ \gamma_{xy} \end{array} \right\} = \begin{bmatrix} \dfrac{\partial}{\partial x} & 0 \\ 0 & \dfrac{\partial}{\partial y} \\ \dfrac{\partial}{\partial y} & \dfrac{\partial}{\partial x} \end{bmatrix} \left\{ \begin{array}{c} u \\ v \end{array} \right\} = -z \begin{bmatrix} \dfrac{\partial}{\partial x} & 0 & 0 \\ 0 & \dfrac{\partial}{\partial y} & 0 \\ \dfrac{\partial}{\partial y} & \dfrac{\partial}{\partial x} & 0 \end{bmatrix} \left\{ \begin{array}{c} \theta_x \\ \theta_y \\ w \end{array} \right\} = -zB_b d \tag{3.195}
$$

where $B_b = \begin{bmatrix} \dfrac{\partial H_1}{\partial x} & 0 & 0 & \cdots & \dfrac{\partial H_{nn}}{\partial x} & 0 & 0 \\ 0 & \dfrac{\partial H_1}{\partial y} & 0 & \cdots & 0 & \dfrac{\partial H_{nn}}{\partial y} & 0 \\ \dfrac{\partial H_1}{\partial y} & \dfrac{\partial H_1}{\partial x} & 0 & \cdots & \dfrac{\partial H_{nn}}{\partial y} & \dfrac{\partial H_{nn}}{\partial x} & 0 \end{bmatrix}$

and $d = \{\theta_{x,1} \quad \theta_{y,1} \quad w_1 \quad \cdots \quad \theta_{x,nn} \quad \theta_{y,nn} \quad w_{nn}\}^T$. From Equation (3.192) the transverse shearing strains are expressed in terms of the nodal displacements as

$$
\left\{ \begin{array}{c} \gamma_{xz} \\ \gamma_{yz} \end{array} \right\} = \begin{bmatrix} -1 & 0 & \dfrac{\partial}{\partial x} \\ 0 & -1 & \dfrac{\partial}{\partial y} \end{bmatrix} \left\{ \begin{array}{c} \theta_x \\ \theta_y \\ w \end{array} \right\} = B_s d \tag{3.196}
$$

where $[B_s] = \begin{bmatrix} -H_1 & 0 & \dfrac{\partial H_1}{\partial x} & \cdots & -H_{nn} & 0 & \dfrac{\partial H_{nn}}{\partial x} \\ 0 & -H_1 & \dfrac{\partial H_1}{\partial y} & \cdots & 0 & -H_{nn} & \dfrac{\partial H_{nn}}{\partial y} \end{bmatrix}$

According to the Reissner–Mindlin theory, the stiffness matrix of the effective model of the cracked plate element can be written as

$$
k_e = \frac{h^3}{12} \iint_A B_b^T TCT^T B_b dA + \kappa h \iint_A B_s^T SDS^T B_s dA \tag{3.197}
$$

where h denotes the thickness of the plate; A denotes the plane area of the plate element; B_b and B_s are the strain–displacement relationship matrices for the bending and transverse shear strains, respectively; and κ is the shear energy correction factor of 5/6. Thus, the sensitivity of the elemental stiffness matrix with respect to the crack variables can be obtained as

$$\frac{\partial k_e}{\partial \alpha} = \frac{h^3}{12} \iint_A B_b^T T \begin{bmatrix} \dfrac{E\alpha(\alpha - 2\beta v^2)}{(\alpha - \beta v^2)^2} & \dfrac{-E\beta^2 v^3}{(\alpha - \beta v^2)^2} & 0 \\[3mm] \dfrac{-E\beta^2 v^3}{(\alpha - \beta v^2)^2} & \dfrac{-E\beta^2 v^2}{(\alpha - \beta v^2)^2} & 0 \\[3mm] 0 & 0 & 0 \end{bmatrix} T^T B_b dA, \tag{3.198a}$$

$$\frac{\partial k_e}{\partial \beta} = \frac{h^3}{12} \iint_A B_b^T T \begin{bmatrix} \dfrac{E\alpha^2 v^2}{(\alpha - \beta v^2)^2} & \dfrac{E\alpha^2 v}{(\alpha - \beta v^2)^2} & 0 \\[3mm] \dfrac{E\alpha^2 v}{(\alpha - \beta v^2)^2} & \dfrac{E\alpha^2}{(\alpha - \beta v^2)^2} & 0 \\[3mm] 0 & 0 & 0 \end{bmatrix} T^T B_b dA, \tag{3.198b}$$

$$\frac{\partial k_e}{\partial \chi} = \frac{h^3}{12} \iint_A B_b^T T \begin{bmatrix} 0 & 0 & 0 \\ 0 & 0 & 0 \\ 0 & 0 & \dfrac{E}{2(1+v)} \end{bmatrix} T^T B_b dA, \tag{3.198c}$$

$$\frac{\partial k_e}{\partial \delta_{13}} = \mu h \iint_A B_s^T S \begin{bmatrix} \dfrac{E}{2(1+v)} & 0 \\ 0 & 0 \end{bmatrix} S^T B_s dA, \tag{3.198d}$$

$$\frac{\partial k_e}{\partial \delta_{23}} = \mu h \iint_A B_s^T S \begin{bmatrix} 0 & 0 \\ 0 & \dfrac{E}{2(1+v)} \end{bmatrix} S^T B_s dA, \tag{3.198e}$$

$$\frac{\partial k_e}{\partial \xi} = \frac{h^3}{12} \iint_A B_b^T \frac{\partial \overline{C}}{\partial \xi} B_b dA + \kappa t \iint_A B_s^T \frac{\partial \overline{D}}{\partial \xi} B_s dA \tag{3.198f}$$

The vector of damage variables, $\Delta p_i = \{\alpha_i \quad \beta_i \quad \chi_i \quad \delta_{13,i} \quad \delta_{23,i} \quad \xi_i\}^T (i = 1, \cdots, ne)$, of the ith plate element characterizing a thick plate element with an inclined through crack can be solved with the penalty function method based on the sensitivity of the Uniform Load Surface Curvature discussed in Section 6.6.3 (Wu and Law, 2004a).

Experimental verification

Three aluminium plate specimens with the same dimensions of $600 \, \text{mm} \times 500 \, \text{mm} \times 3 \, \text{mm}$ are named as Plates A, B and C. The experimental set-up for testing is shown in Figure 3.39, with the perpendicular crack replaced by an inclined crack. A rectangular mesh of 7×6 measuring points is outlined on each plate. The experimental procedure is the same as that for the thin plate given in Section 3.2.7.

Artificial cracks are cut in each specimen as shown in Figure 3.43(a) for cases of different crack lengths and orientations, and the scheme of crack cutting is listed in Table 3.4. The crack in the first state is 0.5mm wide, and it is 0.3 mm wide for the other two states.

As the thickness of the test plates is very small relative to the other dimensions of the plates, the transverse shear deformations are neglected, so that the vector

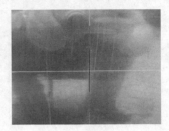

Figure 3.43(a)   Artificial cracks (State I)

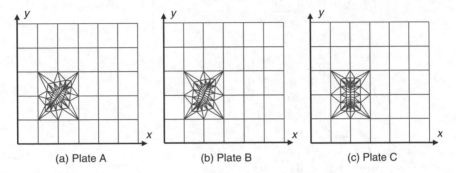

| (a) Plate A | (b) Plate B | (c) Plate C |

Figure 3.43(b)   Finite element meshes to model the cracked plates A, B and C

Table 3.4 Scheme of artificial crack cutting for the three specimens

Crack information		Plate A	Plate B	Plate C
Centre (x, y)		(200,200)	(200,200)	(200,200)
Orientation (ξ)		45°	60°	90°
Length (mm)	State I	80	80	80
	State II	140	120	120
	State III	200	160	160

of damage variables consists of only four components for each plate element, i.e. $p_i = \{\alpha_i \quad \beta_i \quad \chi_i \quad \xi_i\}^T (i = 1, \cdots, ne)$. The initial finite element model of the intact plate with one nine-node plate element and 26 four-node elements is shown in Figure 3.44, with the parameter vectors for all the elements initialized to unity. The crack identification procedure adopted here is to solve an over-determined set of equations characterized by a sensitivity matrix of Uniform Load Surface curvature (Wu and Law, 2007) with $3Q \times 4ne$ dimensions, where $Q = 7 \times 6$ and $ne = 27$. The ULS curvatures for both the test plate and initial finite element model are estimated from the modal data of the first five modes. The identified crack parameters for each specimen with different crack states are plotted in Figure 3.45.

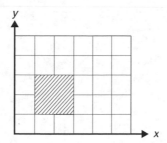

Figure 3.44 Simplified finite element model of the plate with a inclined crack

Besides testing the three plate specimens with different crack states, three refined finite element models were also constructed by OPENFEM to model the cracked aluminium plates, as shown in Figure 3.43(b) for a study similar to that for the thin plate in Section 3.2.7. The resulting crack parameters are shown separately in Figure 3.45 with different symbols as denoted in the legends, and they are curve-fitted as shown.

The test results are consistent with the numerical results, and both sets of results exhibit the same trend of changing with the crack propagation. Parameter α decays very slowly and changes little, which means the stiffness reduction in the direction of crack extension is limited. Parameter β drops sharply with the extension of the crack indicating a remarkable reduction in the stiffness normal to the crack line. There are two phases for χ, which represent the in-plane twisting stiffness of the cracked plate element, in the cases when the crack angle $\xi > 45°$. Firstly, the parameter drops evenly when the relative crack length is less than 0.43 of the element dimension, and then the twisting stiffness degrades abruptly as the crack propagates towards the element edges.

3.2.9 *Model of thick plate reinforced with Fibre-Reinforced-Plastic*

Delamination is one of the most common types of damage in laminated FRP composites due to their relatively weak inter-laminar strengths. For a concrete beam or plate reinforced with the FRP bonding technique for enhancing the ultimate failure loads and the desirable failure behaviours, the connecting layer between the FRP plate and concrete host is more vulnerable to delamination damage due to the distinctly different ductility of the FRP and concrete materials. Although the fracture mechanics of the FRP-concrete debonding are different to that of the inter-laminar delamination, which occurred inside the FRP composite, and many experimental studies show that the FRP-concrete debonding failure scenarios may be very varied (Teng et al., 2002), their effects on the static and dynamic behaviours of the structure are similar in the design consideration. The presence of delamination may significantly affect the structural integrity; and hence severely reduce the stiffness and strength of the structure statically, which may, in some cases, lead to catastrophic failure.

Delamination mainly involves a separation of the two types of materials, and its effects on the structural mass are usually small and thus are usually neglected. Diaz Valdes and Soutis (1999a; 1999b) studied the effect of delamination on the modal

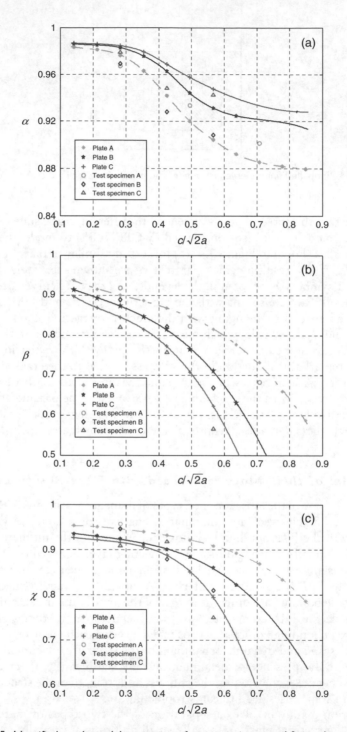

Figure 3.45 Identified crack model parameters from experiment and finite element model

frequencies of existing laminated composite beams. Modal frequencies were obtained by a novel method known as resonant ultrasound spectroscopy. It was demonstrated that changes of the modal frequencies after delamination, when compared with those of an intact specimen, give a good indication on the presence and magnitude of damage. With the help of the modal strain energy distribution in a beam, Griffin and Sun (1991) presented a delamination identification method based on changes in the damping ratio of a delaminated structure. Luo and Hanagud (1995) carried out a comparative study on the sensitivity of four quantitative indices: frequencies, mode shapes, damping ratios and delamination coefficients. It is clear from the above works that the frequency information alone is insufficient to provide a reliable prediction of delamination, and that the delamination coefficients give a better indication of the presence of delamination among the four indices.

Finite element models of delaminated beams and plates were also developed for an accurate prediction of the effects of delamination. They are based on Euler beam theory (Majumdar and Suryanarayan, 1988), engineering beam theory (Tracy and Pardoen, 1989) and Timoshenko beam theory and the Galerkin method (Shen and Grady, 1992). A finite element model using layer-wise theory was also developed for laminated composite plates (Barbero and Reddy, 1991), which can be used to study the multi-delamination problem in the thickness of the plate. When incorporated with artificial intelligence computation techniques of neural networks (Islam and Craig, 1994; Chaudhry and Ganino, 1994; Okafor et al., 1996,) or genetic algorithms (Krawczuk and Ostachowicz, 2002), these models enable real-time non-destructive delamination detection based on vibration measurements.

In the consideration of the dynamic behaviour of the delaminated beams (or plates), the delaminated members can be assumed to deform independently without taking into account the mutual contact and friction effects. This kind of modelling is known as 'free mode' analysis (Wang et al., 1982). In order to take into account the possible dynamic contact between delaminated layers in an approximate way, a 'constrained mode' has also been proposed, in which the delaminated layers are assumed to have the same transverse displacement but are allowed to slide over each other (Majumdar and Suryanarayan, 1988). More refined models may adopt contact elements or contact conditions in the finite element modelling to avoid overlapping between the upper and lower portions of a delaminated plate and the dynamic analysis must be performed by time integration techniques (Kwon and Aygunes, 1996).

None of the above models is suitable for structural damage identification, which is an inverse problem mathematically. A new finite element mesh is usually required for each possible damage case to study the location and magnitude of a particular delamination damage. Perel and Palazotto (2002) developed a finite element model of a delaminated beam particularly for the damage identification procedure. Three parameters characterizing the location and magnitude of delamination are selected and estimated by minimizing the discrepancy between the computed and measured responses.

A finite element formulation of the FRP-bonded concrete plate with this type of delamination fault is presented in the next section for the non-destructive evaluation from vibration measurement. An adhesive interface where possible debonding could occur is introduced between the FRP and the concrete plates. A scalar damage parameter characterizing the extent of delamination is incorporated into the formulation of the

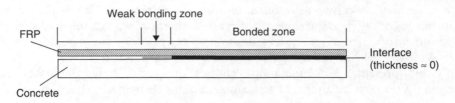

Figure 3.46 Propagation of delamination from the free end of FRP-bonded concrete beams

finite element model that is compatible with the vibration-based damage identification procedure.

3.2.9.1   Damage-detection-oriented model of delamination of fibre-reinforced plastic and thick plate

Assumptions and governing equations of the interface layer
An FRP-bonded concrete plate, as shown in Figure 3.46, can be modelled as a laminate comprising two plies: the concrete host and the FRP plate, between which an interface is introduced, where the delamination is likely to occur. The interface behaves as an entity with large stiffness but negligible thickness. Its physical justification at a microscopic level is the existence of the thin resin-rich layer between the plies. When the loading level increases, or the fatigue durability of the structure decreases during its service life, delamination damage starts and gradually develops at the interface. From the viewpoint of micro-mechanics, there is an interim state between the intact and the delaminated interface states. This damage zone containing micro-defects is called the Weak Bonding Zone (WBZ) in this model. Macro delamination forms when these micro-defects grow and coalesce. To represent these micro-defects in the context of continuum damage mechanics, a damage parameter is required to describe the macroscopic effects of these distributed micro-defects.

The parameter p_b is therefore introduced to characterize the growth of the micro-cracks or the macro-bonding condition of the interface

$$p_b = \begin{cases} 0 & \text{delaminated} \\ > 0 \text{ and } < 1 & \text{weak bonding} \\ 1 & \text{intact} \end{cases} \tag{3.199}$$

The adhesive layer is assumed to carry only constant shear and peel strains over its thickness in both perfect bonding and weak bonding conditions. For the plate segment with delamination at the interface, it is assumed that there is no stress transferring between the concrete host and the FRP layer. In addition, contact and friction between the two debonded surfaces are not considered for simplicity. Based on the well-known constant shear and peel strain assumption (Tong and Steven, 1999), the peel and shear stress in the interface layer are given by Tong et al. (2001) in the form of

$$\sigma_b = \frac{p_b E_{ad}(1 - \nu_{ad})}{(1 - 2\nu_{ad})(1 + \nu_{ad})h_{ad}}(w_f - w_c) \tag{3.200}$$

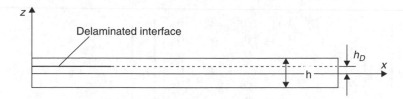

Figure 3.47 Diagram of the FRP-bonded concrete beam (interface thickness = 0)

$$\tau_b = \frac{p_b E_{ad}}{2(1 + v_{ad})h_{ad}}\gamma \tag{3.201}$$

where h, E, v and w denote the thickness, the elastic modulus, the Poisson ratio and the transverse displacement, respectively; and the subscripts f, ad and c represent the lower surface of the FRP plate, the adhesive layer and the upper surface of the concrete host, respectively. The shear strain, γ_{xz}, takes the form

$$\gamma_{xz} = \frac{1}{2}\left(\frac{\partial w_c}{\partial x} + \frac{\partial w_f}{\partial x}\right) + \frac{1}{2h_{ad}}\left[\left(\frac{h}{2} + h_D\right)\frac{\partial w_c}{\partial x} + \left(\frac{h}{2} - h_D\right)\frac{\partial w_f}{\partial x}\right] + \frac{u_f - u_c}{h_{ad}} \tag{3.202}$$

where h denotes the thickness of the whole FRP-bonded concrete plate; h_D denotes the distance between the interface and the mid-plane of the plate, as shown in Figure 3.47; and u represents the longitudinal displacement along the x-axis. The transverse shear strain γ_{yz}, can be similarly written as Equation (3.202).

Basic equations for the three-dimensional anisotropic material
The FRP composite sheet is made of the typical anisotropic material. This section presents the basic equations for the three-dimensional anisotropic material as the basis of further formulations. These equations include:
   The strain–displacement equations

$$\varepsilon_{xx} = \frac{\partial u}{\partial x}, \quad \varepsilon_{yy} = \frac{\partial v}{\partial y}, \quad \varepsilon_{zz} = \frac{\partial w}{\partial z}, \quad \varepsilon_{xy} = \frac{1}{2}\left(\frac{\partial u}{\partial y} + \frac{\partial v}{\partial x}\right),$$

$$\varepsilon_{xz} = \frac{1}{2}\left(\frac{\partial u}{\partial z} + \frac{\partial w}{\partial x}\right), \quad \varepsilon_{yz} = \frac{1}{2}\left(\frac{\partial v}{\partial z} + \frac{\partial w}{\partial y}\right) \tag{3.203}$$

The equations of motion

$$\begin{cases} \sigma_{xx,x} + \sigma_{xy,y} + \sigma_{xz,z} = \rho\ddot{u} \\ \sigma_{yx,x} + \sigma_{yy,y} + \sigma_{yz,z} = \rho\ddot{v} \\ \sigma_{zx,x} + \sigma_{zy,y} + \sigma_{zz,z} = \rho\ddot{w} \end{cases} \tag{3.204}$$

The constitutive relations in the principal coordinate system

$$
\begin{Bmatrix} \sigma_1 \\ \sigma_2 \\ \sigma_3 \\ \sigma_4 \\ \sigma_5 \\ \sigma_6 \end{Bmatrix} = \begin{bmatrix} C_{11} & C_{12} & C_{13} & 0 & 0 & 0 \\ C_{12} & C_{22} & C_{23} & 0 & 0 & 0 \\ C_{13} & C_{23} & C_{33} & 0 & 0 & 0 \\ 0 & 0 & 0 & C_{44} & 0 & 0 \\ 0 & 0 & 0 & 0 & C_{55} & 0 \\ 0 & 0 & 0 & 0 & 0 & C_{66} \end{bmatrix} \begin{Bmatrix} \varepsilon_1 \\ \varepsilon_2 \\ \varepsilon_3 \\ \varepsilon_4 \\ \varepsilon_5 \\ \varepsilon_6 \end{Bmatrix} \tag{3.205}
$$

where $\Delta = \dfrac{1 - v_{12}v_{21} - v_{23}v_{32} - v_{13}v_{31} - 2v_{12}v_{23}v_{31}}{E_1 E_2 E_3}$

$$
C_{11} = \frac{1 - v_{23}v_{32}}{E_2 E_3 \Delta}, \quad C_{12} = \frac{v_{21} + v_{23}v_{31}}{E_1 E_3 \Delta}, \quad C_{13} = \frac{v_{31} + v_{21}v_{32}}{E_1 E_2 \Delta},
$$

$$
C_{23} = \frac{v_{32} + v_{12}v_{31}}{E_1 E_2 \Delta}, \quad C_{22} = \frac{1 - v_{13}v_{31}}{E_1 E_3 \Delta}, \quad C_{33} = \frac{1 - v_{12}v_{21}}{E_1 E_2 \Delta},
$$

$$
C_{44} = G_{23}, \quad C_{55} = G_{31}, \quad C_{66} = G_{12}
$$

The corresponding equations of the concrete host can be similarly written, except that the coefficients in Equation (3.205) should be replaced by the constitutive relations of isotropic material where

$$
C_{11} = C_{22} = C_{33} = \frac{E}{1 - v^2}, \quad C_{12} = C_{13} = C_{23} = \frac{vE}{1 - v^2},
$$

$$
C_{44} = C_{55} = C_{66} = G \tag{3.206}
$$

Strain–stress formulation of the delaminated plate

Considering the FRP-bonded plate with no externally applied loads on the upper and lower surfaces, the stress boundary conditions at the upper and lower surfaces are given as

$$
\begin{cases} \sigma_{zz} = 0 \\ \sigma_{xz} = 0 \quad \text{at } z = \pm\dfrac{h}{2} \\ \sigma_{yz} = 0 \end{cases} \tag{3.207}
$$

According to classical plate theory, it is assumed that the thickness of the plate does not change during deformation such that

$$
\varepsilon_{zz} = 0 \tag{3.208}
$$

In the case of a thin plate, the normal stress in the z-axis can be assumed equal to zero

$$
\sigma_{zz} = 0 \tag{3.209}
$$

and the third row of the constitutive relations in Equation (3.205) can be discarded. However, in this formulation, the peel stress at the interface layer has to be taken into account, and the strain ε_{zz} could be very small but not equal to zero. Furthermore, by taking into account the boundary conditions in Equation (3.207), the strains along the z-axis in the concrete host and the FRP sheet take the following forms

$$\varepsilon_{zz}^{(1)}(x,y,z) = \phi^{(1)}(x,y)\left(1 + \frac{2}{h}z\right), \quad \varepsilon_{zz}^{(2)}(x,y,z) = \phi^{(2)}(x,y)\left(1 - \frac{2}{h}z\right) \quad (3.210)$$

where $\phi(x,y)$ is an unknown function that characterizes the strain distribution in the mid-plane of the plate, and the superscripts (1) and (2) represent terms in the concrete and FRP sheet, respectively.

On further inspection of the stiffness coefficients in Equation (3.205), the strain ε_{zz} is found to be very small due to the small dimension of the plate in the z-direction compared with the other two dimensions, the Poisson ratios, ν_{31} and ν_{32}, are also quite small compared with ν_{21}. Therefore, it can be assumed that the coefficients C_{13} and C_{23} are negligible and the peel stress and strain in the z-axis are independently related by a stiffness scalar. The resulting constitutive equations take the form

$$\begin{Bmatrix} \sigma_1 \\ \sigma_2 \\ \sigma_3 \\ \sigma_4 \\ \sigma_5 \\ \sigma_6 \end{Bmatrix} = \begin{bmatrix} Q_{11} & Q_{12} & 0 & 0 & 0 & 0 \\ Q_{12} & Q_{22} & 0 & 0 & 0 & 0 \\ 0 & 0 & Q_{33} & 0 & 0 & 0 \\ 0 & 0 & 0 & Q_{44} & 0 & 0 \\ 0 & 0 & 0 & 0 & Q_{55} & 0 \\ 0 & 0 & 0 & 0 & 0 & Q_{66} \end{bmatrix} \begin{Bmatrix} \varepsilon_1 \\ \varepsilon_2 \\ \varepsilon_3 \\ \varepsilon_4 \\ \varepsilon_5 \\ \varepsilon_6 \end{Bmatrix} \quad (3.211)$$

where $Q_{11} = \dfrac{E_1^2}{E_1 - \nu_{12}^2 E_2}$, $Q_{12} = \dfrac{\nu_{12}E_1 E_2}{E_1 - \nu_{12}^2 E_2}$, $Q_{22} = \dfrac{E_1 E_2}{E_1 - \nu_{12}^2 E_2}$,

$$Q_{33} = E_3, \quad Q_{44} = G_{23}, \quad Q_{55} = G_{13}, \quad Q_{66} = G_{12}$$

The constitutive relations are expressed in another coordinate system by rotating about the z-axis counter-clockwise through an angle of θ with respect to the principal coordinate system of the material. Thus, the stress–strain relationships can be rewritten as

$$\begin{Bmatrix} \sigma_{xx} \\ \sigma_{yy} \\ \sigma_{zz} \\ \sigma_{yz} \\ \sigma_{xz} \\ \sigma_{xy} \end{Bmatrix} = \begin{bmatrix} \overline{Q}_{11} & \overline{Q}_{12} & 0 & 0 & 0 & \overline{Q}_{16} \\ \overline{Q}_{12} & \overline{Q}_{22} & 0 & 0 & 0 & \overline{Q}_{26} \\ 0 & 0 & \overline{Q}_{33} & 0 & 0 & 0 \\ 0 & 0 & 0 & \overline{Q}_{44} & \overline{Q}_{45} & 0 \\ 0 & 0 & 0 & \overline{Q}_{45} & \overline{Q}_{55} & 0 \\ \overline{Q}_{16} & \overline{Q}_{26} & 0 & 0 & 0 & \overline{Q}_{66} \end{bmatrix} \begin{Bmatrix} \varepsilon_{xx} \\ \varepsilon_{yy} \\ \varepsilon_{zz} \\ \varepsilon_{yz} \\ \varepsilon_{xz} \\ \varepsilon_{xy} \end{Bmatrix} \quad (3.212)$$

where $\overline{Q}_{11} = Q_{11}c^4 + 2(Q_{12} + 2Q_{66})s^2c^2 + Q_{22}s^4,$

$\overline{Q}_{12} = (Q_{11} + Q_{22} - 4Q_{66})s^2c^2 + Q_{12}(s^4 + c^4),$

$\overline{Q}_{16} = (Q_{11} - Q_{12} - 2Q_{66})sc^3 + (Q_{12} - Q_{22} + 2Q_{66})s^3c,$

$\overline{Q}_{26} = (Q_{11} - Q_{12} - 2Q_{66})s^3c + (Q_{12} - Q_{22} + 2Q_{66})sc^3,$

$\overline{Q}_{22} = Q_{11}s^4 + 2(Q_{12} + 2Q_{66})s^2c^2 + Q_{22}c^4,$

$\overline{Q}_{33} = Q_{33}, \quad \overline{Q}_{44} = Q_{44}c^2 + Q_{55}s^2, \quad \overline{Q}_{45} = (Q_{55} - Q_{44})cs,$

$\overline{Q}_{55} = Q_{55}c^2 + Q_{44}s^2, \overline{Q}_{66} = (Q_{11} + Q_{22} - 2Q_{12} - 2Q_{66})s^2c^2 + Q_{66}(s^4 + c^4),$

$c = \cos\theta; \quad s = \sin\theta.$

Considering the strain–displacement relationship, $\varepsilon_{zz} = \partial w/\partial z$, the transverse displacement, $w(x, y, z)$, can be obtained from Equation (3.210) as

$$w^{(1)}(x,y,z) = \phi^{(1)}(x,y)\int_{-h/2}^{z}\left(1 + \frac{2}{h}z\right)dz = \phi^{(1)}(x,y)\left(\frac{z^2}{h} + z + \frac{h}{4}\right),$$

$$w^{(2)}(x,y,z) = \phi^{(2)}(x,y)\int_{z}^{h/2}\left(1 - \frac{2}{h}z\right)dz = \phi^{(2)}(x,y)\left(\frac{z^2}{h} - z + \frac{h}{4}\right) \quad (3.213)$$

Therefore, the transverse deflection at the upper surface of the concrete is given by $w_c = w^{(1)}(x, y, h_d)$ and the deflection at the lower surface of the FRP is given by $w_f = w^{(2)}(x, y, h_d)$. Substituting them into Equation (3.200) gives

$$\phi^{(2)}(x,y)\left(\frac{h_D^2}{h} - h_D + \frac{h}{4}\right) - \phi^{(1)}(x,y)\left(\frac{h_D^2}{h} + h_D + \frac{h}{4}\right) = \frac{\sigma_b(1 - 2v_{ad})(1 + v_{ad})h_{ad}}{p_bE_{ad}(1 - v_{ad})}$$

$$(3.214)$$

By writing the constant peel stress at the interface layer as $\sigma_b = Q_{33}\varepsilon_{zz}^{(1)}\big|_{z=h_D} = Q_{33}\phi^{(1)}$ $(x, y)\left(1 + \frac{2h_D}{h}\right)$ and substituting it into Equation (3.214), $\phi^{(2)}$ can be expressed as a function of $\phi^{(1)}$

$$\phi^{(2)}(x,y) = \left[\frac{Q_{33}(1 + \frac{2h_D}{h})(1 - 2v_{ad})(1 + v_{ad})h_{ad}}{p_bE_{ad}(1 - v_{ad})} + \frac{\frac{h_D^2}{h} + h_D + \frac{h}{4}}{\frac{h_D^2}{h} - h_D + \frac{h}{4}}\right]\phi^{(1)}(x,y)$$

$$(3.215)$$

Equation (3.215) links the strain distributions in the two plates enabling the subsequent mathematical derivation of delamination formulation. Taking the longitudinal displacement along the x-axis at the mid-plane of concrete host as $u_0^{(1)}(x,y) = u^{(1)}(x,y,z)\big|_{z=\frac{h_D}{2}-\frac{h}{4}}$, the longitudinal displacement of the concrete can be written as

$$u^{(1)}(x,y,z) = u_0^{(1)}(x,y) + \int_{\frac{h_D}{2}-\frac{h}{4}}^{z}\left(2\varepsilon_{xz}^{(1)} - \frac{\partial w^{(1)}}{\partial x}\right)dz \quad (3.216)$$

Similarly, the longitudinal displacement of the FRP sheet takes the form

$$u^{(2)}(x,y,z) = u_0^{(2)}(x,y) + \int_{\frac{h_D}{2}+\frac{h}{4}}^{z} \left(2\varepsilon_{xz}^{(2)} - \frac{\partial w^{(2)}}{\partial x} \right) dz \tag{3.217}$$

where $u_0^{(2)}(x,y) = u^{(2)}(x,y,z)\big|_{z=\frac{h_D}{2}+\frac{h}{4}}$.

Inspecting the elements in rows 4 and 5 in the constitutive Equation (3.212) in conjunction with the boundary conditions in Equation (3.207), it can be concluded that

$$\begin{cases} \varepsilon_{xz} = 0 \\ \varepsilon_{yz} = 0 \end{cases} \text{at } z = \pm \frac{h}{2} \tag{3.218}$$

Therefore, the transverse shear strain $\varepsilon_{xz}^{(1)}(x,y,z)$ in the concrete host and $\varepsilon_{xz}^{(2)}(x,y,z)$ in the FRP sheet can be assumed to take the form

$$\varepsilon_{xz}^{(1)}(x,y,z) = \varphi^{(1)}(x,y) \left(1 + \frac{2}{h}z \right), \quad \varepsilon_{xz}^{(2)}(x,y,z) = \varphi^{(2)}(x,y) \left(1 - \frac{2}{h}z \right) \tag{3.219}$$

where $\varphi(x,y)$ is an unknown function similar to $\phi(x,y)$ that characterizes the strain distribution at the mid-plane of the plate. Substituting Equations (3.213) and (3.219) into Equations (3.216) and (3.217), the displacement along the x-axis can be expressed as

$$u^{(1)}(x,y,z) = u_0^{(1)}(x,y) + \int_{\frac{h_D}{2}-\frac{h}{4}}^{z} \left[2\varphi^{(1)} \left(1 + \frac{2}{h}z \right) - \frac{\partial \phi^{(1)}}{\partial x} \left(\frac{z^2}{h} + z + \frac{h}{4} \right) \right] dz$$

$$= u_0^{(1)}(x,y) + 2\varphi^{(1)} \left(\frac{z^2}{h} + z + \frac{3h}{16} - \frac{h_D}{4} - \frac{h_D^2 h}{4} \right)$$

$$+ \frac{\partial \phi^{(1)}}{\partial x} \frac{(2h_D - h - 4z)(16z^2 + 20hz + 8h_Dz + 7h^2 + 8h_Dh + 4h_D^2)}{192h}$$

or in the matrix form

$$u^{(1)}(x,y,z) = \begin{Bmatrix} 1 \\ z \\ z^2 \\ z^3 \end{Bmatrix} \begin{bmatrix} 1 & \dfrac{\partial}{\partial x} \dfrac{(8h_D^3 - 7h^3 + 6h^2 h_D + 12h_D^2 h)}{192h} & \dfrac{3h}{8} - \dfrac{h_D}{2} - \dfrac{h_D^2 h}{2} \\ 0 & -\dfrac{h}{4}\dfrac{\partial}{\partial x} & 2 \\ 0 & -\dfrac{h}{2}\dfrac{\partial}{\partial x} & \dfrac{2}{h} \\ 0 & -\dfrac{1}{3h}\dfrac{\partial}{\partial x} & 0 \end{bmatrix} \begin{Bmatrix} u_0^{(1)} \\ \phi^{(1)} \\ \varphi^{(1)} \end{Bmatrix}$$

$$\tag{3.220}$$

Similarly, for the FRP sheet

$$
u^{(2)}(x, y, z) = u_0^{(2)}(x, y) + \int_{\frac{h_D}{2} + \frac{h}{4}}^{z} \left[2\varphi^{(2)} \left(1 - \frac{2}{h}z \right) - \frac{\partial \phi^{(2)}}{\partial x} \left(\frac{z^2}{h} - z + \frac{h}{4} \right) \right] dz
$$

$$
= u_0^{(2)}(x, y) + 2\varphi^{(2)} \left(-\frac{z^2}{h} + z - \frac{3h}{16} - \frac{h_D}{4} + \frac{h_D^2 h}{4} \right)
$$

$$
+ \frac{\partial \phi^{(2)}}{\partial x} \frac{(4z - 2h_D - h)(16z^2 - 20hz + 8h_Dz + 7h^2 - 8h_Dh + 4h_D^2)}{192h}
$$

or

$$
u^{(2)}(x, y, z) = \left\{ \begin{array}{c} 1 \\ z \\ z^2 \\ z^3 \end{array} \right\} \left[\begin{array}{ccc} 1 & \dfrac{\partial}{\partial x} \dfrac{(-8h_D^3 - 7h^3 - 6h^2 h_D + 12h_D^2 h)}{192h} & \dfrac{h_D^2 h}{2} - \dfrac{3h}{8} - \dfrac{h_D}{2} \\ 0 & \dfrac{h}{4} \dfrac{\partial}{\partial x} & 2 \\ 0 & -\dfrac{h}{2} \dfrac{\partial}{\partial x} & -\dfrac{2}{h} \\ 0 & \dfrac{1}{3h} \dfrac{\partial}{\partial x} & 0 \end{array} \right] \left\{ \begin{array}{c} u_0^{(2)} \\ \phi^{(2)} \\ \varphi^{(2)} \end{array} \right\}
$$

$$(3.221)$$

Further with the use of the strain–displacement relation, $\varepsilon_{xx} = \partial u / \partial x$, the strains $\varepsilon_{xx}^{(1)}$ and $\varepsilon_{xx}^{(2)}$ can be found in terms of the unknown characterizing functions u_0, ϕ and φ as

$$
\varepsilon_{xx}^{(1)} = \left\{ \begin{array}{c} 1 \\ z \\ z^2 \\ z^3 \end{array} \right\} \left[\begin{array}{ccc} \dfrac{\partial}{\partial x} & \dfrac{\partial^2}{\partial x^2} \dfrac{(8h_D^3 - 7h^3 + 6h^2 h_D + 12h_D^2 h)}{192h} & \left(\dfrac{3h}{8} - \dfrac{h_D}{2} - \dfrac{h_D^2 h}{2} \right) \dfrac{\partial}{\partial x} \\ 0 & -\dfrac{h}{4} \dfrac{\partial^2}{\partial x^2} & 2\dfrac{\partial}{\partial x} \\ 0 & -\dfrac{h}{2} \dfrac{\partial^2}{\partial x^2} & \dfrac{2}{h} \dfrac{\partial}{\partial x} \\ 0 & -\dfrac{1}{3h} \dfrac{\partial^2}{\partial x^2} & 0 \end{array} \right] \left\{ \begin{array}{c} u_0^{(1)} \\ \phi^{(1)} \\ \varphi^{(1)} \end{array} \right\}
$$

$$(3.222)$$

and

$$
\varepsilon_{xx}^{(2)} = \left\{ \begin{array}{c} 1 \\ z \\ z^2 \\ z^3 \end{array} \right\} \left[\begin{array}{ccc} \dfrac{\partial}{\partial x} & \dfrac{\partial^2}{\partial x^2} \dfrac{(-8h_D^3 - 7h^3 - 6h^2 h_D + 12h_D^2 h)}{192h} & \left(\dfrac{h_D^2 h}{2} - \dfrac{3h}{8} - \dfrac{h_D}{2} \right) \dfrac{\partial}{\partial x} \\ 0 & \dfrac{h}{4} \dfrac{\partial^2}{\partial x^2} & 2\dfrac{\partial}{\partial x} \\ 0 & -\dfrac{h}{2} \dfrac{\partial^2}{\partial x^2} & -\dfrac{2}{h} \dfrac{\partial}{\partial x} \\ 0 & \dfrac{1}{3h} \dfrac{\partial^2}{\partial x^2} & 0 \end{array} \right] \left\{ \begin{array}{c} u_0^{(2)} \\ \phi^{(2)} \\ \varphi^{(2)} \end{array} \right\}
$$

$$(3.223)$$

The displacements at the delamination surfaces of the FRP and concrete sheets, u_f and u_c in Equation (3.202), respectively, are written as

$$u_c = u^{(1)}(x, y, h_D) = u_0^{(1)}(x, y) + \int_{\frac{h_D}{2} - \frac{h}{4}}^{h_D} \left(2\varepsilon_{xz}^{(1)} - \frac{\partial w^{(1)}}{\partial x} \right) dz,$$

$$u_f = u^{(2)}(x, y, h_D) = u_0^{(2)}(x, y) + \int_{\frac{h_D}{2} + \frac{h}{4}}^{h_D} \left(2\varepsilon_{xz}^{(2)} - \frac{\partial w^{(2)}}{\partial x} \right) dz \qquad (3.224)$$

Substituting Equations (3.213), (3.219) and (3.224) into Equation (3.201), and expressing $u_0^{(2)}(x, y)$ as a function of $u_0^{(1)}(x, y)$

$$u_0^{(2)}(x, y) = F(\phi^{(1)}, \varphi^{(1)}, \varphi^{(2)}) \cdot u_0^{(1)}(x, y) \qquad (3.225)$$

where $F(\phi^{(1)}, \varphi^{(1)}, \varphi^{(2)})$ is a general function of the characterizing functions $\phi^{(1)}(x, y)$, $\varphi^{(1)}(x, y)$ and $\varphi^{(2)}(x, y)$. This gives the relationship between the displacements along the x-axis of the two plates, which is similar to Equation (3.215) for the out-of-plane displacements. Similar to Equations (3.216) to (3.225), the formulations of the displacements along the y-axis can be obtained as

$$v^{(1)}(x, y, z) = v_0^{(1)}(x, y) + \int_{\frac{h_D}{2} - \frac{h}{4}}^{z} \left(2\varepsilon_{yz}^{(1)} - \frac{\partial w^{(1)}}{\partial y} \right) dz,$$

$$v^{(2)}(x, y, z) = v_0^{(2)}(x, y) + \int_{\frac{h_D}{2} + \frac{h}{4}}^{z} \left(2\varepsilon_{yz}^{(2)} - \frac{\partial w^{(2)}}{\partial y} \right) dz \qquad (3.226)$$

and the transverse strains

$$\varepsilon_{yz}^{(1)}(x, y, z) = \psi^{(1)}(x, y) \left(1 + \frac{2}{h} z \right), \qquad \varepsilon_{yz}^{(2)}(x, y, z) = \psi^{(2)}(x, y) \left(1 - \frac{2}{h} z \right) \qquad (3.227)$$

It should be noted that the coefficients C_{13} and C_{23} in Equation (3.205) would not be negligible in the case of a thick plate, and they need to be included in the above formulation.

Finite element formulation of the delaminated plate
The virtual work principle of the delaminated plate element can be written in the form

$$\iiint_{V_{Con}} \rho^{(1)} (\ddot{u}^{(1)} \delta u^{(1)} + \ddot{v}^{(1)} \delta v^{(1)} + \ddot{w}^{(1)} \delta w^{(1)}) dV$$

$$+ \iiint_{V_{FRP}} \rho^{(2)} (\ddot{u}^{(2)} \delta u^{(2)} + \ddot{v}^{(2)} \delta v^{(2)} + \ddot{w}^{(2)} \delta w^{(2)}) dV$$

$$+ \iiint\limits_{V_{Con}} (\sigma_{xx}^{(1)}\delta\varepsilon_{xx}^{(1)} + \sigma_{xz}^{(1)}2\delta\varepsilon_{xz}^{(1)} + \sigma_{yy}^{(1)}\delta\varepsilon_{yy}^{(1)} + \sigma_{yz}^{(1)}2\delta\varepsilon_{yz}^{(1)} + \sigma_{zz}^{(1)}\delta\varepsilon_{zz}^{(1)} + \sigma_{xy}^{(1)}2\delta\varepsilon_{xy}^{(1)})dV$$

$$+ \iiint\limits_{V_{FRP}} (\sigma_{xx}^{(2)}\delta\varepsilon_{xx}^{(2)} + \sigma_{xz}^{(2)}2\delta\varepsilon_{xz}^{(2)} + \sigma_{yy}^{(2)}\delta\varepsilon_{yy}^{(2)} + \sigma_{yz}^{(2)}2\delta\varepsilon_{yz}^{(2)} + \sigma_{zz}^{(2)}\delta\varepsilon_{zz}^{(2)} + \sigma_{xy}^{(2)}2\delta\varepsilon_{xy}^{(2)})dV = 0$$

$$(3.228)$$

in which the volume and mass of the interface layer are ignored. In order to derive the finite element formulations, the equation of motion in Equation (3.204) and the constitutive relations in Equation (3.212) are substituted into Equation (3.228) with the unknown characterizing functions, $\phi^{(1)}(x,y)$, $u_0^{(1)}(x,y)$, $v_0^{(1)}(x,y)$, $\varphi^{(1)}(x,y)$, $\varphi^{(2)}(x,y)$, $\psi^{(1)}(x,y)$ and $\psi^{(2)}(x,y)$ represented by piecewise interpolation polynomials.

Table 3.5 gives the maximum order of derivatives of the unknown functions required in this finite element model. If the virtual work principle contains spatial derivatives of a characterizing function with a highest order of m, then the chosen interpolation polynomial has to satisfy the following conditions: (a) it must be a complete polynomial of degree m or higher; (b) the polynomial and all its derivatives up to order $m-1$ must be continuous across the element boundaries. The Hermit polynomials of the third degree satisfying the above requirements are therefore chosen to interpolate the function $\phi^{(1)}(x,y)$ as

$$\phi^{(1)}(x,y) = [N]_{1\times16}\{\overline{\phi}^{(1)}\}_{16\times1} \qquad (3.229)$$

where

$$N_1 = H_1(\xi)H_1(\eta), \quad N_2 = H_3(\xi)H_1(\eta), \quad N_3 = H_1(\xi)H_3(\eta), \quad N_4 = H_3(\xi)H_3(\eta),$$

$$N_5 = H_2(\xi)H_1(\eta), \quad N_6 = H_4(\xi)H_1(\eta), \quad N_7 = H_2(\xi)H_3(\eta), \quad N_8 = H_4(\xi)H_3(\eta),$$

$$N_9 = H_1(\xi)H_2(\eta), \quad N_{10} = H_3(\xi)H_2(\eta), \quad N_{11} = H_1(\xi)H_4(\eta), \quad N_{12} = H_3(\xi)H_4(\eta),$$

$$N_{13} = H_2(\xi)H_2(\eta), \quad N_{14} = H_4(\xi)H_2(\eta), \quad N_{15} = H_2(\xi)H_4(\eta), \quad N_{16} = H_4(\xi)H_4(\eta),$$

$$H_1(\xi) = 1 - 3\xi^2 + 2\xi^3, \quad H_2(\xi) = 3\xi^2 - 2\xi^3, \quad H_3(\xi) = \xi - 2\xi^2 + \xi^3, \quad H_4(\xi) = \xi^3 - \xi^2$$

and $\{\overline{\phi}^{(1)}\} = \{\{\overline{\phi}_1^{(1)}\} \quad \{\overline{\phi}_2^{(1)}\} \quad \{\overline{\phi}_3^{(1)}\} \quad \{\overline{\phi}_4^{(1)}\}\}^T$

Table 3.5 The required maximum order of the characterizing functions in the virtual work formulation

Characterizing functions	Maximum order of derivative
$\phi^{(1)}$	2
$u_0^{(1)}$	1
$v_0^{(1)}$	1
$\varphi^{(1)}, \varphi^{(2)}$	1
$\psi^{(1)}, \psi^{(2)}$	1

where $\{\overline{\phi}_i^{(1)}\} = \{\overline{\phi}_i^{(1)} \quad \partial\overline{\phi}_i^{(1)}/\partial x \quad \partial\overline{\phi}_i^{(1)}/\partial y \quad \partial^2\overline{\phi}_i^{(1)}/\partial x\partial y\}^T$ and i denotes the node number.

Following the same rules, the two-dimensional Lagrange polynomials of the first degree are chosen to interpolate the other six characterizing functions (take $\varphi^{(1)}(x, y)$ for example):

$$\varphi^{(1)}(x, y) = [M]_{1 \times 4}\{\overline{\varphi}^{(1)}\}_{4 \times 1} \tag{3.230}$$

where

$$M_1 = \frac{1}{4}(1 - \xi)(1 - \eta), \quad M_2 = \frac{1}{4}(1 + \xi)(1 - \eta),$$

$$M_3 = \frac{1}{4}(1 + \xi)(1 + \eta), \quad M_4 = \frac{1}{4}(1 - \xi)(1 + \eta),$$

and $\{\overline{\varphi}^{(1)}\} = \{\overline{\varphi}_1^{(1)} \quad \overline{\varphi}_2^{(1)} \quad \overline{\varphi}_3^{(1)} \quad \overline{\varphi}_4^{(1)}\}^T$.

Further, a general nodal displacement vector, $\{\overline{w}\}_{10 \times 1}$, is introduced, the components of which are defined as

$$\overline{w}_1 = \overline{\phi}^{(1)}, \quad \overline{w}_2 = \frac{\partial\overline{\phi}^{(1)}}{\partial x}, \quad \overline{w}_3 = \frac{\partial\overline{\phi}^{(1)}}{\partial y}, \quad \overline{w}_4 = \frac{\partial^2\overline{\phi}^{(1)}}{\partial x\partial y}, \quad \overline{w}_5 = \overline{u}_0^{(1)}, \quad \overline{w}_6 = \overline{v}_0^{(1)},$$

$$\overline{w}_7 = \overline{\varphi}^{(1)}, \quad \overline{w}_8 = \overline{\varphi}^{(2)}, \quad \overline{w}_9 = \overline{\psi}^{(1)}, \overline{w}_{10} = \overline{\psi}^{(2)}$$

By substituting the polynomial approximations of the unknown functions in Equations (3.229) and (3.230) into the virtual work principle expression in Equation (3.228), the finite element formulation of the delaminated plate model can be obtained in terms of the general nodal displacement vector

$$\{\delta\overline{w}\}^T \left([m]\{\ddot{\overline{w}}\} + [k]\{\overline{w}\}\right) = 0 \tag{3.231}$$

where $[m]_{40 \times 40}$ and $[k]_{40 \times 40}$ are the elemental mass and stiffness matrices of the four-node rectangular plate element, respectively.

Verification of the finite element model

An example of a cantilever FRP-bonded plate is considered. The plate consists of a 30 mm thick concrete host and a bonded 6mm thick T300/5208 carbon fibre reinforced polymer layer, and it has the in-plane dimensions 1.6m × 0.8 m. The fibre orientation of the FRP layer is parallel to the x-axis, which is normal to the support, as shown in Figure 3.48. Table 3.6 summarizes the material properties of the components of the structure. The cantilever plate is separately represented by two finite element models, in which one is based on the above-derived formulations with the bonding parameters $p_b = 1$ for each element and the other finite element model consists of two layers of eight-node solid elements, with one layer for the concrete host and the other for the FRP. The two models have the same in-plane meshing of 20 × 10, as shown in Figure 3.48. It is assumed that the delamination starts from the free end over the full width of the plate. To simulate the delamination damage with the present model, the

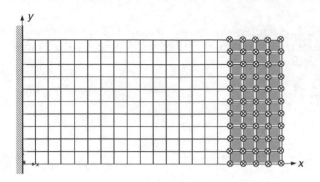

Figure 3.48 Finite element model of the plate with delamination at the free end

Table 3.6 Material properties of the cantilever plate

Properties	Concrete	T300/5208 (CFRP)
Mass density (kg/m^3)	2402	1600
Elastic modulus E_1(Pa)	2.482E10	1.81E11
Elastic modulus E_2(Pa)	–	1.03E10
Poisson ratio ν_{21}	0.2	0.28
Torsional modulus G_{12}	1.034E10	7.17E9

Table 3.7 Natural frequencies (Hz) of the plate with delamination at the free end

Mode order	Natural frequency (Hz)				Percentage reduction (%) due to delamination	
	Solid element model		Present model		Solid element model	Present model
	Intact	Delaminated	Intact	Delaminated		
1	15.087	15.078	15.135	14.833	0.06%	2.00%
2	37.135	36.074	37.576	36.668	2.86%	2.42%
3	94.397	93.264	93.318	92.563	1.20%	0.81%
4	135.71	126.89	136.18	124.96	6.50%	8.24%
5	–	163.88	–	156.76	–	–
6	185.17	170.61	186.46	173.76	7.86%	6.81%
7	–	184.95	–	180.67	–	–
8	264.33	252.39	262.31	245.05	4.52%	6.58%

bonding parameter could be set to a tiny but non-vanishing value ($p_b = 1 \times 10^{-5}$ in this example) for the delaminated elements. For the two-layer solid element model, the delamination is simulated by un-joining the common joints at the concrete-FRP interface. The un-joined nodes are shown as $\otimes$ in Figure 3.48.

Modal analysis is performed separately on the two finite element models for the intact and delaminated plate. The first eight natural frequencies are listed in Table 3.7. The results from the present model are consistent with those from the two-layer solid

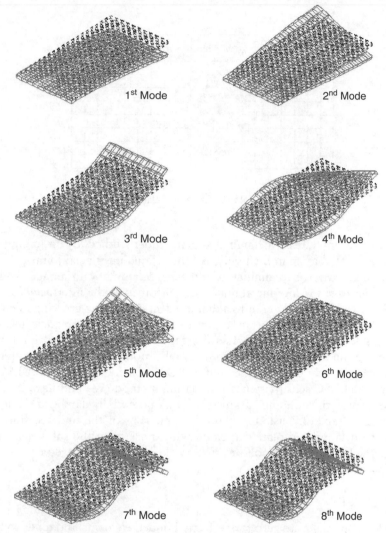

Figure 3.49 Identified mode shapes of the cantilevered delaminated plate (- - - - original plate; ____ deformed plate)

element model for both the intact and delaminated states. Both models successfully predict the 'delamination modes' without sliding and sticking (friction) between the two component plates, which are caused by the delamination and shown as the 5th and 7th modes in Figure 3.49.

Verification of the delamination detection

A new damage scenario with the previously studied cantilever plate, as shown in Figure 3.50, is considered. The dark-grey area represents the delaminated zone, and the surrounding grey area denotes the weak-bonding zone. The two-layer solid element

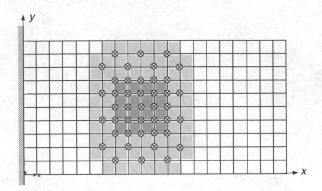

Figure 3.50 Finite element model of the plate with delamination at mid-span

model is used to simulate the delaminated plate under studied, by un-joining the inter-facial joints denoted by $\otimes$ in the figure. All the related interfacial joints are un-joined in the delaminated zone to simulate the macro delamination damage, while in the weak-bonding zone the interfacial nodes are un-joined at the interfacial joints with a staggered pattern to simulate the weakening effect due to the growing micro-defects. It is assumed that the free vibration response of the delaminated plate is 'measured' by 'accelerators' located at all nodal points along the z-direction on the FRP side to extract the frequencies and mode shapes. The first eight modes from the 'measurement' are used to approximately estimate the Uniform Load Surface curvature (ULSC) and its sensitivity. The ULSC sensitivity-based updating method (Wu and Law, 2007) is then applied to identify the bonding parameters for all potentially damaged elements in the present model. Table 3.7 lists the natural frequencies of the updated finite element model of the delaminated plate compared with those from the solid element model. To denote the local delamination severity, a debonding index based on the identified bonding parameters is defined as

$$DI = -\log p_b \qquad\qquad (3.232)$$

The index approaches to zero for perfectly bonded elements and tends to be a large positive scalar when the interface is completely debonded. Figure 3.51 shows the delamination index map of the updated plate model. It is seen that the completely debonded zone and the weak-bonding zone can be distinctly localized separately with different index values. The debonding index corresponding to the completely delami-nated zone attains a value of 3.0 to 5.0, whereas the index for the weak-bonding zone is mostly not larger than 1.0.

3.3 Conclusions

This chapter gives the formulation of a group of local damages in beam and plate structures which have been specifically developed for the purpose of structural damage detection. They are called the group of Damage-Detection-Oriented models, and are different from the usual models for the direct analysis for the behaviour of structures

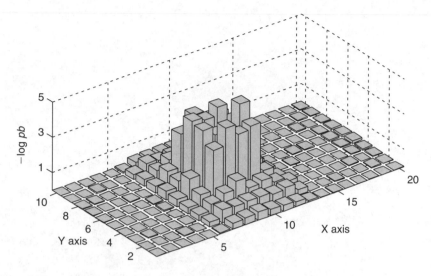

Figure 3.51 Identified delamination index map of the plate with delamination at mid-span

under load. A damage index has been included in each of these models, which is suitable to be included in any type of parametric identification algorithm for the condition assessment of structure. The system identification technique which goes along with these damage models is referred to in Chapters 6 to 8 of this book for further illustration of the complete process of condition assessment of a structure.

Chapter 4

Model reduction

4.1 Introduction

It is computationally expensive and sometimes difficult to determine the natural frequencies and mode shapes for a structure with a large number of elements using the full finite element model. For the inverse problems in structural analysis, the dynamic responses of a structure required for the system identification are usually measured, and it is difficult and impractical to obtain the responses at all the degrees-of-freedom (DOFs) of the structure. It is desirable to reduce the size of the system matrices to have a more manageable and economic solution for both the direct and inverse problems from only the responses of the measured DOFs. Such a reduction is usually referred to as the 'model condensation method'. Another technique, called the 'model expansion method', to solve the problem of limited measured DOFs of a structure and the large number of DOFs of a structure is also presented in this chapter.

4.2 Static condensation

The oldest and perhaps the most popular model condensation method is the Guyan/Irons method (Guyan, 1965; Irons, 1965), which is applicable to both the static and dynamic problems. Since the dynamic effect is ignored in the condensation, this method is usually referred to as the 'static condensation'.

Considering a system with a static force, F_m, applied at some selected DOFs, its stiffness matrix and response vector can be partitioned as

$$\begin{bmatrix} K_{mm} & K_{ms} \\ K_{sm} & K_{ss} \end{bmatrix} \begin{Bmatrix} x_m \\ x_s \end{Bmatrix} = \begin{Bmatrix} F_m \\ 0 \end{Bmatrix} \tag{4.1}$$

where m and s denote the master (selected) DOFs and slave (truncated) DOFs, respectively. Combining the upper and lower parts of Equation (4.1), the relationship between the selected response x_m and the truncated response x_s is

$$x_s = -K_{ss}^{-1} K_{sm} x_m \tag{4.2}$$

The response vector can be described using the selected response x_m only as

$$x = \begin{Bmatrix} x_m \\ x_s \end{Bmatrix} = \begin{bmatrix} I \\ -K_{ss}^{-1}K_{sm} \end{bmatrix} x_m = T_s x_m \tag{4.3}$$

where $T_s = \begin{bmatrix} I \\ -K_{ss}^{-1}K_{sm} \end{bmatrix}$ is the static transformation matrix between the full state vector x and the master coordinates x_m. The condensed system matrix, M_{Rs}, K_{Rs}, can then be expressed as

$$M_{Rs} = T_s^T M T_s, \quad K_{Rs} = T_s^T K T_s \tag{4.4}$$

where M and K are the system matrices before reduction. The eigen-solution for the condensed system can be denoted as

$$(K_{Rs} - \omega^2 M_{Rs})\phi_m = 0 \tag{4.5}$$

where ω^2 and ϕ_m are the eigenvalues and eigenvectors of the condensed system, respectively.

It should be noted that the incorrect selection of the master DOFs may result in a singularity in the eigenvalue problem. Different selection schemes (Shan and Raymund, 1982; Matta, 1987) have been developed to improve the accuracy of the eigenvalue problem. These schemes are also applicable to other methods referred to in this chapter.

It is noted that any frequency-response function generated from the reduced matrices in Equation (4.4) is exact only at zero frequency. In a dynamic system, the inertia effect increases with increasing frequency, and the static condensation method may lead to a significant error. Therefore, dynamic condensation methods have subsequently been developed to avoid this deficiency.

4.3   Dynamic condensation

A reduction method including the dynamic effects, known as the 'dynamic condensation method', is discussed in this section (Kuhar and Stahle, 1974; Miller, 1980). Inertia terms that were omitted in Equation (4.1) are included to give

$$\begin{bmatrix} M_{mm} & M_{ms} \\ M_{sm} & M_{ss} \end{bmatrix} \begin{Bmatrix} \ddot{x}_m \\ \ddot{x}_s \end{Bmatrix} + \begin{bmatrix} K_{mm} & K_{ms} \\ K_{sm} & K_{ss} \end{bmatrix} \begin{Bmatrix} x_m \\ x_s \end{Bmatrix} = 0 \tag{4.6}$$

Thus, the eigenvalue problem becomes

$$\left(-\omega^2 \begin{bmatrix} M_{mm} & M_{ms} \\ M_{sm} & M_{ss} \end{bmatrix} + \begin{bmatrix} K_{mm} & K_{ms} \\ K_{sm} & K_{ss} \end{bmatrix} \right) \begin{Bmatrix} x_m \\ x_s \end{Bmatrix} = 0 \tag{4.7}$$

The slave vector of responses, x_s, may again be solved in terms of x_m as

$$\begin{aligned} x_s &= -[K_{ss} - \omega^2 M_{ss}]^{-1}[K_{sm} - \omega^2 M_{sm}]x_m \\ &= T_d x_m \end{aligned} \tag{4.8}$$

where $T_d = -[K_{ss} - \omega^2 M_{ss}]^{-1}[K_{sm} - \omega^2 M_{sm}]$ is the dynamic transformation matrix between the full state vector and the master coordinates. The eigenvalue problem of the reduced system becomes

$$(-\omega^2(M_{mm} - M_{ms}T_d) + (K_{mm} - K_{ms}T_d))x_m = 0 \tag{4.9}$$

Note that the eigenvalue, ω, in T_d is unknown. It can be found by using an iterative process to find the transformation matrix with the set of initial eigenvalues equal to zero. A new eigenvalue vector based on the updated transformation matrix is then calculated from Equation (4.9) and the process is repeated until the eigenvalue no longer changes.

After obtaining the dynamic transformation matrix, the reduced system matrix can be calculated as

$$M_{Rd} = T_d^T M T_d, \quad K_{Rd} = T_d^T K T_d \tag{4.10}$$

To avoid the matrix inversion of T_d in Equation (4.8), the first term on the right-hand-side is expanded into a Taylor series (Gordis, 1992) with the higher order terms larger than ω^2 neglected, so T_d becomes

$$\begin{aligned} T_d &= -[K_{ss} - \omega^2 M_{ss}]^{-1}[K_{sm} - \omega^2 M_{sm}] = -K_{ss}^{-1}[I - \omega^2 M_{ss}K_{ss}^{-1}]^{-1}[K_{sm} - \omega^2 M_{sm}] \\ &= -K_{ss}^{-1}[I + \omega^2 M_{ss}K_{ss}^{-1}][K_{sm} - \omega^2 M_{sm}] \\ &= -K_{ss}^{-1}[K_{sm} + \omega^2(M_{ss}K_{ss}^{-1}K_{sm} - M_{sm})] \end{aligned} \tag{4.11}$$

Substituting Equations (4.8) and (4.11) into Equation (4.7) and neglecting the higher order terms, gives

$$\omega^2(M_{mm} - M_{ms}K_{ss}^{-1}K_{sm} - K_{ms}K_{ss}^{-1}M_{ss}K_{ss}^{-1}K_{sm})x_m = (K_{mm} - K_{ms}K_{ss}^{-1}K_{sm})x_m \tag{4.12}$$

Thus, the eigenvalue can be determined directly from Equation (4.12) without a requirement of iteration. Actually, this eigenvalue is the same as the one calculated from the reduced model using the Guyan/Irons method, which can be expressed as

$$\omega^2 M_{Rs}\phi_m = K_{Rs}\phi_m \tag{4.13}$$

where M_{Rs} and K_{Rs} are already shown in Equation (4.4). Using the relationship in Equation (4.13), the dynamic transformation matrix T_d can be modified as

$$T_{id} = -K_{ss}^{-1}[K_{sm} + (M_{ss}K_{ss}^{-1}K_{sm} - M_{sm})M_R^{-1}K_R] \tag{4.14}$$

where T_{id} is the transformation matrix referred to in Equation (4.10). This model condensation method is called the Improved Reduction System (IRS) method.

This IRS method was originally proposed by O'Callahan (1989) with a different formulation but giving the same transformation matrix. The following gives the formulation developed by O'Callahan for further comparison.

Considering a vector of static force, F, acting on all the DOFs of a structure and with the same partitioning as in the Guyan/Irons method, Equation (4.1) becomes

$$\begin{bmatrix} K_{mm} & K_{ms} \\ K_{sm} & K_{ss} \end{bmatrix} \begin{Bmatrix} x_m \\ x_s \end{Bmatrix} = \begin{Bmatrix} F_m \\ F_s \end{Bmatrix} \tag{4.15}$$

where F_s is the applied force on the slave DOFs. The truncated set of equation becomes

$$K_{sm}x_m + K_{ss}x_s = F_s \tag{4.16}$$

Solving for x_s in Equation (4.16), gives

$$x_s = -K_{ss}^{-1}K_{sm}x_m + K_{ss}^{-1}F_s \tag{4.17}$$

Thus, the full state vector, x, can be expressed as

$$x = T_s x_m + x_{Fd} \tag{4.18}$$

where T_s is demonstrated in Equation (4.3) and x_{Fd} is the displacement adjustment from the truncated distributed force defined as

$$x_{Fd} = \begin{Bmatrix} 0 \\ K_{ss}^{-1}F_s \end{Bmatrix} = \begin{bmatrix} 0 & 0 \\ 0 & K_{ss}^{-1} \end{bmatrix} F = K_s^{-1}F \tag{4.19}$$

According to Equations (4.13) and (4.15), the full space modal vector, ϕ, can be expressed by ϕ_m using the same static reduction, that is $\phi = T_s\phi_m$, and x_{Fd} can be obtain as

$$x_{Fd} = K_s^{-1}MT_s\phi_m\omega^2 = K_{ss}^{-1}M_{ss}T_sM_R^{-1}K_R\phi_m \tag{4.20}$$

then

$$x = (T_s + K_{ss}^{-1}M_{ss}T_sM_R^{-1}K_R)\phi_m = T_{IRS}\phi_m \tag{4.21}$$

where

$$T_{IRS} = -K_{ss}^{-1}[K_{sm} + (M_{ss}K_{ss}^{-1}K_{sm} - M_{sm})M_R^{-1}K_R] \tag{4.22}$$

Note that T_{IRS} in Equation (4.22) is identical to T_{id} in Equation (4.14).

The mass and stiffness matrices obtained from using the transformation matrix in Equation (4.22) or Equation (4.14) exhibit improved system characteristics compared to those obtained from static condensation with the inertia effects included. The mass and stiffness matrices for the improved reduced system are thus

$$M_{RI} = T_{IRS}^T M T_{IRS}, \quad K_{RI} = T_{IRS}^T K T_{IRS} \tag{4.23}$$

The quality of the IRS results is relatively insensitive to the number and location of the selected DOFs compared with the methods mentioned in previous sections.

Other dynamic condensation methods and iterative methods involving eigenvalue analysis have also been developed. An accurate method of dynamic condensation with a simplified computation for sub-structures was developed by Leung (1978; 1979). An improved dynamic condensation approach was developed by Suarez and Singh (1992) for structural eigenvalue analysis, which can use the kept and reduced DOFs separately or together to calculate the total eigen-solution of the system.

4.4 Iterative condensation

Improvements have been proposed by other researchers which modify the method into an iterative process. Such a method is called the Iterative IRS method (IIRS). There are two types of IIRS method. The first one was proposed by Blair (1991). Since the transformation matrix for the IRS method makes use of the reduced mass and stiffness matrices obtained from the Guyan/Irons method, an improved transformation matrix can be constructed using the newly approximated matrices described in Equation (4.23) via iterations with Equation (4.22) as

$$T_{IRS,i+1} = -K_{ss}^{-1}[K_{sm} + (M_{ss}T_s - M_{sm})M_{RI,i}^{-1}K_{RI,i}] \tag{4.24}$$

where i is the number of iterations. Equation (4.23) becomes

$$M_{RI,i+1} = T_{IRS,i+1}^T M T_{IRS,i+1}, \quad K_{RI,i+1} = T_{IRS,i+1}^T K T_{IRS,i+1} \tag{4.25}$$

The second type of IIRS method (Friswell et al., 1995) further incorporates the transformation matrix from the last iteration in the computation by modifying Equation (4.24) as

$$T_{IRS,i+1} = -K_{ss}^{-1}[K_{sm} + (M_{ss}T_{IRS,i} - M_{sm})M_{RI,i}^{-1}K_{RI,i}] \tag{4.26}$$

The two types of iterative method are quite similar and both of them show good accuracy for model reduction. Another iterative scheme for dynamic condensation was also introduced by Qu and Fu (2000) and the full proof of its convergence was given. Two criteria for the computation convergence were introduced with improved computational efficiency.

4.5 Moving force identification using the improved reduced system

4.5.1 Theory of moving force identification

The equation of motion for the vehicle-bridge system can be written as (Law et al., 2004a)

$$M_b\ddot{R} + C_b\dot{R} + K_bR = H_cP_{int} \tag{4.27}$$

where M_b, C_b and K_b are the mass, damping and stiffness matrices of the bridge, respectively; $\ddot{R}, \dot{R}$ and R are the nodal acceleration, velocity and displacement vectors of the bridge, respectively; and H_cP_{int} is the equivalent nodal load vector of the bridge-vehicle interaction force with

$$H_c = \begin{Bmatrix} 0 & \cdots & 0 & \cdots & H_1 & \cdots & 0 \\ 0 & \cdots & H_2 & \cdots & 0 & \cdots & 0 \end{Bmatrix}^T$$

H_c is an $NN \times N_p$ matrix with zero entries, except at the DOFs corresponding to the nodal displacements of the beam elements on which the load is acting; and NN is the number of DOFs of the bridge after considering the boundary condition.

From Equation (4.27), the strain at a point x and at time t can be written as

$$\varepsilon(x, t) = -z \frac{\partial^2 H(x) \, R(t)}{\partial x^2} \tag{4.28}$$

where z represents the distance from the neutral axis of the beam to the bottom surface.

The Improved Reduced System reduction scheme (O'Callahan, 1989) is adopted to condense the unmeasured DOFs to the measured DOFs of the bridge deck. All the measured DOFs are designated as the master DOFs and denoted by $R_m(t)$. The remaining structural DOFs are called the slave DOFs, and are denoted by $R_s(t)$. The response vector of the bridge is then partitioned as

$$R(t) = \begin{Bmatrix} R_m(t) \\ R_s(t) \end{Bmatrix} \tag{4.29}$$

Accordingly, the bridge mass, damping, stiffness and shape function matrices are also partitioned as

$$M_b = \begin{bmatrix} M_{mm} & M_{ms} \\ M_{sm} & M_{ss} \end{bmatrix}, \quad C_b = \begin{bmatrix} C_{mm} & C_{ms} \\ C_{sm} & C_{ss} \end{bmatrix}, \quad K_b = \begin{bmatrix} K_{mm} & K_{ms} \\ K_{sm} & K_{ss} \end{bmatrix},$$

$$H = \begin{bmatrix} H_{mm} \\ H_{ss} \end{bmatrix}^T, \quad H_c = \begin{bmatrix} H_{cmm} \\ H_{css} \end{bmatrix} \tag{4.30}$$

The total response matrix of the system, $R(t)$, can then be represented by a partitioned matrix related to the master set of DOFs $R_m(t)$ multiplied by the transformation matrix as

$$R(t) = W \, R_m(t), \tag{4.31}$$

where $W = W_s + W_i, \quad W_s = \begin{bmatrix} I \\ -K_{ss}^{-1} K_{sm} \end{bmatrix}$

and $W_i = \begin{bmatrix} I \\ K_{ss}^{-1}(M_{sm} - M_{ss}K_{ss}^{-1}K_{sm})(W_s^T M_b W_s)^{-1}(W_s^T K_b W_s) \end{bmatrix}$.

where I is the identity matrix. Substituting Equations (4.30) and (4.31) into Equation (4.27) and pre-multiplying W^T on both sides, gives

$$U = H_{cr} \, P_{int} \tag{4.32}$$

and $U = M_r \ddot{R}_m + C_r \dot{R}_m + K_r R_m$
where $M_r = W^T M_b W, C_r = W^T C_b W, K_r = W^T K_b W$, and $H_{cr} = W^T H_c$.

Substituting Equation (4.31) and $H = \begin{bmatrix} H_{mm} \\ H_{ss} \end{bmatrix}^T$ into Equation (4.28), matrices $R_m, \dot{R}_m$ and $\ddot{R}_m$ can be obtained by the generalized orthogonal function method.

The moving forces obtained from Equation (4.32) using a straightforward least-squares solution would be unbound. A regularization technique can be used to solve the ill-posed problem in the form of minimizing the function.

$$J(P, \lambda) = \| B\,P_{int} - U \|^2 + \lambda \| P_{int} \|^2 \qquad (4.33)$$

where λ is the non-negative regularization parameter.

4.5.2 Numerical example

The vehicle-bridge system is shown in Figure 4.1. The bridge deck is modelled as a simply supported beam with the physical and material parameters as shown in Table 4.1. The vehicle is modelled as two-axle loads at 4.26 m spacing, moving at a constant speed. The two axle loads are:

$$P_1(t) = 6268(1.0 + 0.1\sin(10^\pi t) + 0.05\sin(40^\pi t))\,\text{kg};$$

$$P_2(t) = 12332(1.0 - 0.1\sin(10^\pi t) + 0.05\sin(50^\pi t))\,\text{kg}.$$

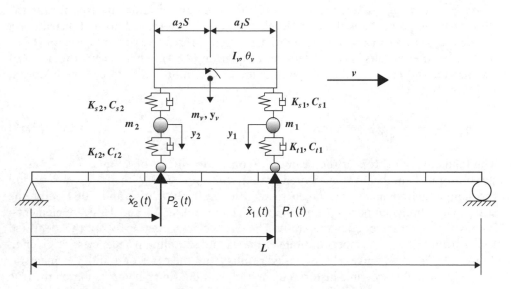

Figure 4.1 Vehicle-bridge system

Table 4.1 Parameters of vehicle-bridge system

Bridge		Vehicle	
$L = 30\,\text{m}$	$I_v = 1.47 \times 10^5\,\text{kgm}^2$	$m_1 = 1500\,\text{kg}$	$m_2 = 1000\,\text{kg}$
$EI = 2.5 \times 10^{10}\,\text{Nm}^2$	$m_v = 17735\,\text{kg}$	$k_{s1} = 2.47 \times 10^6\,\text{N/m}$	$k_{s2} = 4.23 \times 10^6\,\text{N/m}$
$\rho A = 5.0 \times 10^3\,\text{kg/m}$	$S = 4.27\,\text{m}$	$k_{t1} = 3.74 \times 10^6\,\text{N/m}$	$k_{t2} = 4.60 \times 10^6\,\text{N/m}$
$\xi = 0.02$ for all modes	$a1 = 0.519$	$c_{s1} = 3.00 \times 10^4\,\text{N/m/s}$	$c_{s2} = 4.00 \times 10^4\,\text{N/m/s}$
$z = 1.0\,\text{m}$	$a2 = 0.481$	$c_{t1} = 3.90 \times 10^3\,\text{N/m/s}$	$c_{t2} = 4.30 \times 10^3\,\text{N/m/s}$

Table 4.2 The percentage error of the identified forces for different discretization schemes and sampling rates

Sampling frequency (Hz)	Number of elements							
	4		8		12		16	
	Axle-1	Axle-2	Axle-1	Axle-2	Axle-1	Axle-2	Axle-1	Axle-2
100	20.93	18.36	21.11	17.49	20.41	17.64	21.46	17.66
200	15.02	13.12	11.42	9.19	11.50	9.20	11.50	9.12
300	14.31	12.57	11.09	8.99	10.99	8.92	10.98	8.90
400	14.64	12.92	10.92	9.03	10.94	9.04	10.95	8.94
500	14.68	12.92	10.95	9.03	10.97	9.04	10.97	8.94

The calculated responses are polluted with white noise to simulate the polluted measurement as

$$\varepsilon = \varepsilon_{calculated}(1 + E_p * N_{oise})$$

where ε_j and $\varepsilon_{j\,calculated}$ are the vectors of measured and calculated responses at the jth measuring point; E_p is the noise level; and N_{oise} is a standard normal distribution vector with zero mean and unit standard deviation. The relative percentage error (RPE) in the identified results is calculated from Equation (4.34), where $\|\bullet\|$ is the norm of the matrix; and $P_{identified}$ and P_{true} are the identified and the true force time histories, respectively.

$$RPE = \frac{\|P_{identified} - P_{true}\|}{\|P_{true}\|} \times 100\% \tag{4.34}$$

The loads move on top of the beam at a constant velocity of 15 m/s. The strain at the bottom of the beam is measured at $L/4$, $L/2$ and $3L/4$ with three sensors, and the sampling frequencies are taken to be 100, 200, 300, 400 and 500 Hz for the study. The simply supported beam is discretized into 4, 8, 12 and 16 beam elements. No noise effect is included in this study. The relative percentage errors of the identified forces from different numbers of finite elements and sampling frequencies are given in Table 4.2. Figure 4.2 gives the identified results from the cases of 4 and 8 elements.

The identified force–time histories from both 4 and 8 finite elements match the true forces very well in the middle half of the duration. The forces from the four-element case have large fluctuations after the entry of the second load and before the exit of the first load, while those from eight-element case have slight fluctuations at these moments. Table 4.2 also shows that the relative percentage errors of the identified forces from the four-element case are much larger than those from the other discretization cases. These indicate that discretizing the beam into eight elements would be sufficient for an accurate identification.

The sampling frequency is shown not to have any significant effect when it is larger or equal to 200 Hz. It is noted that the first five modes are included in the measured responses with this sampling frequency indicating that the higher modes do not make a significant contribution to the identification accuracy.

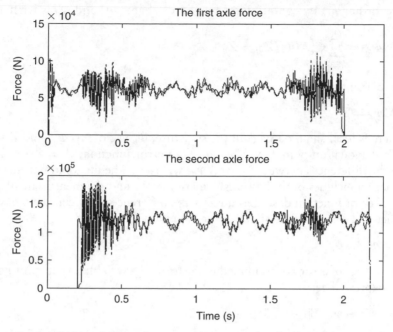

Figure 4.2 Identified results from 3 measured points and without noise (___ true force, — 8 elements, ...4 elements)

4.6 Structural damage detection using incomplete modal data

4.6.1 *Mode shape expansion*

The measured modes are incomplete in practice because of the limited number of sensors, and because the rotational DOFs are usually difficult to measure. Law et al. (1998) have presented a mode shape expansion by which the mode shape of a limited number of measured DOFs is expanded to the full dimension of the finite element model. The mode shape expansion preserves the connectivity of the structure in the final expanded mode shape.

If there is a small change in the stiffness of the structure, each perturbed mode shape can be described as a linear combination of the original mode shapes as

$$\Phi_d = \begin{bmatrix} \Phi_{dc} \\ \Phi_{df} \end{bmatrix} = \Phi_u Z = \begin{bmatrix} \Phi_{uc} \\ \Phi_{uf} \end{bmatrix} Z \qquad (4.35)$$

where the subscript c denotes the master DOFs of the experimental model; the subscript f denotes the number of additional DOFs to be expanded to form the complete mode shape; matrix Z is the transformation matrix; and Φ_d and Φ_u are the mode shape matrices of the damaged and the undamaged structures, respectively. The cross-orthogonality condition for two mode shapes Φ_u and Φ_d is written as

$$\Phi_{ui}^T M \Phi_{dj} = \sum_{k=1}^{m} \Phi_{ui}^T M \Phi_{uk} Z_{kj} = Z_{ij} \tag{4.36}$$

and in a matrix form as

$$\Phi_u^T M \Phi_d = Z \tag{4.37}$$

When there is local small damage in the structure, Φ_u is close to Φ_d; and Z should be close to a diagonal unity matrix. Therefore, an error function, $\Delta_1 = I - Z$, is used to describe the differences between Φ_u and Φ_d, where I is the diagonal unit matrix. This error function includes both the transformation error and the measurement errors.

Another error function describes directly the differences between the measured and the predicted damaged mode shapes from Equation (4.35) as

$$\Delta_2 = \Phi_{dc} - \Phi_{uc} Z \tag{4.38}$$

The two types of error are combined in the form of a weighted Frobenius norm error function expressed as

$$\Psi = \|\Phi_{dc} - \Phi_{uc} Z\|_F^2 + W\|I - Z\|_F^2 \tag{4.39}$$

where W is the weight coefficient. By differentiating this error function with respect to each element of the matrix Z and setting it equal to zero, the transformation matrix Z is obtained as

$$Z = (\Phi_{uc}^T \Phi_{uc} + WI)^{-1}(\Phi_{uc}^T \Phi_{dc} + WI) \tag{4.40}$$

Note that Φ_{uf} is not measured and it is practical to use the analytical mode shapes instead. When Φ_{uc} is not available, i.e. there is no measured information from the undamaged model, the analytical values can also be used in the expansion.

Substituting Equation (4.40) into Equation (4.35) yields the additional DOFs required to form the complete damaged mode shape,

$$\Phi_{df} = \Phi_{uf} Z \tag{4.41}$$

This method does not require an accurate analytical model but it should have the correct connectivity assumptions for the structure with appropriate types of finite elements.

The expansion result can be adjusted by weighing the accuracy of the analytical model and the modal analysis model with the weight coefficient, W. If the modal analysis model is reliable and accurate, less weight is given to the analytical information by making the weight W less than 1.0. Otherwise the value of W is made larger than 1.0. Note that when W equals zero, the expansion method converges to the SEREP method

4.6.2 Application

The frame structure (2.82m high and 1.41m wide) is shown in Figure 4.3, with details of the geometrical and physical information and the finite element model. The structure

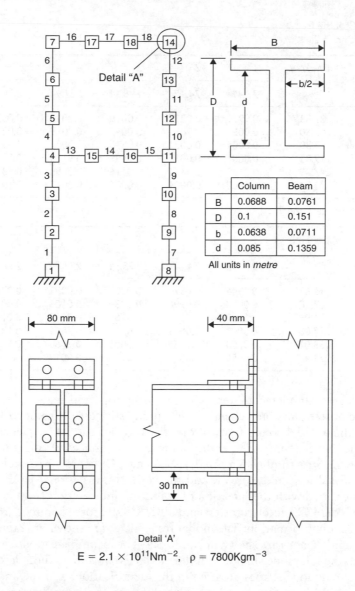

The table within the figure:

	Column	Beam
B	0.0688	0.0761
D	0.1	0.151
b	0.0638	0.0711
d	0.085	0.1359

All units in *metre*

Detail 'A'

$E = 2.1 \times 10^{11} \text{Nm}^{-2}, \quad \rho = 7800 \text{Kgm}^{-3}$

Figure 4.3 Finite element model of two-storey frame structure

is modelled by 18 two-dimensional beam elements of equal length. The beams are connected to the column horizontally by top- and seat-angles and double web-angles with nuts and bolts. Details of this type of semi-rigid beam-column connection are also shown in Figure 4.3. The initial rotational stiffness of the beam-column connection is approximately 3.0×10^6 Nm/rad from static tests. The bottoms of the columns are fully welded to base plates which are welded to the rigid floor, and the same rotational stiffness as for the beam-column connection is assumed for these supports in the finite element modelling.

Table 4.3 MAC of the frame structure
(a) Damage at node 7

Analy. Freq. Damaged Freq.	1	2	3	4	5	6
	22.49	74.17	198.51	221.77	261.32	280.17
1 21.83	0.9979	0.0295	0.0019	0.0023	0.0325	0.0034
2 67.50	0.0053	0.9834	0.0047	0.0019	0.0340	0.0002
3 193.30	0.0004	0.0022	0.9535	0.0002	0.0377	0.0006
4 219.83	0.0001	0.0015	0.0689	0.9173	0.0221	0.0103
5 247.09	0.0145	0.0131	0.0116	0.0598	0.8359	0.1291
6 278.24	0.0000	0.0001	0.0006	0.0017	0.1021	0.7513

(b) Damage at nodes 4 and 7

Analy. Freq. Damaged Freq.	1	2	3	4	5	6
	22.49	74.17	198.51	221.77	261.32	280.17
1 18.67	0.9946	0.0373	0.0006	0.0082	0.0206	0.0011
2 67.10	0.0024	0.9796	0.0030	0.0106	0.0032	0.0002
3 191.99	0.0007	0.0032	0.9788	0.0018	0.0115	0.0049
4 217.08	0.0001	0.0001	0.0513	0.7016	0.1336	0.0702
5 235.02	0.0203	0.0153	0.0001	0.2766	0.6690	0.0910
6 273.42	0.0050	0.0017	0.0126	0.0019	0.1182	0.5678

The damage is simulated by removing both the top- and seat-angles at the joint. This release of restraints only changes the bending stiffness of the horizontal member but retains the original connectivity of the structure. Six modal frequencies and mode shapes of the frame before and after the 'damage' is introduced are measured and determined using the Structural Modal Analysis Package. Only the horizontal translational DOFs at nodes 2 to 7 and 9 to 14, and the vertical translational DOFs at nodes 15 to 18 are measured with the intention of collecting information from all parts of the structures. B&K 4370 accelerometers and a B&K 8202 force hammer are used to collect the vibration information. The modal frequencies are shown in Table 4.3 and the theoretical mode shapes are shown in Figure 4.4. The incomplete mode shapes are then expanded to the full mode shapes with the modal expansion method described above.

The measured mode shapes are used in the identification, and the analytical mode shapes are used in the mode shape expansion of the incomplete measurement. Any comparison of the modal data obtained in the two different states requires correct matching of the modes. This is done by calculating the Modal Assurance Criteria (MAC) between the analytical mode shapes and the measured mode shapes. A larger MAC value indicates greater correlation in the modes of the two states. The MAC were calculated between the analytical and experimental mode shapes of the frame, and it is larger than or equal to 0.98 for the first two modes in the two damage cases as shown in Table 4.3.

Two damage cases were studied. Case 1 is with damage at node 7. Case 2 is with the same damage at both nodes 7 and 11. The damage location was successfully identified with the first two test mode shapes after the mode shape expansion with 15 analytical modes were included, and the results are shown in Figures 4.5 and 4.6. The damage

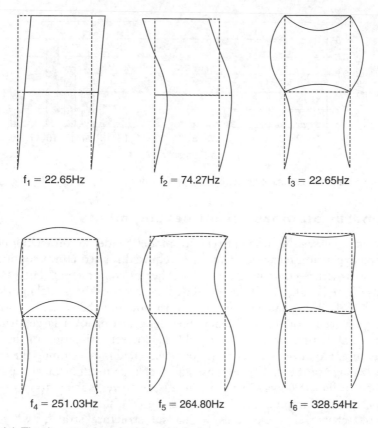

$f_1 = 22.65Hz$ $f_2 = 74.27Hz$ $f_3 = 22.65Hz$

$f_4 = 251.03Hz$ $f_5 = 264.80Hz$ $f_6 = 328.54Hz$

Figure 4.4 The theoretical mode shapes of the frame structure

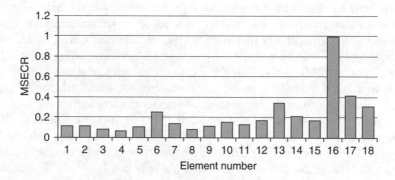

Figure 4.5 Identified results for case (a) from two modes

at node 7 was located as damage in element 16 in Figure 4.5, as this beam element is connected to the column at node 7. Damages at nodes 7 and 11 were identified in Figure 4.6 as damages in elements 15 and 16, which are the two beam elements connected to the column at nodes 11 and 7, respectively.

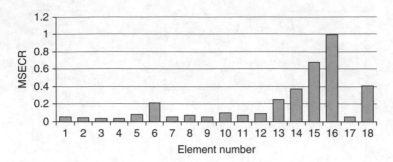

Figure 4.6 Identified results for case (b) from two modes

4.7 Remarks on more recent developments

This chapter discusses the different types of model condensation techniques and the mode-shape expansion technique which are required for structural condition assessment with a structure containing a large number of components. Their use without constraints (Terrell et al., 2007) changes the connectivity of the model matrices leading to a spread of the local damage information throughout the structure in the inverse problem. As a result of this deficiency, the identification of a large structure with many structural components does not yield the correct estimation of the conditions of the structure. However, Guo et al. (2006) succeeded in developing the relationship between a change in the global response and a change in an elemental parameter of a sub-structure via the Guyan method which is, however, subject to error due to the missing inertia and damping forces with the slave DOFs.

Some researchers treat the problem with a 'sub-structure' strategy, where the interface responses are measured; as input acting on the sub-structures (Gontier and Bensaibi, 1995; Yun and Lee, 1997; Yang and Huang, 2006). However, it is not always possible to obtain all the interface measurements, particularly with rotational responses. Some researchers reconstructed these forces based on the frequency response function or the transmissibility formulation (Devriendt and Fontul, 2005; Sjövall and Abrahamsson, 2007). Others proposed parameter identification of the sub-structures (Koh and Shankar, 2003) based on the receptance theory and the genetic algorithm approach without the need for interface measurements, and with modal parameters (Yun and Bahng, 2000) as the input to neural networks to identify the stiffness of the sub-structures. There are also limited studies in the time domain. Koh et al (2003) gave a good report on the sub-structural identification based on his SSI approach. Yuen and Katafygiotis (2006) proposed a probability-based method with the Bayes' theorem to identify local damages in sub-structures of a shear frame building. The authors have also reconstructed the interaction forces between sub-structures directly from the measured responses (Law et al., 2008) and the local anomalies are then identified via a sensitivity-based method. However, the required measured information is too much to be acceptable for practical use.

Chapter 5

Damage detection from static measurement

5.1 Introduction

Damage detection methods can be classified into two major categories depending on the nature of the experimental data: dynamic identification methods that use dynamic test data and static identification methods that use static test data. Compared with the static identification techniques, the dynamic methods have been more fully developed (Doebling et al., 1998b). To obtain a good estimation of the damage parameters, many difficulties inherent in the dynamic identification methods have to be overcome, such as the damping and mass changes due to the damage and the accurate measurements of higher vibration modes and frequencies. These may be very difficult with the experimental data analysis. However, the static identification method is usually simpler, since the static equilibrium equation only involves the stiffness properties of a structure. Also, static testing is comparatively cheaper and many advanced techniques have been developed recently for the static measurement. Accurate deformation or strain of the structure can be obtained rapidly and economically. Therefore, this group of methods attracts much attention from the engineering industry.

This chapter gives a brief description of this group of methods. Basically, the static parameter estimation is based on measured deformations induced by static loads, such as a truck on a bridge deck, a mass on a building or an actuator-induced static load on a structure. Section 5.2 gives the output error functions for system identification. Section 5.3 describes several techniques to detect the local damage based on static deflection profiles. Finally, the limitations of these methods are discussed in Section 5.4.

5.2 Constrained minimization

Damage can be defined as a structural change affecting the performance of the structure. The damage results in a loss of functionality. For a structural system, the loss of functionality means a reduction in the load-carrying capacity or a reduction in its ability to control motion under imposed loads. With this definition of damage, changes in certain properties of the structure between two time-separated inferences must be considered. Static-based methods allow damage identification by measuring changes in the static structural response. The measured quantities are typically displacements or strains under environmental and applied loads.

5.2.1 *Output error function*

The first method to describe the changes in the static response is the output error function. This is the difference between the analytical and measured static responses. The parameters are identified by minimizing an objective function formulated by the output error function (Hjelmstad and Shin, 1997) defined as the difference between the computed and measured displacements. A recursive quadratic programming algorithm was used to solve the constrained nonlinear optimization problem. Sanayei et al. (1997) presented the output error function based on the static displacements and strains. Both static displacement and strain measurements are used to estimate the element stiffness parameters. The method is extended to identify the element properties of a truss structure using static strains from a series of concentrated forces (Liu and Chian, 1997), and the effect of measurement errors on the identified results was studied using the perturbation technique. Shenton and Hu (2006) presented a static procedure based on the idea that the dead load is redistributed when damage occurs in the structure. The static strain measurements due to dead load only are used for the damage identification.

5.2.1.1 *Displacement output error function*

Consider a discrete structural system with n_d degrees-of-freedom (DOFs), having a stiffness matrix, K, characterized by n_p constitutive parameters, **p**. Let the structure be separately subject to n_f static load cases. For a linear elastic structure, the force–displacement relationship based on the finite element model is

$$\mathbf{f}_i = \mathbf{K}\mathbf{u}_i \tag{5.1}$$

where $\mathbf{f}_i$ is the vector of applied forces for the ith load case and $\mathbf{u}_i$ is the corresponding displacement vector.

The output error function is defined as the difference between the analytical and measured displacements for the ith load case as follows

$$e_i(\mathbf{p}) = \mathbf{K}^{-1}\mathbf{f}_i - \mathbf{u}_i \tag{5.2}$$

where $\mathbf{K}^{-1}$ is the inverse of the stiffness matrix.

Since the displacements are taken only at a subset of DOFs, matrix $\mathbf{u}_i$ may be partitioned into $\mathbf{u}_{ai}$ and $\mathbf{u}_{bi}$, which are the measured and unmeasured displacements, respectively. Giving

$$\begin{bmatrix} \mathbf{f}_{ai} \\ \mathbf{f}_{bi} \end{bmatrix} = \begin{bmatrix} \mathbf{K}_{aa} & \mathbf{K}_{ab} \\ \mathbf{K}_{ba} & \mathbf{K}_{bb} \end{bmatrix} \begin{bmatrix} \mathbf{u}_{ai} \\ \mathbf{u}_{bi} \end{bmatrix} \tag{5.3}$$

Using the static condensation

$$\mathbf{f}_{ai} = (\mathbf{K}_{aa} - \mathbf{K}_{ab}\mathbf{K}_{bb}^{-1}\mathbf{K}_{ba})\mathbf{u}_{ai} + \mathbf{K}_{ab}\mathbf{K}_{bb}^{-1}\mathbf{f}_{bi} \tag{5.4}$$

With the applied forces and measured responses, the output error function can be defined as

$$e_i(\mathbf{p}) = (\mathbf{K}_{aa} - \mathbf{K}_{ab}\mathbf{K}_{bb}^{-1}\mathbf{K}_{ba})^{-1}(\mathbf{f}_{ai}^m - \mathbf{K}_{ab}\mathbf{K}_{bb}^{-1}\mathbf{f}_{bi}^m) - \mathbf{u}_{ai}^m \tag{5.5}$$

where $\mathbf{u}_{ai}^m, (\mathbf{f}_{ai}^m, \mathbf{f}_{bi}^m)$ are the displacements at the measured DOFs and the vector of applied forces for the ith load case.

For all the load cases, the output error function can be written in matrix form as

$$\mathbf{e(p)} = (\mathbf{K}_{aa} - \mathbf{K}_{ab}\mathbf{K}_{bb}^{-1}\mathbf{K}_{ba})^{-1}(\mathbf{F}_a^m - \mathbf{K}_{ab}\mathbf{K}_{bb}^{-1}\mathbf{F}_b^m) - \mathbf{U}_a^m \qquad (5.6)$$

where $\mathbf{e(p)} = \{e_i(\mathbf{p}), i = 1, 2, \ldots n_f\}^T$, $\mathbf{F}_a^m = \{\mathbf{f}_{ai}^m, i = 1, 2, \ldots n_f\}^T$, $\mathbf{F}_b^m = \{\mathbf{f}_{bi}^m, i = 1, 2, \ldots n_f\}^T$ and $\mathbf{U}_a^m = \{\mathbf{u}_{ai}^m, i = 1, 2, \ldots n_f\}^T$ are obtained from the test data.

If the stiffness parameters are unchanged, then $\mathbf{e(p)}$ will be zero. Otherwise, it will not be zero. To adjust the parameter, $\mathbf{p}$, with a small increment, a first-order Taylor series expansion is used to linearize the vector, $\mathbf{e(p)}$, that is a nonlinear function of the parameters

$$e(\mathbf{p} + \Delta\mathbf{p}) = e(\mathbf{p}) + S(\mathbf{p})\Delta\mathbf{p} \qquad (5.7)$$

where $S(\mathbf{p}) = \frac{\partial e(\mathbf{p})}{\partial \mathbf{p}}$ is the sensitivity matrix.

A constrained nonlinear optimization problem is solved for the optimal parameters by minimizing the objective function of the output error as

$$\underset{\alpha \in R^{n_\alpha}}{\text{Minimize}} \; J(p) = e(\mathbf{p} + \Delta\mathbf{p})^T \mathbf{W} e(\mathbf{p} + \Delta\mathbf{p}) \qquad (5.8)$$

where $\mathbf{W}$ is a weighting factor. The unknown parameters, $\mathbf{p}$, are obtained from Equation (5.8).

5.2.1.2 Strain output error function

Equation (5.1) shows the relationship between the forces and displacements based on the finite element model. To utilize strain measurements, a mapping matrix between the displacements and strains can be derived from the geometric relationship of nodal displacements to elemental strains (Sanayei and Saletnik, 1996). The mapping matrix is defined as

$$\boldsymbol{\varepsilon} = BU \qquad (5.9)$$

where $\boldsymbol{\varepsilon}$ is the vector of strains and B is the corresponding mapping matrix. Substituting Equation (5.1) into Equation (5.9), gives the static equation for a constrained structural system as

$$\boldsymbol{\varepsilon}_i = BK^{-1}\mathbf{f}_i \qquad (5.10)$$

Similar to Equation (5.2), the strain output error function can be defined as

$$e_i(\mathbf{p}) = BK^{-1}\mathbf{f}_i - \boldsymbol{\varepsilon}_i \qquad (5.11)$$

Similar to Equation (5.3), the strains from multiple load tests can be grouped into

$$\begin{bmatrix} \boldsymbol{\varepsilon}_{ai} \\ \boldsymbol{\varepsilon}_{bi} \end{bmatrix} = \begin{bmatrix} B_a \\ B_b \end{bmatrix} K^{-1}\mathbf{f}_i \qquad (5.12)$$

and the strain output error function is written as

$$e_i(\mathbf{p}) = \mathbf{B}_a \mathbf{K}^{-1} \mathbf{f}_i^m - \mathbf{\varepsilon}_{ai}^m \tag{5.13}$$

where $\mathbf{\varepsilon}_{ai}^m, \mathbf{f}_i^m$ are the strains at the measured DOFs and the vector of applied forces for the ith load case, respectively. Combining all the load cases, Equation (5.13) can be rewritten as

$$e(\mathbf{p}) = \mathbf{B}_a \mathbf{K}^{-1} \mathbf{F}^m - \mathbf{\varepsilon}_a^m \tag{5.14}$$

where $\mathbf{F}^m = \{\mathbf{f}_i^m, i = 1, 2, \ldots, n_f\}^T, \mathbf{\varepsilon}_a^m = \{\mathbf{\varepsilon}_{ai}^m, i = 1, 2, \ldots, n_f\}^T$.

Similar to Equation (5.8), a constrained nonlinear optimization problem can be constructed and solved for the optimal parameters.

5.2.2   Damage detection from the static response changes

The changes in the static response of a structure are characterized as a set of nonlinear simultaneous equations that relates the changes in the static response to the location and severity of damage. A structural component is considered to be damaged if any of its stiffness parameters has been reduced. The static loads applied at a subset of DOFs and the measured static response at another subset of DOFs are used to detect the damage in structural elements. These subsets can be overlapped, partially overlapped or independent. Though structural damage is usually associated with a non-linear type of behaviour, this method uses small magnitude loads that cause structures to behave only in their elastic range. Bakhtiari-Nejad et al. (2005) presented an algorithm for damage identification based on the changes in the static displacements. Zhu et al. (2007) studied the effect of the load carried by a reinforced concrete beam on the assessment result of a crack damage, and they concluded that accurate assessment can only be obtained when a load close to the one that creates the damage is used in the identification. Zhu and Law (2007b) introduced the scalar damage parameters to characterize the changes in the interface between concrete and steel. The damage is identified using changes in the static responses of the beam.

The force–displacement relationship of a damaged structure can be expressed as

$$F = \tilde{K}\tilde{U} = (K - \Delta K)(\overline{U} - \Delta U) \tag{5.15}$$

where $\mathbf{F} = \mathbf{\Phi}\mathbf{P}$ is the force vector, in which $P = \{P_1, P_2, \ldots, P_{Np}\}^T$ is the vector of static loads; $\Phi = \{\Phi_1 \Phi_2 \ldots \Phi_l \ldots \Phi_{Np}\}$ is a $2(N+1) \times N_p$ shape function matrix; N_p is the number of loads; $\overline{U}$ and $\tilde{U}$ are the vectors of nodal deformation without and with damage, respectively; and $\Delta U = \overline{U} - \tilde{U}$ is the vector of deformation difference due to the damage. When under the same applied load, Equation (5.15) indicates qualitatively that a reduction in the stiffness matrix corresponds to an increase in the vector of deformations.

Vector ΔU can be estimated from Equation (5.15) as

$$\Delta U = -K^{-1} \Delta K K^{-1} F + K^{-1} \Delta K \Delta U \approx -K^{-1} \Delta K K^{-1} F \tag{5.16}$$

by neglecting the second-order terms. Substituting the force–displacement relationship of the intact structure and the stiffness matrix of the damage structure into Equation (5.16), the analytical vector of deformation due to damage is obtained as

$$\Delta U \approx \sum_{i=1}^{N} \alpha_i K^{-1} A_i^T K_i A_i K^{-1} \Phi P = \sum_{i=1}^{N} \alpha_i \hat{U}_i \tag{5.17}$$

where $\hat{U}_i = K^{-1} A_i^T K_i A_i K^{-1} \Phi P$ and $\mathbf{A}_i$ is the transformation matrix for the ith element for assembling the global stiffness matrix from the constituting elemental stiffness matrix.

For N_s measuring points at $\{x_s, s = 1, 2, \ldots, N_s\}$, Equation (5.17) can be written as

$$\mathbf{U}_s = \Phi_s U \tag{5.18}$$

where $\mathbf{U}_s = \{u_1, u_2, \ldots, u_{N_s}\}^T$ and $\Phi_s = \{\phi_1, \phi_2, \ldots, \phi_{N_s}\}^T$.

The error between the vectors of difference between the calculated and measured deformations of the structure is obtained from Equations (5.16) and (5.18) as

$$e(\alpha) = \Phi_s \Delta U - \Delta U_s \tag{5.19}$$

where ΔU_s is the vector of differences between the measured displacement of structures with and without damage.

The algorithm to identify the damage is based on minimizing the least-squares error function in Equation (5.19) as

$$\text{Minimize } J(\alpha) = \frac{1}{2}\|e(\alpha)\|^2 = \frac{1}{2}\left\|\sum_{i=1}^{N} \alpha_i \Phi_s \hat{U}_i - \Delta U_s\right\|^2$$

$$\text{subject to } 0 \le \alpha_i \le 1, \qquad (i = 1, 2, \ldots, N) \tag{5.20}$$

This model can be cast into the following quadratic programming problem (Banan and Hjelmstad, 1994) for determining the damage indices

$$\text{Minimize } J(\alpha) = \frac{1}{2}\alpha^T K^T \Phi_s^T \Phi_s \alpha K - \Delta U_s^T \Phi_s \alpha K + \frac{1}{2}\Delta U_s^T \Delta U_s$$

$$\text{subject to } 0 \le \alpha_i \le 1, \qquad (i = 1, 2, \ldots, N) \tag{5.21}$$

where $\alpha = \{\alpha_i\}^T$. The algorithm presented by Goldfarb and Idnani (1983) is used to solve this quadratic programming problem. Details of the iterative algorithm used to solve the nonlinear optimization problem are as follows:

1) Calculate the matrix $\Phi_s = \{\phi_1, \phi_2, \ldots, \phi_{N_s}\}^T$, $\Phi = \{\Phi_1 \Phi_2 \ldots \Phi_l \ldots \Phi_{Np}\}$.
2) Initially assume that there is no damage in the beam, i.e. $\alpha_0 = \{0, 0, \ldots, 0\}^T$.
3) When the deformation of the intact structure under a given load is not available, the baseline deformation vector of the structure without damage $\overline{U}_s$ is calculated and used instead of the measured deformation vector from the intact structure,

and the initial vector of measured deformation difference from the intact and damage structures is obtained as $\Delta U_{s0} = \overline{U}_s - U_s$.

4) Identify the damage index α_j using Equation (5.15). $j = 1$ for the first cycle of iteration.

5) Calculate ΔU using Equation (5.16) with the updated $\alpha = \alpha_j + \alpha_0$ and $U_{Re\,constructed} = \overline{U}_s - \Phi_s \Delta U$.

6) Calculate the following criteria of convergence:

$$
\begin{cases}
Error1 = \dfrac{\| U_s - U_{Re\,constructed} \|}{\| U_s \|} \times 100\% \\[2ex]
Error2 = \dfrac{\| \alpha_{j+1} - \alpha_j \|}{\| \alpha_j \|} \times 100\%
\end{cases}
\tag{5.22}
$$

α_j, α_{j+1} are the identified damage indices in two successive iterations. Convergence is achieved when both errors are less than the pre-defined tolerance values.

7) When the computed error does not converge, calculate $\Delta U_s = \overline{U}_{Re\,constructed} - U_s$ and $\alpha_0 = \alpha$. Repeat Steps 4 to 6 until convergence is reached.

5.2.3 Damage detection from combined static and dynamic measurements

In a finite element model, the characteristics of a structure are defined in terms of the stiffness, damping and mass matrices. Any variations in these matrices affect the dynamic response of the structure. Some researchers have studied algorithms using both the static and dynamic test data. A modified error function for the statistical parameter estimation was presented (Oh and Jung, 1998) to combine the curvature or slope of the mode shapes and the static displacements. This combination gives rise to a promising damage detection and assessment algorithm. Later Wang et al. (2001) used both the static displacements and changes in natural frequencies in a structural damage identification algorithm.

The ith eigenvalue, λ_i, and the corresponding eigenvector, φ_i, of an n-DOFs system are obtained by solving the characteristic equation

$$
K\varphi_i = \lambda_i M\varphi_i
\tag{5.23}
$$

where K is the structural stiffness matrix and M is the structural mass matrix. Combining Equations (5.1) and (5.23), the relationship between the modal parameters, static responses and structural parameters is expressed in terms of a first-order of Taylor series expansion as

$$
\begin{Bmatrix} \Lambda \\ \Phi \\ \Gamma \end{Bmatrix} = \begin{Bmatrix} \Lambda \\ \Phi \\ \Gamma \end{Bmatrix}_{P_p} + S(P - P_p)
\tag{5.24}
$$

where P is the vector of structural parameters; vector P_p is the prior estimate of the structural parameters; the vector with subscript P_p denotes the responses of the system when the parameters $P = P_p$; Λ is the vector of measured eigenvalues; and Φ and Γ are vectors of dynamic and static error responses, respectively. Let n_r be the total number of responses for the target structure, and S be the partial derivative or sensitivity matrix of the eigenvalues and dynamic and static error responses. It can be written as

$$S = \begin{bmatrix} \dfrac{\partial \Lambda}{\partial p_j} & \dfrac{\partial \Phi}{\partial p_j} & \dfrac{\partial \Gamma}{\partial p_j} \end{bmatrix}^T \tag{5.25}$$

The different types of responses in Equation (5.24) are combined to form the error function. To prevent the case where the contribution of a special response type is reduced due to a relatively small magnitude of the response, an adjusted dynamic response should be defined. Therefore, when the combined data of static and modal tests are used in the identification procedure, the measured mode shapes are weighted so that the maximum values of each type of response are the same.

It is noted from Equation (5.24) that the number of structural parameters is not the same as the number of measured responses, such that the inverse matrix of S does not always exist. Therefore, an estimate T on such an inverse can be defined as

$$\Delta p = T \Delta R \tag{5.26}$$

where $\Delta R = \begin{bmatrix} \Lambda \\ \Phi \\ \Gamma \end{bmatrix} - \begin{bmatrix} \Lambda \\ \Phi \\ \Gamma \end{bmatrix}_{p_p}$.

The objective of this procedure is to find the best unbiased estimator, Δp, based on measured values of Λ, Φ and Γ and prior estimates of the structural parameters, p_p. A scalar performance error function in the Bayesian estimation theory is defined for the derivation of the statistical system identification formula as

$$E = (R - R_k)^T W_r (R - R_k) + (P_k^* - P_k)^T W_p (P_k^* - P_k) \tag{5.27}$$

where W_r and W_p are the weighting matrices for the measured responses and the structural parameters, respectively, which can be assumed as the inverse of the covariance matrices of the errors on the measured responses and the initial structural parameters, respectively; R is the measured responses; R_k, P_k are predicted vectors of responses and structural parameters, respectively, at the k-th step; and P_k^* is the vector of objective parameters.

Other error functions can also be used to estimate the structural parameters. The changes in curvatures are local in nature and hence can be used to detect and locate the damage in the structure. The differences in the curvatures of mode shapes can be used to form the error response vector as well as for estimating the uncertainty of the structural parameters. Certainly, the slope of mode shapes is also another possible candidate for formulating the error response vectors.

5.3 Variation of static deflection profile with damage

5.3.1 The static deflection profile

In Section 5.2, an equivalent scalar parameter is used to describe the variation in the stiffness parameter of a structural element. This approach can determine changes in structural element stiffness, including failure in an element. The identified cross-sectional properties of the structural elements can be taken as parameters for assessment and to determine the load-carrying capacity of the structure. In a real case, the presence of damage in a continuous body is represented by a reduction in some physical parameters of the material in an appropriate constitutive model. In the case of damage caused by the presence of a crack, it is well known that, besides the stress concentration occurring at the crack tip, there is a zone, adjacent to the crack, denoted as 'ineffective' in view of its low stress level. If a straight beam is subjected to concentrated damage, such as a crack or a saw cut at a certain cross-section, the presence of the ineffective zone in the crack vicinity can be accounted for by a loss of the flexural stiffness in the Euler–Bernoulli beam theory.

The variation in a stiffness parameter can also be modelled as a distribution function in the structural element, such as modelling the concentrated damages as Dirac's delta distribution in the flexural stiffness (Di Paola and Bilello, 2004; Caddemi and Greco, 2006; Buda and Caddemi, 2007). Basically, the presence of damage can be represented through a variation of its bending stiffness as

$$EI(x, \mathbf{x}_d, \boldsymbol{\beta}_d) = EI \times [(1 - d(x, \mathbf{x}_d, \boldsymbol{\beta}_d))] \tag{5.28}$$

where EI is the bending stiffness of the undamaged beam; $d(x, \mathbf{x}_d, \boldsymbol{\beta}_d)$ is the damage distribution function; while $\mathbf{x}_d, \boldsymbol{\beta}_d$ are q-dimensional vectors (q is the number of damage events) of damage locations and extent, respectively. The upper bound of the damage distribution function, $d(x, \mathbf{x}_d, \boldsymbol{\beta}_d) = 0$, corresponds to the undamaged case while the lower bound, $d(x, \mathbf{x}_d, \boldsymbol{\beta}_d) = 1$, indicates a local failure due to damage. For a concentrated crack, $d(x, x_d, \beta_d) = \beta_d \delta(x - x_d)$, $\delta(x - x_d)$ is a Dirac's delta distribution centred at x_d.

The static equations of an Euler–Bernoulli beam subjected to the external load $f(x)$ are easily obtained in the form

$$
\begin{aligned}
Q'(x) &= -f(x) \\
M'(x) &= Q(x) \\
\varphi(x) &= -u'(x) \\
\chi'(x) &= \varphi'(x) \\
\chi(x) &= M(x)/EI(x, \mathbf{x}_d, \boldsymbol{\beta}_d)
\end{aligned}
\tag{5.29}
$$

where $f(x)$ is the external vertical load; $Q(x)$ and $M(x)$ are the shear force and the bending moment, respectively; $u(x), \varphi(x), \chi(x)$ are the deflection, slope and curvature functions, respectively; and the prime denotes differentiation with respect to the spatial coordinate x, spanning from 0 to the length L of the beam. The first four equations in Equation (5.29) are the equilibrium and compatibility equations, and the final one is the constitutive equation relating curvature and bending moment through the spatial variable flexural stiffness $EI(x)$.

Combining the equilibrium, the compatibility and the constitutive equations yields the following fourth-order differential governing equation in the beam deflection theory as

$$[EI(x, \mathbf{x}_d, \boldsymbol{\beta}_d)u''(x)]'' = f(x) \tag{5.30}$$

With the presence of a local damage, the fourth-order governing differential equation is obtained as

$$\{EI[1 - \beta_d\delta(x - x_d)]u''(x)\}'' = f(x) \tag{5.31}$$

The deflection function, $u(x)$, can be obtained by integrating the governing equation of the Euler–Bernoulli beam as

$$u(x) = c_4 + c_3 x + \frac{c_1}{2EI}[x^2 + 2\frac{\beta_d}{1 - \beta_a A}(x - x_d)U(x - x_d)] + \frac{C_2}{6EI}$$

$$\times [x^3 + 6\frac{\beta_d}{1 - \beta_d A}U(x - x_d)] + \frac{f^{[4]}(x)}{EI}$$

$$+ \frac{\beta_d}{1 - \beta_d A}\frac{f^{[2]}(x_d)}{EI}(x - x_d)U(x - x_d) \tag{5.32}$$

where $U(x - x_d)$ is the unit step distribution, also called the Heaviside function. It represents the formal primitive of the Dirac's delta with discontinuity at x_d, $U(x - x_d) = 0$ for $x < x_d$ and $U(x - x_d) = 1$ for $x \geq x_d$. Constants c_1, c_2, c_3 and c_4 can be obtained by the enforcement of boundary conditions; $f^{[k]}(x)$ denotes a primitive of order of the external load function, $f(x)$; and A is a constant defined as $A\delta(x - x_d) = \delta(x - x_d)\delta(x - x_d)$ (Bagarello, 1995).

According to the explicit solutions of the response in terms of the crack position and intensity, the damage parameters can be identified by means of the following minimization problem

$$J(\mathbf{x}_d, \boldsymbol{\beta}_d) = \frac{1}{2}\sum_{i=1}^{n_{lc}}\sum_{j=1}^{nm} w_i[u_i(x_j, \mathbf{x}_d, \boldsymbol{\beta}_d) - u_i^e(x_j)]^2 \tag{5.33}$$

or

$$J(\mathbf{x}_d, \boldsymbol{\beta}_d) = \frac{1}{2}\sum_{i=1}^{n_{lc}}\sum_{j=1}^{nm} w_i[\Delta u_i(x_j, \mathbf{x}_d, \boldsymbol{\beta}_d) - \Delta u_i^e(x_j)]^2 \tag{5.34}$$

where w_i is the weight associated with the ith load case; n_{lc} is the number of load cases; nm is the number of measurements; $u_i(x_j, \mathbf{x}_d, \boldsymbol{\beta}_d), u_i^e(x_j)$ are the theoretical and experimental measured deflection at nm different cross-sections for n_{lc} load cases, respectively; and $\Delta u_i(x_j, \mathbf{x}_d, \boldsymbol{\beta}_d), \Delta u_i^e(x_j)$ are the theoretical and experimental measured variations of a structural response due to damage, respectively.

5.3.2 *Spatial wavelet transform*

From the above discussion, the closed form deflection can be written in terms of the damage intensities and positions. The local damage is treated as an abnormal flexural stiffness that induces some singularity in the deflection function.

An important feature of the wavelet transform is its ability to characterize the local irregularity of a function and its sensitivity to small changes in the structure. A crack in a structure introduces singularities to the dynamic displacement mode shapes and the static deflection profile. These small changes cannot be identified directly from the structural response, but they may be observed in the wavelet transforms since local abnormalities in the signal would result in large wavelet coefficients in the proximity of the damage. The applicability of the wavelet damage detection techniques depends on the measurement precision and the sensor spacing.

With the development of modern sensor technology, the measurement of the beam displacements at a large number of spatially distributed points can be obtained by processing digital photographs of the beams (Rucka and Wilde, 2006). The high-resolution camera equipped with a metric optic matrix that allows precise mapping of 3D objects into 2D digital photographs is also a means to obtain high quality data. Recently, the operational deflection of a bridge deck under moving loads is used for damage assessment (Zhu and Law, 2006). The study shows that the response at a single point from the passage of a very slow moving load is equivalent to the displacement of the whole bridge deck loaded at the measuring point. This method will be studied in detail in Section 8.2.

5.4 Application

5.4.1 *Damage assessment of concrete beams*

5.4.1.1 *Effect of measurement noise*

A numerical example with a four-metre long simply supported uniform rectangular concrete beam is studied. It has a 300 mm high and 200 mm wide cross-section and 3.8 m clear span. There are three 20 mm diameter mild steel bars at the bottom of the beam corresponding to a steel percentage of 1.57%, and two 6 mm diameter steel bars at the top of the beam section. Mild steel links with a 6 mm diameter are provided at 195 mm spacing over the whole beam length. The density, tensile strength, elastic modulus and Poisson's Ratio of concrete are 2351.4 kg/m^3, 3.77 MPa, 30.2 GPa and 0.16, respectively. The elastic modulus and yield stress of the mild steel bars are 181.53 GPa and 300.07 MPa, respectively.

One per cent white noise is added to the calculated displacements of the beam to simulate the polluted measurements with

$$U_s = U_{calculated}(1 + E_p \cdot N_{noise}) \qquad (5.35)$$

where U_s and $U_{calculated}$ are the polluted and the original 'measured' displacements; E_p is the noise level; and N_{noise} is a standard normally distributed vector with zero mean and unit standard deviation, and it is generated independently for each component of the measured displacement.

The beam is discretized into eight elements. A static load of 5000N is applied at $1/2L$. Seven displacement measurements evenly distributed along the beam are used for the identification. The local damage in the beam is modelled with the damage distribution function (Wahab et al., 1999) with damage index α_i for the ith beam element with $0.0 \geq \alpha_i \geq 1.0$. The method described in Section 5.2.2 is adopted for the identification. The Monte Carlo method is used in the simulations and one hundred sets of simulated results are obtained.

Figures 5.1(a) and 5.1(c) show the relationship between the mean values of the identified results and the number of simulations when the damage index, α_3, is 0.1 and 0.3, respectively. Figures 5.1(b) and 5.1(d) show their histograms for 100 simulations compared with the corresponding normal distributions. The mean values of the identified damage indices on Figures 5.1(a) and 5.1(c) converge to a constant value when the number of simulations is larger than 80. Figures 5.1(b) and 5.1(d) show that the histograms of the identified results are close to the normal distribution. These results indicate that the estimated damage indices have approximately normal distributions, if the displacement measurement noise follows a normal distribution. More simulated results would give a clearer indication of distribution with the damage indices.

Figure 5.2 shows the range of identified results from 1% noise polluted static responses when the damage index, α_3, is 0.1 and 0.3, respectively. In practice, static measurements are more accurate than dynamic measurements, and 1% simulated noise pollution is considered good enough to represent the random error with real static

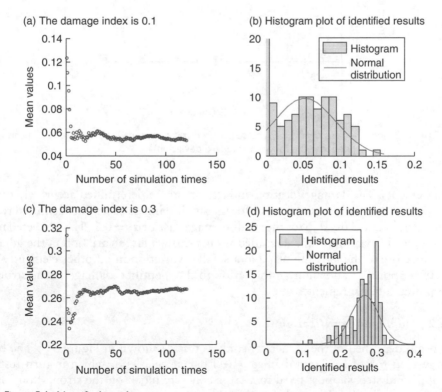

Figure 5.1 Identified results

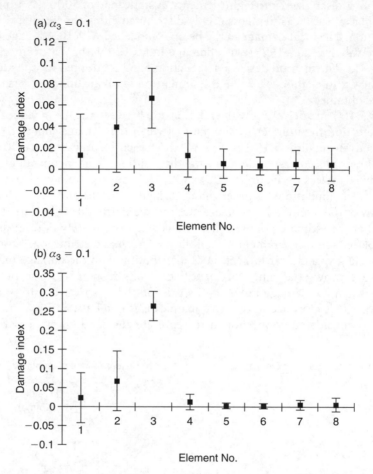

Figure 5.2 Identified results from 1% noise polluted static responses (■ denotes the mean value), (|————| denotes the range of standard deviation)

measurements. The damage location and extent can be determined accurately for the large-damage case of $\alpha_3 = 0.3$ but not for the small-damage case. The identified results for elements 2 and 3 overlap for the small-damage case of $\alpha_3 = 0.1$. The predicted mean and standard deviation of the identified results show a large variation in the adjacent elements close to the left support, and a smaller variation in all other elements. This statistical approach adds more information to the identified damage of the structure with polluted measurement.

5.4.1.2   Damage identification

The beam described in the numerical study is test as shown in Figure 5.3. The beam ends rest on top of a 50 mm diameter steel bar at each end, which in turn rests on top of a solid steel support fixed to a large concrete block on the strong floor of the laboratory. A piece of thin rubber pad is placed between the steel bar and the bottom

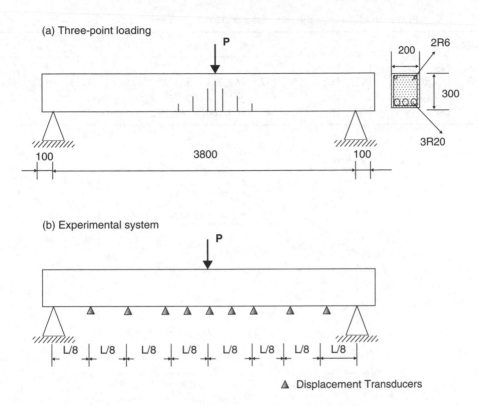

(a) Three-point loading

P

200   2R6

300

3R20

100   3800   100

(b) Experimental system

P

L/8 | L/8 | L/8 | L/8 | L/8 | L/8 | L/8 | L/8

▲ Displacement Transducers

Figure 5.3 Reinforced concrete beam and sensor locations

Table 5.1 Load level in different loading stages

Load Stage	1	2	3	4	5	6	7	8
Assessment load (kN)	10	17	25	35	45	50	55	60
Maximum load (kN)	15	25	35	45	50	55	60	67

of the concrete beam for level adjustment. The vertical stiffness of the rubber pad was measured as 39.41 kN/mm and it is used to modify the measured displacements.

The initial flexural rigidity of the concrete beam is estimated by direct calculation using the measured material property and the geometrical dimensions of the beam. The beam carries no crack and therefore the steel bars inside are not considered contributing as a composite component of the beam. The beam was incrementally loaded at the mid-span to create crack damage using three-point loading. Eight loading cycles starting from 0.0 kN to a pre-specified load level as shown in Table 5.1 were conducted, each with a subsequent unloading stage. The beam was subsequently loaded to yield at 75 kN after the eighth load stage. The crack locations and lengths were monitored in addition to the displacement measurements and they are shown in Table 5.2. The beam is divided into 8 and 16 finite elements separately for the study. The first crack appeared

Table 5.2 Cracks in each element at the end of the final loading stage

Element No.		4	5	6	7	8	9	10	11	12	13
front	No. of cracks	1	1	1	1	2	2	1	2	1	1
view	Length (mm)	162	162	150	177	163, 211	211, 192	181	141, 137	163	171
back	No. of cracks	–	1	1	1	2	2	–	–	–	–
view	Length (mm)	–	161	170	145	180, 160	160, 202	–	–	–	–

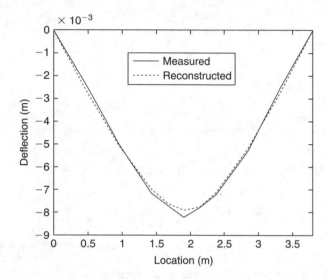

Figure 5.4 Comparison of deflections at the mid-span at 60 kN

in elements 8 and 9 at 17 kN in the 16 finite element model. Nine displacement trans-ducers were located at the bottom of the beam to measure the deflection under load, as shown in Figure 5.3. The static responses at all the nine measurement points were used in the damage identification. The INV300 data acquisition system was used to record data from all 10 channels, including the applied load, with a sampling rate of 200.12 Hz.

Figure 5.4 compares the measured and reconstructed deflection at the mid-span of the beam at 60 kN in the final load stage. The two curves are very close to each other and nearly symmetrical about the mid-span of the beam. This is consistent with the small error (results not shown) between the two curves calculated by the following formula.

$$Error = \frac{\|\mathbf{U}_s - \mathbf{U}_{Re\,constructed}\|}{\|\mathbf{U}_s\|} \times 100\% \tag{5.36}$$

5.4.1.3   Damage evolution under load

The displacement in a subsequent loading stage with an assessment load close to the maximum load of the previous stage is used to identify the damage. For example, the

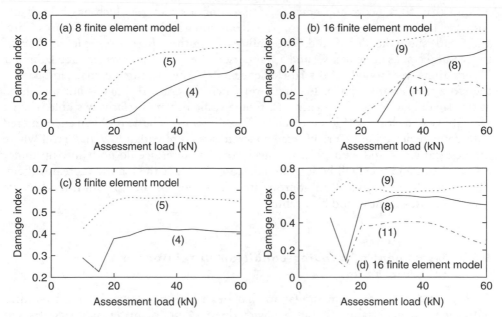

Figure 5.5 Identified damage indices (a), (b) evolution with load and (c), (d) after the final loading stage

displacement at an assessment load of 45 kN in the sixth loading stage is used to assess the damage created by the same load at the fifth loading stage. This is to simulate the real practice of assessment at a second loading cycle instead of the same load cycle. The assessment load for each load stage is shown in Table 5.1.

Elements 4 and 5 are identified to have damage in the case of the model with 8 finite elements, while elements 8, 9 and 11 are identified to have damage in the model with 16 finite elements. The identified damage indices are plotted in Figures 5.5(a) and 5.5(b), indicating a nonlinear increase with an increase in the applied load. The element numbers are marked as (•) beside the curves. The curve flattens at around 35 kN for element 5 and 45 kN for element 4. This is because part of the deformation does not recover after unloading in the previous load cycle and the damage index would be an underestimation of the true value. It is noted that the damage is more or less symmetrical about the mid-span, and the crack damage in the right half of the beam is slightly more severe than the left half, as seen in Table 5.2. It should be noted that Figures 5.5(a) and 5.5(b) are damage evolution diagrams of the damaged elements with load. Each data point on the curve corresponds to the damage created by the associated load. After the beam was loaded up to 60 kN in the eighth loading stage, the damage indices in elements 4 and 5 become 0.41 and 0.55, respectively. Similar discussions apply to the damages in the 16-element model.

5.4.1.4   *Damage identification–Simulating practical assessment*

The crack damage in the beam after the eighth loading stage was again assessed with loads at different load levels. The beam was unloaded and then loaded again with the assessment loads of 10, 15, 20, 25, 30, 35, 40, 45, 50, 55 and 60 kN in turn. This is equivalent to the common practice of assessing a structure which has been badly damaged under extreme loading with a smaller loading in the assessment. Figures 5.5(c) and 5.5(d) show the identified damage indices using the different static loads. Part of the flexural cracks in elements 4 and 5 was closed after the beam was unloaded. Similar observations were found for the 16-element model. The identified damage indices are therefore smaller at a smaller loading level when the crack damage is not mobilized at this small load level. The identified damage value becomes relatively stable when a load above and close to 25 kN is reached. This illustrates a basic problem encountered with damage detection of a reinforced concrete structure with a low load level where the flexural crack and damage in the steel–concrete interface do not show up under a small load, and they will be fully mobilized only under the load that creates them. Therefore, the inclusion of an appropriate proportion of the operation load in the condition assessment would be essential.

5.4.2 Assessment of bonding condition in reinforced concrete beams

A reinforced concrete beam similar to that described in Section 5.4.1.1 is studied with the same material properties but with three 16 mm diameter mild steel bars at the bottom of the beam instead of three 20 mm diameter bars. The first six natural frequencies of the beam without damage are 30.5, 121.9, 274.4, 488.3, 764.8 and 1015.6 Hz. The three steel bars are at the same distance from the centroidal axis of the section and are assumed to experience the same bond slippage under loading. The bonding stiffness of 9.05 MPa/mm is used in the calculation. The steel bar and concrete bonding model described in Section 3.2.4 is used in this study. The method described in Section 5.2.2 is adopted for the identification.

The above simply supported beam with one static load, $p(t) = 5000\,\text{N}$, at $2/3L$ from the left support is studied. The static responses under static load are used in the identification and the measuring points are distributed evenly along the beam. In the simulation, white noise is added to the calculated strain and displacements of the beam to simulate the polluted measurements with

$$\begin{cases} \mathbf{u} = \mathbf{u}_{calculated}(1 + E_P \cdot N_{noise}) \\ \varepsilon = \varepsilon_{calculated}(1 + E_P \cdot N_{noise}) \end{cases} \tag{5.37}$$

where $\mathbf{u}$, ε and $\mathbf{u}_{calculated}$, $\varepsilon_{calculated}$ are the polluted and the original strains and displacements, respectively; E_p is the noise level; and N_{noise} is a standard normally distributed vector with zero mean and unit standard deviation.

5.4.2.1 Local beam damage identification

The simply supported beam is divided into eight equal finite elements, and damage is in the fourth element with $\boldsymbol{\alpha}_B = \{0, 0, 0, 0.1, 0, 0, 0, 0\}^T$. A different number of sensors, distributed evenly along the beam, is used to measure the static displacement responses

Table 5.3 Identified local beam damage indices ($\alpha_{Be}^{identified}$) from a different number of sensors

Measurements	Displacements			Strains		
Noise level	1%	5%	10%	1%	5%	10%
1	0.039/60.93	0.040/60.42	0.040/59.77	0.068/31.66	0.069/30.76	0.070/29.64
2	0.003/97.27	0.003/97.33	0.003/94.40	0.067/33.17	0.068/32.24	0.069/31.07
3	0.100/0.05	0.100/0.14	0.100/0.27	0.085/15.40	0.086/14.32	0.087/12.96
4	0.100/0.15	0.099/0.77	0.099/1.54	0.100/0.19	0.101/0.97	0.102/1.94
5	0.100/0.22	0.099/1.08	0.098/2.16	0.100/0.14	0.101/1.15	0.103/2.61
6	0.100/0.12	0.099/0.58	0.099/1.16	0.100/0.10	0.100/0.19	0.100/0.36
7	0.100/0.09	0.100/0.45	0.099/0.90	0.100/0.03	0.100/0.17	0.100/0.34
8	0.100/0.17	0.099/0.83	0.098/1.67	0.100/0.11	0.101/0.55	0.101/1.10
9	0.099/0.68	0.097/3.38	0.093/6.76	0.100/0.11	0.101/0.56	0.101/1.12

Number of Sensors (row label for rows 1–9)

Note: •/• denotes the identified damage index/percentage error of identification.

under the static load; 1, 5 and 10% noise are added to simulate the measured responses. The errors in the identified damage indices are defined as

$$Error = \frac{\sum\limits_{i=1}^{Nd} \left| \alpha_{Bei}^{true} - \alpha_{Bei}^{identified} \right|}{\sum\limits_{i=1}^{Nd} \alpha_{Bei}^{true}} \times 100\% \tag{5.38}$$

where $\alpha_{Bei}^{true}, \alpha_{Bei}^{identified}$ are the true and identified damage indices of the ith damage element and Nd is the number of damage elements.

Table 5.3 shows the identified damage indices from using a different number of sensors. Figure 5.6(a) shows the identified damage indices from measurements of seven sensors, and Figure 5.6(b) shows the results when the static load is applied at different locations with seven sensors.

The identified damage indices shown in Figure 5.6 are very close to the true values, and they are insensitive to the noise in the measurements. The results show that the method based on static measurement is effective to determine the damage location, and the damage extent can be estimated accurately. Acceptable results highlighted in Table 5.3 are obtained when the number of displacement measurements is not less than three or the number of strain measurements is not less than four. It should be noted that the load position has no effect on the detection results as long as it is not close to the supports which may lead to low signal-to-noise ratio in the measurement.

5.4.2.2 Identification of local bonding

The parameters are the same as those used above and the bonding damage is in the third element with $\alpha_b = \{0, 0, 0.1, 0, 0, 0, 0, 0\}^T$. Table 5.4 shows the identified damage indices from using a different number of sensors, and Figure 5.7 shows the identified damage indices from seven sensors with different measurement noise.

Figure 5.7 shows that the identified bonding damage indices are very close to the true value. The identified bonding damage indices from displacement measurements

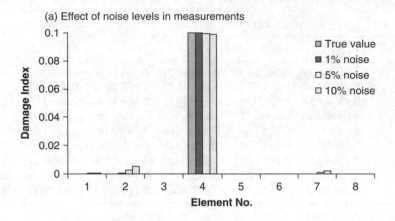

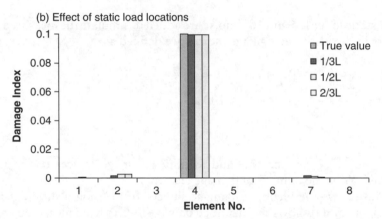

Figure 5.6 Identified damage indices (a) from measurements with different noise levels; (b) with different static load positions

Table 5.4 Identified local bonding damage indices $\alpha_{be}^{identified}$ from a different number of sensors

Measurements	Displacements			Strains		
Noise level	1%	5%	10%	1%	5%	10%
1	0.000/100.00	0.000/100.00	0.000/100.00	0.001/99.05	0.003/97.47	0.002/98.07
2	0.092/7.70	0.093/6.75	0.094/5.57	0.087/12.80	0.088/11.69	0.090/10.31
3	0.102/2.10	0.104/3.71	0.106/5.73	0.102/1.87	0.103/2.85	0.105/4.47
4	0.102/1.85	0.104/2.57	0.106/5.69	0.097/2.63	0.099/1.24	0.100/0.42
5	0.102/1.82	0.105/5.08	0.106/5.87	0.095/5.06	0.097/3.04	0.099/1.15
6	0.104/3.46	0.111/11.36	0.119/18.49	0.103/3.16	0.105/5.06	0.107/6.47
7	0.102/2.10	0.108/7.48	0.115/15.06	0.100/0.29	0.102/1.45	0.103/2.91
8	0.104/4.29	0.117/16.70	0.127/26.64	0.103/2.46	0.105/5.41	0.108/8.14
9	0.102/1.48	0.103/2.80	0.106/5.47	0.101/0.81	0.104/3.92	0.108/7.82

Note: •/• denotes the identified damage index/percentage error of identification.

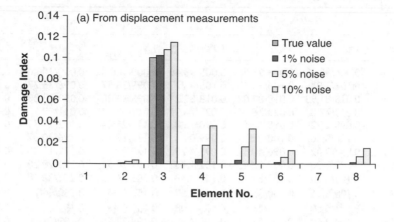

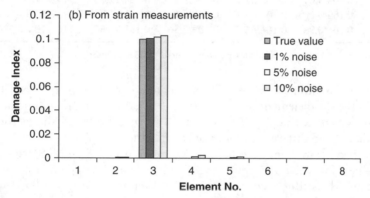

Figure 5.7 Identified bonding damage indices from displacement and strain measurements

increased with the noise level. However, the identified bonding damage indices from strain measurements are less sensitive to the noise level than the displacement measurements. This shows that the present method is effective to determine the bonding damage location and the bonding damage extent can be estimated accurately.

The highlighted acceptable results in Table 5.4 are obtained when the number of sensors is not less than three for both displacement and strain measurements. However, an increase in the number of sensors does not, in general, give a more accurate identified result as observed in the cases of 5 to 8 sensors. The identified bonding damage indices from strain measurements are much closer to the true value than those from displacement measurements. A comparison between Tables 5.3 and 5.4 shows that errors for the identification of bonding damage is larger than those in the identification of beam damage for the same test and beam configurations.

5.4.2.3 *Simultaneous identification of local bonding and beam damages*

The effects of these two types of damage are assumed to be independent. However, they may be coupled with real damage and they cannot be identified using existing damage models or the models described in Section 3.2.4. The method of identification

Table 5.5 Identified damage indices from strain measurements

Damage Index	$\alpha_{Be}^{identified}$			$\alpha_{be}^{identified}$		
Noise level	1%	5%	10%	1%	5%	10%
1	0.004/95.62	0.004/95.56	0.005/95.49	0.000/100.00	0.000/100.00	0.000/100.00
2	0.103/2.94	0.104/3.75	0.105/4.77	0.095/36.57	0.096/35.77	0.098/34.77
3	0.018/81.93	0.018/82.08	0.018/82.27	0.098/34.80	0.099/33.88	0.101/32.74
4	0.023/77.03	0.023/76.96	0.023/76.92	0.011/92.94	0.011/92.82	0.011/92.66
5	0.098/2.22	0.098/1.94	0.098/1.68	0.111/25.84	0.112/25.31	0.113/24.76
6	0.100/0.66	0.100/0.37	0.099/0.66	0.177/17.73	0.196/30.67	0.199/32.96
7	0.097/3.02	0.096/4.18	0.095/4.79	0.179/19.51	0.179/19.38	0.180/19.75
8	0.098/2.20	0.099/0.64	0.102/1.45	0.193/28.92	0.192/28.29	0.194/28.97
9	0.103/2.70	0.103/3.44	0.105/4.48	0.205/36.33	0.207/38.18	0.212/41.37
10	0.100/0.06	0.100/0.29	0.099/0.59	0.152/1.10	0.158/5.47	0.166/10.95
11	0.101/0.50	0.103/2.54	0.105/5.08	0.151/0.49	0.151/0.51	0.152/1.01
12	0.104/3.69	0.105/5.09	0.105/4.81	0.184/22.75	0.198/31.90	0.189/25.96
13	0.100/0.20	0.101/1.02	0.102/2.03	0.152/1.35	0.159/6.25	0.168/12.45
14	0.101/0.86	0.104/4.28	0.109/8.56	0.153/2.15	0.166/10.73	0.182/21.46

(row label, vertical: Number of Sensors)

Note: •/• denotes the identified damage index/percentage error of identification.

is referred to (Zhu and Law, 2004). The simulated damage is as follows. The bonding damage is in the third element with $\alpha_b = \{0, 0, 0.15, 0, 0, 0, 0, 0\}^T$ and beam damage is in the sixth element with $\alpha_B = \{0, 0, 0, 0, 0, 0.1, 0, 0\}^T$. Table 5.5 shows the identified damage indices from different number of strain gauges.

The identified beam damage indices shown in Figures 5.8 and 5.9 are very close to the true values. But the bonding damage is identified as a spread zone in the beam. This is because the stiffness contribution from bonding is very small compared with the stiffness from the beam itself. The order of damage stiffness from the beam is much larger than that of the bond damage, and this leads the identification algorithm to identify the more significant parameter more accurately than the less significant parameter. This phenomenon occurs with most damage detection algorithms.

The results in Table 5.5 show that the accuracies of both types of damage indices are acceptable when at least ten sensors are used in the measurement, except for the case with twelve sensors. More sensors do not, in general, give good identified results on the bonding damage. This may be due to the noise that accompanies the additional measured information. This again confirms the need to have an optimized selection of sensors for an optimal identification of both types of damage.

5.5   Limitations with static measurements

The main problems with static identification methods are:

1)   There is less information available for the static damage identification method compared to the dynamic ones, which makes it more difficult to obtain accurate results.

2)   The effect of damage may be concealed due to the existence of alternate load paths under the static load. Scenarios with different patterns of loadings have to

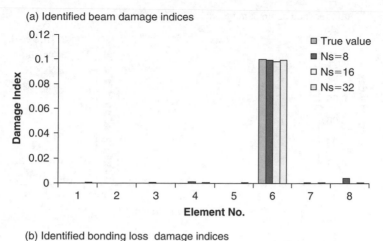

(a) Identified beam damage indices

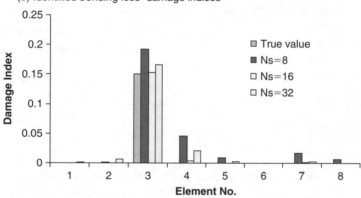

(b) Identified bonding loss damage indices

Figure 5.8 Identified damage indices from strain measurements with 5% noise
(Ns = number of sensors)

be studied to cover all components of a structure with a major contribution to the deformation.

3)   The static-based techniques also face practical challenges when applied to large civil structures. It is nearly impossible to apply a controlled lateral load to a large civil structure. This would require a reaction structure of a size equal to that of the structure with tremendous actuating capability.

4)   Some of the techniques described in this chapter use static displacements in the identification procedure. However, in most cases it is impractical to measure displacements in a large structure, simply because there is no fixed/absolute reference for the measurement.

5.6 Conclusions

The static approach for damage assessment has been shown in this chapter to be feasible but subject to many constraints. The lack of measured data is the main reason

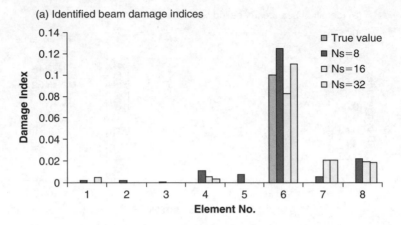

(a) Identified beam damage indices

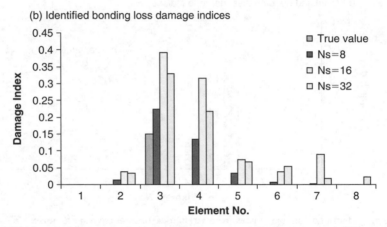

(b) Identified bonding loss damage indices

Figure 5.9   Identified damage indices from displacement measurements with 5% noise
(Ns = number of sensors)

to prevent its application on a large-scale structure with a large number of components. This may be overcome with the high quality data obtained from the high-resolution camera. This group of methods, however, may be suitable to provide a reference set of results for calibrating results obtained from dynamic measurements.

Chapter 6

Damage detection in the frequency domain

6.1 Introduction

Static methods have been discussed in Chapter 5 to identify local damage from static structural responses. Similar to the static methods, the vibration-based technique is also an attractive alternative for structural damage detection, and many researchers have made great efforts to advance knowledge in this area. The basic idea of the vibration-based technique is to measure the dynamic characteristics at some specific stages over the life span of the structure, and use them as a database to assess the condition of the structure.

This chapter discusses this group of vibration-based structural damage assessment techniques. The basic concept of modal analysis is firstly introduced in Section 6.2 and the mathematical theory of the eigenvalue problem is presented in Section 6.3. Based on the sensitivity of the eigenvalue and eigenvector, a non-destructive damage detection algorithm is presented in Section 6.4. To further increase the sensitivity of modal responses to the local change in stiffness or mass properties, the high-order modal parameters, such as the element modal strain energy, modal flexibility and unit load surface are introduced in Section 6.6, and their curvature in the spatial domain are presented in Section 6.7.

6.2 Spatial distributed system

The dynamics of a structure are physically decomposed by frequency and position. This is clearly evidenced by the analytical solution of partial differential equations of continuous systems such as beams and strings. Modal analysis is based on the fact that the vibration response of a linear time-invariant dynamic system can be expressed as the linear combination of a set of simple harmonic motions called the natural modes of vibration. The natural modes of vibration are inherent to a dynamic system and are determined completely by its physical properties (mass, stiffness, damping) and their spatial distributions. Each mode is described in terms of its modal parameters: natural frequency, modal damping factor and the characteristic displacement pattern, called the mode shape. The mode shape may be real or complex with each one corresponding to a natural frequency. The degree of participation of each natural mode in the overall vibration is determined both by the properties of the excitation sources and by the mode shapes of the system.

Modal analysis embraces both theoretical and experimental techniques. The theoretical modal analysis is based on the physical model of a dynamic system comprising its mass, stiffness and damping properties. These properties may be in the form of partial differential equations. An example is the wave equation of a uniform vibratory string established from its mass distribution and elasticity properties. The solution of the equation provides the natural frequencies and mode shapes of the string and its forced vibration responses. However, a more realistic physical model will usually comprise the mass, stiffness and damping properties in terms of their spatial distributions, namely the mass, stiffness and damping matrices. These matrices are incorporated into a set of normal differential equations of motion. The superposition principle of a linear dynamic system enables us to transform these equations into a typical eigenvalue problem, the solution of which provides the modal information of the system. Modern finite element analysis empowers the discretization of almost any linear dynamic structure and hence has greatly enhanced the capacity and scope of theoretical modal analysis. However, the rapid development over the last two decades of data acquisition and processing capabilities has given rise to major advances in the experimental realm of the analysis, which has become known as modal testing.

6.3 The eigenvalue problem

6.3.1 Sensitivity of eigenvalues and eigenvectors

The real symmetric eigenvalue problem associated with linear vibration is defined as

$$\mathbf{K}\phi_j = \lambda_j \mathbf{M}\phi_j \qquad (6.1)$$

where $\mathbf{K}$ and $\mathbf{M}$ are the stiffness and mass matrices, and they are of order n symmetric matrices. $\mathbf{M}$ is positive definite and $\mathbf{K}$ is positive definite or semi-positive definite. Eigenvector, ϕ_j, is the jth mode shape and eigenvalue, λ_j, is the square of the jth natural frequency. The eigenvector, ϕ_j, is typically normalized as

$$\phi_j \mathbf{M}\phi_j = 1 \qquad (6.2)$$

To obtain the derivative of the eigenvalue, Equation (6.1) is differentiated with respect to a system parameter, p, as

$$(\mathbf{K} - \lambda_j \mathbf{M})\frac{\partial \phi_j}{\partial p} = -\left(\frac{\partial \mathbf{K}}{\partial p} - \lambda_j \frac{\partial \mathbf{M}}{\partial p}\right)\phi_j + \frac{\partial \lambda_j}{\partial p}\mathbf{M}\phi_j \qquad (6.3)$$

Pre-multiplying both sides of Equation (6.3) with ϕ_j^T, the eigenvalue derivative is given by

$$\frac{\partial \lambda_j}{\partial p} = \phi_j^T \left(\frac{\partial \mathbf{K}}{\partial p} - \lambda_j \frac{\partial \mathbf{M}}{\partial p}\right)\phi_j \qquad (6.4)$$

Since the matrix, $(\mathbf{K} - \lambda_j \mathbf{M})$, is singular, the eigenvector derivative, $\partial \phi_j / \partial p$, cannot be found directly from Equation (6.3). Nelson (1976) proposed an algorithm expressing the eigenvector derivative in terms of a particular solution, v, and a homogeneous solution, $c\phi_j$, as

$$\frac{\partial \phi_j}{\partial p} = v + c\phi_j \tag{6.5}$$

where c is an undetermined coefficient. The particular solution is found by identifying the component of the jth eigenvector with the largest absolute value and constraining its derivative to zero. The undetermined coefficient can be obtained by substituting Equation (6.5) into the derivative of Equation (6.2) to give

$$c_j = -\mathbf{v}^T \mathbf{M} \phi_j - 0.5\phi_j^T \frac{\partial \mathbf{M}}{\partial p} \phi_j \tag{6.6}$$

Nelson's method is powerful for computing the eigenvector derivatives of general real matrices with distinct eigenvalues, because it only requires knowledge of the eigenpairs that are to be differentiated. However, the algorithm of the method is lengthy and complicated. For a large-scale system, there may be thousands of degrees-of-freedom (DOFs). In this case, only a few of the first lower modes are calculated, while the other higher modes are truncated. Assuming that the first k modes are available with the other $(n-k)$ modes truncated, an eigenvector derivative can be written as (Lee and Jung, 1997a)

$$\frac{\partial \phi_j}{\partial p} = \sum_{i=1}^{k} c\phi_i \tag{6.7}$$

where $c_i = \begin{cases} \dfrac{1}{\lambda_i - \lambda_j}\phi_i^T f_j, & i \neq j \\ h_j, & i = j \end{cases}$, $\quad f_j = -\left(\dfrac{\partial \mathbf{K}}{\partial p} - \lambda_j \dfrac{\partial \mathbf{M}}{\partial p}\right)\phi_j + \dfrac{\partial \lambda_j}{\partial p}\mathbf{M}\phi_j \quad$ and $h_j = -0.5\phi_j^T \mathbf{M}\phi_j$.

A pseudostatic term is added to the expansion of Equation (6.7) to obtain the eigenvector derivative as (Wang, 1985)

$$\frac{\partial \phi_j}{\partial p} = \mathbf{K}^{-1}\left[-\left(\frac{\partial \mathbf{K}}{\partial p} - \lambda_j \frac{\partial \mathbf{M}}{\partial p}\right)\phi_j + \frac{\partial \lambda_j}{\partial p}\mathbf{M}\phi_j\right] + \sum_i d_i \phi_i \tag{6.8}$$

Rudisill (1974) proposed an exact solution method solving an unsymmetric linear algebraic equation with an additional condition. Rewriting Equation (6.1) into

$$(\mathbf{K} - \lambda_j \mathbf{M})\phi_j = 0 \tag{6.9}$$

rearranging Equation (6.3)

$$(\mathbf{K} - \lambda_j \mathbf{M})\frac{\partial \phi_j}{\partial p} - \frac{\partial \lambda_j}{\partial p}\mathbf{M}\phi_j = -\left(\frac{\partial \mathbf{K}}{\partial p} - \lambda_j \frac{\partial \mathbf{M}}{\partial p}\right)\phi_j \tag{6.10}$$

and imposing the constraint on the length of the eigenvectors,

$$\phi_j^T \phi_j = 1 \tag{6.11}$$

Differentiating Equation (6.11) with respect to the parameter gives

$$\phi_j^T \frac{\partial \phi_j}{\partial p} = 0 \tag{6.12}$$

Equations (6.9) and (6.12) can be written as a single matrix equation as follows

$$\begin{bmatrix} \phi_j^T & 0 \\ \mathbf{K} - \lambda_j \mathbf{M} & -\mathbf{M}\phi_j \end{bmatrix} \begin{Bmatrix} \dfrac{\partial \phi_j}{\partial p} \\ \dfrac{\partial \lambda_j}{\partial p} \end{Bmatrix} = \begin{Bmatrix} 0 \\ -\left(\dfrac{\partial \mathbf{K}}{\partial p} - \lambda_j \dfrac{\partial \mathbf{M}}{\partial p} \right) \end{Bmatrix} \phi_j \tag{6.13}$$

The derivatives $\partial \phi_j / \partial p$ and $\partial \lambda_j / \partial p$ can be found by solving Equation (6.13). The algorithm of this method is simple and compact. However, the demand on computer storage and CPU time is high, since the method has to deal with the unsymmetric coefficient matrix to find the eigenpair derivative. For the symmetric coefficient matrix, an equation can be obtained similar to Equation (6.10) by differentiating the mass normalized Equation (6.2) with respect to the parameter, and rearranging to give

$$\phi_j^T \mathbf{M} \frac{\partial \phi_j}{\partial p} + 0.5 \phi_j^T \frac{\partial \mathbf{M}}{\partial p} \phi_j = 0 \tag{6.14}$$

Combining Equations (6.10) and (6.14) as a single matrix equation, the derivatives $\partial \phi_j / \partial p$ and $\partial \lambda_j / \partial p$ can be found as

$$\begin{bmatrix} \mathbf{K} - \lambda_j \mathbf{M} & -\mathbf{M}\phi_j \\ -\phi_j^T \mathbf{M} & 0 \end{bmatrix} \begin{Bmatrix} \dfrac{\partial \phi_j}{\partial p} \\ \dfrac{\partial \lambda_j}{\partial p} \end{Bmatrix} = \begin{Bmatrix} -\left(\dfrac{\partial \mathbf{K}}{\partial p} - \lambda_j \dfrac{\partial \mathbf{M}}{\partial p} \right) \phi_j \\ 0.5 \phi_j^T \dfrac{\partial \mathbf{M}}{\partial p} \phi_j \end{Bmatrix} \tag{6.15}$$

6.3.2 System with close or repeated eigenvalues

When there are multiple eigenvalues and a parameter is perturbed, the eigenvectors will be split into as many as m (multiplicity of multiple eigenvalues) distinct eigenvectors. Consider the following eigenvalue problem, in which ψ, of order $(n \times m)$, is a matrix of eigenvectors with multiple eigenvalues:

$$\mathbf{K}\psi = \mathbf{M}\psi\Lambda \tag{6.16}$$

Here $\Lambda = \lambda I$ and $\psi^T M \psi = I$, and λ is the eigenvalue for the eigenspace spanned by the column of ψ. λ is an eigenvalue of multiplicity m. Adjacent eigenvectors can be expressed in terms of ψ by an orthogonal transformation such as

$$Z = \psi \, \Gamma \tag{6.17}$$

where Γ is an orthogonal transformation matrix of the order $(m \times m)$ with $\Gamma^T \Gamma = I$. The columns of Z are the adjacent eigenvectors for which a derivative can be defined. The adjacent eigenvectors satisfy the mass-orthogonality condition as

$$Z^T M Z = \Gamma^T \psi^T M \psi \, \Gamma = \Gamma^T \Gamma = I \qquad (6.18)$$

The next step is to find Γ such that the derivative of eigenvectors exists and then to find Z, $\partial \Lambda / \partial p$ and $\partial Z / \partial p$. $\partial \Lambda / \partial p$ is obtained as follows:

$$\frac{\partial \Lambda}{\partial p} = diag\left(\frac{\partial \lambda_1}{\partial p}, \frac{\partial \lambda_2}{\partial p}, \cdots, \frac{\partial \lambda_m}{\partial p}\right) \qquad (6.19)$$

Consider the following eigenvalue problem to find Z, $\partial \Lambda / \partial p$ and $\partial Z / \partial p$.

$$KZ = MZ\Lambda \qquad (6.20)$$

Differentiating Equation (6.20) with respect to a parameter and rearranging, gives,

$$(K - \lambda M)\frac{\partial Z}{\partial p} = -\left(\frac{\partial K}{\partial p} - \lambda \frac{\partial M}{\partial p}\right) Z + MZ\frac{\partial \Lambda}{\partial p} \qquad (6.21)$$

Pre-multiplying by ψ^T and substituting Equation (6.17) into Equation (6.21), gives

$$\left[\psi^T \left(\frac{\partial K}{\partial p} - \lambda \frac{\partial M}{\partial p}\right) \psi\right] \Gamma \equiv D\Gamma = \Gamma\frac{\partial \Lambda}{\partial p} \qquad (6.22)$$

The eigenvalue derivative, $\partial \Lambda / \partial p$, and the orthogonal transformation matrix, Γ, can be obtained by solving Equation (6.22), and the adjacent eigenvectors, Z, from Equation (6.17).

If the adjacent eigenvectors are calculated, the following method can be used for the case of repeated eigenvalues. Consider the following eigenvalue problem with the normalization condition $Z^T M Z = I$ and

$$KZ = MZ\Lambda \qquad (6.23)$$

Here $\Lambda = diag(\lambda_j, \lambda_{j+1}, \cdots, \lambda_{j+m-1})$ and $\lambda_j = \lambda_{j+1} = \cdots = \lambda_{j+m-1} = \lambda$. Rewriting Equation (6.21), gives

$$(K - \lambda M)\frac{\partial Z}{\partial p} = F \qquad (6.24)$$

where $F = \left(\lambda \frac{\partial M}{\partial p} - \frac{\partial K}{\partial p}\right) Z + MZ\frac{\partial \Lambda}{\partial p}$, $\dfrac{\partial Z}{\partial p} = \left\{\dfrac{\partial z_1}{\partial p}, \dfrac{\partial z_2}{\partial p}, \cdots, \dfrac{\partial z_m}{\partial p}\right\}^T$ and $F = \{f_1, f_2, \ldots, f_m\}^T$. Consider a shift, $\Delta \lambda$, in Equation (6.24) and let $A_R \dfrac{\partial Z}{\partial p} = [K - (\lambda - \Delta \lambda)M]$, then

$$A_R\frac{\partial Z}{\partial p} = F + \Delta \lambda M\frac{\partial Z}{\partial p} \qquad (6.25)$$

The iteration scheme for the repeated eigenvalue is then given as

$$\frac{\partial \tilde{Z}^{(k+1)}}{\partial p} = A_k^{-1}\left[F + \Delta\lambda M \frac{\partial \tilde{Z}^{(k)}}{\partial p}\right] \tag{6.26}$$

The iteration algorithm can be used to find $\partial Z/\partial p$. The convergence properties of the iteration procedure are discussed in detail by Lee et al. (1997).

6.4 Localization and quantification of damage

The perturbation of the eigenvalue and eigenvector can be expanded into a truncated Taylor series to the first order to have the following linear expression

$$\delta Z = S\delta p \tag{6.27}$$

where δZ is the difference between the measured modal data and the analytical solution; δp is the perturbation in the unknown parameters to be identified; and S is the sensitivity matrix containing the first derivative of the modal parameters with respect to the unknown parameter.
Equation (6.27) can be rewritten as

$$[Z_m - Z_j] = S_j[p_{j+1} - p_j] \tag{6.28}$$

where Z_m is the measured modal parameters used in the updating algorithm; and the subscript j indicates the iteration number at which the sensitivity matrix is computed.

$$p_{j+1} = P_j + S_j^+[Z_m - Z_j] \tag{6.29}$$

where S_j^+ is the pseudo-inverse of S_j.

6.5 Finite element model updating

Let the objective function, f, denote the deviation between the analytical prediction and the real behaviour of a structure. The model updating can be posed as a minimization problem to find a design set, x^*, such that

$$f(x^*) \leq f(x), \quad \forall x \quad \underline{x}_i \leq x_i \leq \bar{x}_i, \quad i = 1, 2, 3, \ldots\ldots n \tag{6.30}$$

where the upper $(\bar{x}_i)$ and lower $(\underline{x}_i)$ bounds on the design variables are required. The objective function in an ordinary least-squares problem is defined as the sum of squared differences

$$f(x) = \sum_{j=1}^{n_r} [z_j(x) - \bar{z}_j]^2 = \sum_{j=1}^{n_r} r_j(x)^2 \tag{6.31}$$

where $z_j(x)$ represents an analytical modal quantity, which is a nonlinear function of the optimization or design variables, $x \in \Re^n$; and $\bar{z}$ refers to the measured modal parameters. In order to obtain a unique solution, the number of residuals, n_r, should be greater than the number, n, of unknown parameters, x.

The updating parameters are the undetermined physical properties of the numerical model. Instead of using the absolute value of each variable, x, its relative variation to the initial value, x_0, is chosen as the dimensionless updating parameter, a. By using the normalized parameters, a, the problem of ill-conditioning due to large relative differences in the parameter magnitudes can be avoided.

$$a^i = -\frac{x^i - x_0^i}{x_0^i} \tag{6.32}$$

$$x^i = x_0^i(1 - a^i) \tag{6.33}$$

The objective of the model updating is to find the value of vector, a^i, in Equation (6.32) such that the error between the measured and analytical modal parameters is minimized. This gives

$$f(a) = \sum_{j=1}^{n_r} r_j(a)^2 \tag{6.34}$$

In general, the residual vector, r, contains the differences between the identified and predicted modal data, such as the eigen-frequencies and the mode shapes.

6.6 Higher order modal parameters and their sensitivity

6.6.1 Elemental modal strain energy

The elemental Modal Strain Energy (MSE) is defined as the product of the elemental stiffness matrix and the second power of its mode shape component. For the jth element and ith mode, the elemental MSE before and after the occurrence of damage are given as

$$MSE_{ij} = \boldsymbol{\Phi}_i^T K_j \boldsymbol{\Phi}_i \quad MSE_{ij}^d = \boldsymbol{\Phi}_i^{d^T} K_j \boldsymbol{\Phi}_i^d \tag{6.35}$$

where MSE_{ij} and MSE_{ij}^d are the jth elemental modal strain energy corresponding to the ith mode shape for the undamaged and damaged states, respectively; K_j is the jth elemental stiffness matrix; and $\boldsymbol{\Phi}_i$ is the ith mode shape. The superscript d denotes the damaged state. Since the damage elements are not known, the undamaged elemental stiffness matrix, K_j, is used instead of the damaged one as an approximation in MSE_{ij}^d.

The elemental Modal Strain Energy Change Ratio (MSECR) has been verified to be a good indicator for damage localization and is defined as (Shi et al., 1998)

$$MSECR_{ij} = \frac{|MSE_{ij}^d - MSE_{ij}|}{MSE_{ij}} \tag{6.36}$$

6.6.1.1 Modal strain energy change sensitivity

Structural damage often causes a loss of stiffness in one or more elements of a structure but not a loss in the mass. In the theoretical development that follows, damage is assumed to affect only the stiffness matrix of the system. When damage occurs in a structure, it can be represented by a small perturbation in the original system. Thus, the stiffness matrix, K^d; the ith modal eigenvalue, λ_i^d; and the ith mode shapes, Φ_i^d, of the damaged system can be expressed as

$$K^d = K + \sum_{j=1}^{L} \Delta K_j = K + \sum_{j=1}^{L} \alpha_j K_j \quad (-1 < \alpha_j \leq 0) \tag{6.37a}$$

$$\lambda_i^d = \lambda_i + \Delta \lambda_i \tag{6.37b}$$

$$\Phi_i^d = \Phi_i + \Delta \Phi_i \tag{6.37c}$$

where α_j is the coefficient defining a fractional reduction in the jth elemental stiffness matrices; and L is the total number of elements in the system. The elemental Modal Strain Energy Change (MSEC) for the jth element in the ith mode is expressed as

$$MSEC_{ij} = \Phi_i^{d^T} K_j \Phi_i^d - \Phi_i^T K_j \Phi_i \tag{6.38}$$

Substituting Equation (6.37c) into Equation (6.38) and neglecting the second-order terms, the $MSEC_j^i$ becomes

$$MSEC_{ij} = 2\Phi_i^T K_j \Delta \Phi_i \tag{6.39}$$

For a small perturbation in an undamped n-DOFs dynamic system, the equation of motion becomes

$$[(K + \Delta K) - (\lambda_i + \Delta \lambda_i)M](\Phi_i + \Delta \Phi_i) = 0 \tag{6.40}$$

Neglecting second-order terms, Equation (6.40) leads to

$$(K - \lambda_i M)\, \Delta \Phi_i = \Delta \lambda_i M \Phi_i - \Delta K \Phi_i \tag{6.41}$$

$\Delta \Phi_i$ in Equation (6.41) can be expressed as a linear combination of mode shapes of the original system (Fox and Kappor, 1968)

$$\Delta \Phi_i = \sum_{k=1}^{n} d_{ik} \Phi_k \tag{6.42}$$

where d_{ik} are scalar factors and n is the total number of modes of the original system. Substituting Equation (6.42) into Equation (6.41), and pre-multiplying Φ_r^T on both sides of Equation (6.41), gives

$$\sum_{k=1}^{n} d_{ik}\Phi_r^T(K - \lambda_i M)\Phi_k = \Delta\lambda_i \Phi_r^T M\Phi_i - \Phi_r^T \Delta K\Phi_i \tag{6.43}$$

With the orthogonal relationship, Equation (6.43) can be simplified into the following when r is not equal to i.

$$d_{ir} = -\frac{\Phi_r^T \Delta K\Phi_i}{\lambda_r - \lambda_i} \quad \text{where } r \neq i \tag{6.44}$$

For the case of $r=i$, d_{rr} equals 0.0 from the orthogonal relationship, $\Phi_i^T M\Phi_i = I$. Therefore (6.42) can be written as

$$\Delta\Phi_i = \sum_{r=1}^{n} -\frac{\Phi_r^T \Delta K\Phi_i}{\lambda_r - \lambda_i}\Phi_r \quad \text{where } r \neq i \tag{6.45}$$

Substituting Equation (6.11) into Equation (6.39) the $MSEC_{ij}$ becomes

$$MSEC_{ij} = 2\Phi_i^T K_j \left(\sum_{r=1}^{n} -\frac{\Phi_r^T \Delta K\Phi_i}{\lambda_r - \lambda_i}\Phi_r\right) \quad \text{where } r \neq i \tag{6.46}$$

Substituting Equation (6.37a) into Equation (6.46), gives

$$MSEC_{ij} = \sum_{p=1}^{L} -2\alpha_p \Phi_i^T K_j \sum_{r=1}^{n} \frac{\Phi_r^T K_p \Phi_i}{\lambda_r - \lambda_i}\Phi_r \quad \text{where } r \neq i \tag{6.47}$$

The term on the left-hand-side of Equation (6.47) is the elemental Modal Strain Energy Change of the jth element in the ith mode, which can be calculated from Equation (6.38) by using the experimental mode shape of the undamaged and damaged states. All the terms on the right-hand-side of Equation (6.47), except α_p, are known information of the undamaged system. Equation (6.47) can be used to quantify the damage magnitude.

If it is assumed that the number of damaged elements to be identified is q, and the number of measured elements for the computation of $MSEC$ in Equation (6.38) is J, Equation (6.47) can be expressed in the following form for the ith mode.

$$\begin{Bmatrix} MSEC_{i1} \\ MSEC_{i2} \\ \cdots \\ MSEC_{iJ} \end{Bmatrix} = \begin{bmatrix} \beta_{11} & \beta_{12} & \cdots & \beta_{1q} \\ \beta_{21} & \beta_{22} & \cdots & \beta_{2q} \\ \cdots & \cdots & \cdots & \cdots \\ \beta_{J1} & \beta_{J2} & \cdots & \beta_{Jq} \end{bmatrix} \begin{Bmatrix} \alpha_1 \\ \alpha_2 \\ \vdots \\ \alpha_q \end{Bmatrix} \tag{6.48}$$

in which the element, $\beta_{st}(s = 1, 2, \ldots, J; t = 1, 2, \ldots, q)$, is the sensitivity coefficient of *MSEC* for the suspected damaged element, and it is given by

$$\beta_{st} = -2\sum_{r=1}^{n^*} \boldsymbol{\Phi}_i^T K_s \frac{\boldsymbol{\Phi}_r^T K_t \boldsymbol{\Phi}_i}{\lambda_r - \lambda_i} \boldsymbol{\Phi}_r \quad \text{where } r \neq i \tag{6.49}$$

where n^* used in the actual calculation is a finite number of modes of the system.

In practice, the group of suspected damaged elements can be determined from any damage localization method in the frequency domain. The number of these elements, q, would be much smaller than the total number of elements, L, in the system. And J is the size of the group of selected elements for the MSEC computation, which can include or exclude the suspected damaged elements with $J \geq q$. When there are m modes to be used to estimate the damage, the number of equations in Equation (6.48) will increase to $m \times J$ to become an over-determined set of equations.

The sensitivity of *MSE* is more informative than the sensitivity of the mode shape for damage identification, as it describes directly the vibration energy change of the element while the mode shape sensitivity describes the change of the mode shape at a DOF. The former is also less subjective to noise effect at a measured DOF than the latter.

In Equation (6.42), n must be equal to the total number of modes of the n-DOFs system in order to obtain an accurate estimation of $\Delta \boldsymbol{\Phi}_i$. But in practice, the number of DOFs of the structure is very large, and only the first few n^* modes of the system are used in the estimation. This introduces a truncation error in the value of $\Delta \boldsymbol{\Phi}_i$ in Equation (6.42) and affects the sensitivity, β_{st}, in Equation (6.49). The error in β_{st} becomes large when the selected modes n^* in Equation (6.42) is small. An improved formulation for the sensitivity has been developed to reduce this truncation error (Shi et al., 2002). The application to the damage assessment of structure with this technique is referred to in Shi et al. (1998; 2000b; 2002).

6.6.2   Modal flexibility

For a structural system with n DOFs, the flexibility matrix can be expressed by superposition of the mass normalized modes, $\{\phi_i\}$, as (Berman and Flannely, 1971)

$$[F] = \sum_{i=1}^{n} \frac{\{\phi_i\}\{\phi_i\}^T}{\omega_i^2} \tag{6.50}$$

where ω_i is the ith natural frequency and $\{\phi_i^T\}M\{\phi_i\} = 1 (i = 1, \cdots, n)$. It can be seen from Equation (6.50) that the modal contribution to the flexibility matrix decreases rapidly as the frequency, ω_i, increases, so the flexibility matrix converges rapidly as the number of contributing lower modes increases. This observation provides a great possibility to approximate closely the flexibility matrix with several lower modes.

In practice, there are only a few, in most cases, two to three, lower vibration modes of a structure that can be obtained with confidence from modal testing. When

r lower modes are available, the modal flexibility matrix of the structure can be approximated as

$$[F] \simeq [F_T] = \sum_{i=1}^{r} \frac{\{\phi_i\}\{\phi_i\}^T}{\omega_i^2} \qquad (6.51)$$

The modal flexibility component, $f_{k,l}$, which represents the displacement at the kth DOF under the unit load at DOF l, is the summation of the products of two related modal coefficients for each available mode.

$$f_{kl} = \sum_{j=1}^{r} \frac{\phi_{lj}\phi_{kj}}{\omega_j^2} \qquad (6.52)$$

If two sets of measurements, one for the intact and another for the damaged structure, are taken and the modal parameters are estimated from the measurements, then the flexibility matrix, F, for the two states can be approximately obtained from Equation (6.51). Considering each column of the flexibility matrix, F, represents a set of nodal displacements due to a unit force at a corresponding DOF, Pandey and Biswas (1994; 1995) simply subtract the flexibility matrix, F_I, of the intact structure from the damaged state flexibility matrix, F_D, to obtain the flexibility changes due to the damage

$$\Delta F = F_D - F_I \qquad (6.53)$$

For each translational DOF j (since it is difficult to measure the rotational DOFs, only the translational DOFs are used in the calculation of flexibility matrices), the maximum absolute value of the elements in the jth column of ΔF is obtained as

$$\delta_j = \max_i |\Delta f_{i,j}| \qquad (6.54)$$

The quantity δ_j is then used as the index to locate damage in structures. However, the maximum absolute value based index is usually replaced with the absolute value of the diagonal element in the corresponding column of the flexibility matrix as

$$\delta_j = |\Delta f_{j,j}| \qquad (6.55)$$

6.6.2.1 Modal flexibility sensitivity

When only r lower modes are available, the modal flexibility matrices in Equation (6.50) can be truncated as

$$[F_T] = \sum_{i=1}^{r} \frac{\{\phi_i\}\{\phi_i\}^T}{\omega_i^2}, \quad \text{and} \quad [F_T^a] = \sum_{i=1}^{r} \frac{\{\phi_i^a\}\{\phi_i^a\}^T}{(\omega_i^a)^2} \qquad (6.56)$$

where $[F_T]$ and $[F_T^a]$, are the measured and analytical Truncated Modal Flexibility (TMF) matrices (Hoyos and Aktan, 1987). They are different from the static flexibility measured from a strain gauge test on a loaded structure. These matrices are symmetric and each coefficient is typically expressed as

$$f_{kl} = \sum_{j=1}^{r} \frac{\phi_{lj}\phi_{kj}}{\omega_j^2}, \quad \text{and} \quad f_{kl}^a = \sum_{j=1}^{r} \frac{\phi_{lj}^a\phi_{kj}^a}{(\omega_j^a)^2} \tag{6.57}$$

where ϕ_{kj} and ϕ_{kj}^a represent the measured and analytical mode shapes of the kth coordinate for the jth mode of the structure, respectively.

The analytical TMF matrix is correlated to the measured TMF matrix by a perturbation matrix, $[\delta F_T]$, contributed by the model error vector $\{\delta\theta\}$, such that

$$[F_T^a] + [\delta F_T] = [F_T] \tag{6.58}$$

Expanding the full TMF matrices and considering the symmetry of them, the above equation can be rewritten in the form of the TMF vectors of dimensions $n \times (n+1)/2$, as

$$\{\delta F_T\} = \{F_T\} - \{F_T^a\} \tag{6.59}$$

Based on the Taylor series expansion to a first-order approximation, the perturbation TMF vector can be expressed as

$$\{\delta F_T\} = \frac{\partial\{F_T^a\}}{\partial\theta_1}\delta\theta_1 + \frac{\partial\{F_T^a\}}{\partial\theta_2}\delta\theta_2 + \cdots + \frac{\partial\{F_T^a\}}{\partial\theta_m}\delta\theta_m = \left[\frac{\partial F_T^a}{\partial\theta}\right]\{\delta\theta\} \tag{6.60}$$

where the subscript m denotes the number of the model error. Substituting Equation (6.60) into Equation (6.59), the inverse problem of model error identification can be formulated into a linear set of equation as

$$\left[\frac{\partial F_T^a}{\partial\theta}\right]\{\delta\theta\} = \{F_T\} - \{F_T^a\} \tag{6.61}$$

The vectors on the right-hand-side of the equation can be obtained from modal measurement and analysis. If the sensitivity matrix of the TMF with respect to the model errors, $[\partial F_T^a/\partial\theta]$, is available, the vector, $\{\delta\theta\}$, which represents the parameter changes correlating the analytical model to the test structure, can be solved directly using least-square techniques or the singular value decomposition method.

Equation (6.61) can also be written as

$$
\begin{bmatrix}
\dfrac{\partial f_{11}}{\partial \theta_1} & \dfrac{\partial f_{11}}{\partial \theta_2} & \cdots & \dfrac{\partial f_{11}}{\partial \theta_m} \\[2mm]
\dfrac{\partial f_{21}}{\partial \theta_1} & \dfrac{\partial f_{21}}{\partial \theta_2} & \cdots & \dfrac{\partial f_{21}}{\partial \theta_m} \\[2mm]
\dfrac{\partial f_{22}}{\partial \theta_1} & \dfrac{\partial f_{22}}{\partial \theta_2} & \cdots & \dfrac{\partial f_{22}}{\partial \theta_m} \\[2mm]
\vdots & \vdots & \vdots & \vdots \\[2mm]
\dfrac{\partial f_{nn}}{\partial \theta_1} & \dfrac{\partial f_{nn}}{\partial \theta_2} & \cdots & \dfrac{\partial f_{nn}}{\partial \theta_m}
\end{bmatrix}
\begin{Bmatrix}
\delta\theta_1 \\ \delta\theta_2 \\ \vdots \\ \delta\theta_m
\end{Bmatrix}
=
\begin{Bmatrix}
f_{11}^x - f_{11}^a \\ f_{21}^x - f_{21}^a \\ f_{22}^x - f_{22}^a \\ \vdots \\ f_{nn}^x - f_{nn}^a
\end{Bmatrix}
\tag{6.62}
$$

where f_{kl}^a and f_{kl}^x are the analytical and measured TMF coefficients, respectively, and $\partial f_{kl}/\partial \theta_i$, $(1 \le k \le n, \; 1 \le l \le k)$ is a TMF sensitivity coefficient with respect to a specific model error. For a large-scale structure in which only the r lower modes, $(r << n)$, were measured, the total number of linear equations in Equation (6.62) would be $n \times (n+1)/2$. However, it should be mentioned that not all the equations are reciprocally independent. It can be found from Equation (6.56) that the rank of the TMF matrix is r (the number of available modes). Without the lost of generality, let us take the first r columns of the flexibility matrix as the basis of constructing the range space $\Re^n$ of modal flexibility, while the remaining columns are linear combinations of them. Thus, the truncated flexibility can be expressed in a matrix form as

$$
[F^T] =
\begin{bmatrix}
f_{1,1} & \otimes & \cdots & \otimes & | & \times \cdots \\
f_{2,1} & f_{2,2} & & \vdots & | & \times \cdots \\
\vdots & \vdots & \ddots & \otimes & | & \times \cdots \\
f_{(n-1),1} & f_{(n-1),2} & \cdots & f_{(n-1),r} & | & \times \cdots \\
f_{n,1} & f_{n,2} & \cdots & f_{n,r} & | & \times \cdots
\end{bmatrix}_{n \times n}
\tag{6.63}
$$

Taking note of the symmetry of Equation (6.63), it can be concluded that there exist $n \times r - r(r-1)/2$ independent terms in the truncated flexibility matrix, leading to the same number of independent equations in Equation (6.62). The condition, $n \times r - r(r-1)/2 \ge m$, is required to ensure an over-determined solution of Equation (6.62).

Taking the derivative of Equation (6.57) with respect to a design parameter θ_i gives the sensitivity coefficients of the modal flexibility as

$$
\frac{\partial f_{kl}}{\partial \theta_i} = \sum_{j=1}^{r} \left[\frac{1}{\omega_j^2} \left(\frac{\partial \phi_{lj}}{\partial \theta_i} \phi_{kj} + \frac{\partial \phi_{kj}}{\partial \theta_i} \phi_{lj} \right) - \frac{2}{\omega_j^3} \frac{\partial \omega_j}{\partial \theta_i} \phi_{lj} \phi_{kj} \right]
\tag{6.64}
$$

The terms $\partial\omega_j/\partial\theta_i$, $\partial\phi_{lj}/\partial\theta_i$ and $\partial\phi_{kj}/\partial\theta_i$ can be computed from Section 6.3. The sensitivity matrix of TMF to a change in the design parameter can then be computed analytically. The application of this technique to the damage assessment of a structure is referred to in Wu and Law (2004b; 2004c).

6.6.3 *Unit load surface*

The modal flexibility matrix of a linear structural system is given in Equation (6.51), while the modal flexibility relating two modal coefficients is given in Equation (6.52). The modal deflection at point k under uniform unit load all over the structure can be approximated as

$$u(k) = \sum_{l=1}^{n} f_{k,l} = \sum_{r=1}^{m} \frac{\phi_r(k) \sum_{l=1}^{n} \phi_r(l)}{\omega_r^2} \tag{6.65}$$

The Uniform Load Surface (ULS) is defined as the deflection vector of the structure under uniform load (Zhang and Aktan, 1998)

$$U_T = \{u(k)\} = F_T \cdot L \tag{6.66}$$

where $L = \{1, \cdots, 1\}_{1 \times n}^T$ is the unit vector representing the uniform load acting on the structure. From Equations (6.52) and (6.65), Zhang and Aktan (1998) observed two features of the ULS comparative to the modal flexibility. Firstly, the ULS is less sensitive to measurement noise than the modal flexibility, because the summation of all the modal coefficients of the corresponding mode, $\sum_{l=1}^{n} \phi_r(l)$ in Equation (6.65), averages out the random error at each measuring point. The second feature is that the ULS converges more rapidly with the lower modes than the modal flexibility. This is also because of the summation of all the modal coefficients of each mode to the ULS in Equation (6.65). Since the modal coefficients of higher modes tend to cancel each other more than those of the lower modes, the lower modes tend to contribute more than the higher modes to the ULS coefficients. This cancelling effect does not exist with the modal flexibility formulation in Equation (6.52). These special properties make the ULS a potentially stable and sensitive damage indicator for structural health monitoring.

6.7 The curvatures

For a continuous structure without damage, the mode shape curvature has a smooth curve shape along the span of the structure. When there is a fault, a change in the curvature in the form of a peak or abrupt slope, appears near the fault position. This local peak position in the curvature can be used as an index to locate local damage in structures. The most practical advantage of the method is that the curvature computation does not refer to the parameters of the intact structure as a baseline. It is also more sensitive to closely distributed damages than the method using mode shape changes. It has been concluded (Pandey et al., 1991; Lu et al., 2002) that the curvature technique is the most efficient to locate these changes in the smooth curves based on the study of damage detection with mode shapes and flexibility for beam-like structures. The following sections describe the finite difference approach, which is commonly adopted for the computation of the curvature from the mode shape. The more sophisticated approach of using polynomial approximation is also given. The gap-smoothing technique is also discussed for the case of incomplete measurement. These techniques are applicable not only to the mode shape information but also to the flexibility and unit load surface as described in last section.

6.7.1 Mode shape curvature

An alternative to using mode shapes to obtain spatial information about vibration changes is using mode shape derivatives, such as curvature. The curvature values are computed from the displacement mode shape using the central difference approximation for mode i and DOF j as

$$\phi_{j,i}'' = \frac{\phi_{j-1,i} - 2\phi_{j,i} + \phi_{j+1,i}}{\Delta l^2} \tag{6.67}$$

where Δl is the distance between the two DOFs considered.

6.7.2 Modal flexibility curvature

Once the flexibility matrix of a structure is estimated from the measured modal parameters, the flexibility curvature can be approximated by a finite difference scheme. In the damage localization of a simply supported reinforced concrete beam (Lu et al., 2002), the flexibility curvature is obtained as

$$fc_i = \frac{f_{i-1,i-1} - 2f_{i,i} + f_{i+1,i+1}}{\Delta l^2} \tag{6.68}$$

where $f_{i,i}$ and Δl are the ith diagonal element of the flexibility matrix and the distance between the two DOFs considered, respectively; and fc_i is the ith item in the flexibility curvature vector.

6.7.3 Unit load surface curvature

Starting from this section, the focus will be on damage detection with plate-like structures. It is assumed that the dynamic response of the plate is acquired by placing sensors in a rectangular grid, such that the mode shapes and the ULS can further be estimated.

An existing curvature-based damage detection technique computes the curvatures using a finite central differentiation procedure. When this technique is incorporated with two-dimensional ULS, the curvatures of the ULS are calculated by a Laplacian operator in each normal direction along the sensor grid as

$$u_{xx}(x_i, y_j) = \frac{u(x_{i+1}, y_j) - 2u(x_i, y_j) + u(x_{i-1}, y_j)}{h_x^2} \tag{6.69a}$$

$$u_{yy}(x_i, y_j) = \frac{u(x_i, y_{j+1}) - 2u(x_i, y_j) + u(x_i, y_{j-1})}{h_y^2} \tag{6.69b}$$

in which the ULS is grouped from a vector into a matrix according to the coordinates of measuring points in the grid, and the grid is assumed to be equally spaced in the x- and y-directions, as shown in Figure 6.1. h_x and h_y are the uniform grid spacings in the corresponding directions.

If two sets of measurements, one from the intact structure and another from the damaged structure, are taken and the modal parameters are estimated from the measurements, the ULS curvature at point (x_i, y_j) for the two states can be obtained using

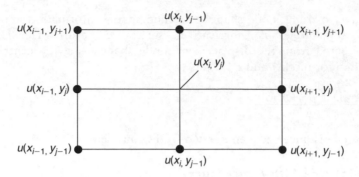

Figure 6.1 Gapped grid for curve fitting at measuring point (x_i, y_j)

Equations (6.52) and (6.69). The presence of the irregularity in the damaged curvature can be detected by subtracting the ULS curvature of the intact state from the curvature of the damaged state. Thereby a map of the damage index can be formulated as

$$d(x_i, y_j) = [\alpha_{xx}|\, u_{xx}^D(x_i, y_j) - u_{xx}(x_i, y_j)| + \alpha_{yy}|\, u_{yy}^D(x_i, y_j) - u_{yy}(x_i, y_j)|]^2 \qquad (6.70)$$

where $|\bullet|$ denotes the absolute value; u_{xx} and u_{yy} are the measured ULS curvature values of the intact structure at the corresponding point in the x- and y-directions, respectively, u_{xx}^D and u_{yy}^D are those of the suspected damaged structure; and α_{xx} and α_{yy} are the weights that can be set from 0.0 to 1.0 to consider the importance of the curvature in the corresponding directions.

If the structure is undamaged when the second set of measurement is carried out, the difference between the two sets of measured ULS curvature would be due to measurement noise only. Therefore values of the damage index map, $d(x_i, y_j)$, slightly oscillate around zero without any distinct peak. In contrast, if the structure is damaged, peaks or slopes will clearly show up at the damaged zone of the plate, as shown later in the numerical study in Section 6.7.6.

6.7.4 Chebyshev polynomial approximation

The accuracy of the central difference method is well known depending on the density of the measurement grid. If the ULS values are estimated on a sparse grid, it will induce a very large error in calculating the curvature from differentiation. The following Chebyshev polynomial in two variables is adopted to model the ULS distribution to avoid this error (Wu and Law, 2004a):

$$u(x, y) = \sum_{i=1}^{N} \sum_{j=1}^{M} c_{ij} T_i(x) T_j(y) \qquad (6.71)$$

where $T_i(x)$ and $T_j(y)$ are the first kind Chebyshev polynomials, and N and M are their orders. To map the standard Chebyshev polynomials from the plane domain of

$\{\xi, \mu\} = [-1, 1] \times [-1, 1]$ to the physical plate domain of $\{x, y\} = [0, L_x] \times [0, L_y]$, two linear transfer functions are defined

$$\xi = 2x/L_x - 1, \quad \mu = 2y/L_y - 1 \tag{6.72}$$

L_x and L_y are the dimensions of the plate in the x- and y-directions, respectively. The Chebyshev polynomials of variable x is then written as

$$T_1(x) = \frac{1}{\sqrt{\pi}}, \quad T_2(x) = \sqrt{\frac{2}{\pi}} \left(\frac{2x}{L_x} - 1 \right),$$

$$T_{i+1}(x) = 2 \left(\frac{2x}{L_x} - 1 \right) T_i(x) - T_{i-1}(x), \quad i = 2, 3, \cdots, N - 1 \tag{6.73}$$

The polynomials of variable y can be formulated similarly.

Without loss of generality, it is assumed that $P = N \times M$ measuring points are set on the rectangular sensor grid so that the ULS can be estimated at these points. Equation (6.71) will be satisfied at all the measuring points, and the Chebyshev polynomial approximation can be written in a matrix form

$$\{u(x_i, y_j)\}_{P \times 1} = [T(x_i)T(y_j)]_{P \times P} \{c_{ij}\}_{P \times 1} \tag{6.74}$$

The coefficient vector, $\{c_{ij}\}$, can then be solved as

$$\{c_{ij}\}_{P \times 1} = [T(x_i)T(y_j)]_{P \times P}^{-1} \{u(x_i, y_j)\}_{P \times 1} \tag{6.75}$$

or obtained by the least-squares method if the number of measuring points $Q > P$.

$$\{c_{ij}\}_{P \times 1} = ([T(x_i)T(y_j)]_{Q \times P}^T [T(x_i)T(y_j)]_{Q \times P})^{-1}$$

$$[T(x_i)T(y_j)]_{Q \times P}^T \{u(x_i, y_j)\}_{Q \times 1} \tag{6.76}$$

It is a better choice to have the measuring points at the Chebyshev zeros $(\hat{x}_i, \hat{y}_j)$, which ensure the convergence for any continuous function that satisfies a Dini–Lipschitz condition (Mason and Handscomb, 2003). The location of these $N \times M$ zeros is given by

$$\hat{x}_i = \left(\cos \frac{(i - 0.5)\pi}{N} + 1 \right) \frac{L_x}{2}, \quad i = 1, \cdots, N \tag{6.77a}$$

$$\hat{y}_j = \left(\cos \frac{(j - 0.5)\pi}{M} + 1 \right) \frac{L_y}{2}, \quad j = 1, \cdots, M \tag{6.77b}$$

The corresponding coefficients can be explicitly obtained as

$$c_{ij} = \frac{\lambda}{P} \sum_{r=1}^{N} \sum_{s=1}^{M} u(\hat{x}_i, \hat{y}_j) \cos\left(\frac{i(r-0.5)\pi}{N}\right) \cos\left(\frac{j(s-0.5)\pi}{M}\right),$$

$$\left\{ \begin{array}{ll} i = 1, \cdots, N; & j = 1, \cdots, M \\ r = 1, \cdots, N; & s = 1, \cdots, M \end{array} \right\} \tag{6.78}$$

where

$$\lambda = \left\{ \begin{array}{lll} 1 & for & i=1, j=1 \\ 2 & for & i=1, j \neq 1 \quad or \quad i \neq 1, j=1 \\ 4 & for & i \neq 1, j \neq 1 \end{array} \right.$$

By making use of the orthogonal property of the Chebyshev polynomial, the curvature of the ULS can then be approximated by the second derivatives of the Chebyshev polynomials in Equation (6.71) as

$$u_{xx}(x,y) = \sum_{i=1}^{N} \sum_{j=1}^{M} c_{ij} \frac{\partial T_i^2(x)}{\partial x^2} T_j(y), \quad u_{yy}(x,y) = \sum_{i=1}^{N} \sum_{j=1}^{M} c_{ij} T_i(x) \frac{\partial T_j^2(y)}{\partial y^2} \tag{6.79a}$$

and

$$u_{xy}(x,y) = \sum_{i=1}^{N} \sum_{j=1}^{M} c_{ij} \frac{\partial T_i(x)}{\partial x} \frac{\partial T_j(y)}{\partial y} \tag{6.79b}$$

Therefore, the formulation of the damage index in Equation (6.70) can be rewritten as

$$d(x_i, y_j) = \left[\alpha_{xx} |u_{xx}^D - u_{xx}| + \alpha_{yy} |u_{yy}^D - u_{yy}| + \alpha_{xy} |u_{xy}^D - u_{xy}| \right]^2 \tag{6.80}$$

It should be noted that the discussions in Sections 6.7.4 and 6.7.5 for a two-dimensional distribution of ULS curvature are applicable for the mode shape curvature and flexibility curvature obtained from the same set of measured mode shape information.

6.7.5   The gap-smoothing technique

Most damage-index methods require the 'footprint', or baseline data set, of the intact structure for comparison, to inspect the change in modal parameters due to damage. Typically the 'footprint' is obtained either from measurements of the undamaged structure, or from a finite element model of the intact structure. An inaccurate finite element model can introduce large model errors, and degrade or even lead to incorrect results in the damage detection. However, most suspected damaged civil structures were constructed several decades ago, and the 'footprint' of the structures in the intact state is not available. To avoid this difficulty, Ratcliffe and Bagaria (1998) proposed the 'gapped-smoothing' technique with modal curvature, which allows for damage detection in a beam structure without prior knowledge of the undamaged state. The 'gapped-smoothing' technique has now been extended and applied to bi-dimensional ULS curvature for the plate structures.

The basic idea of the method is that the ULS curvature of the plate, without any damage, has a smooth surface, and it can be approximated by a cubic polynomial in two variables:

$$\tilde{u}(x,y) = c_0 + c_1 x + c_2 y + c_3 x^2 + c_4 y^2 + c_5 xy + c_6 x^2 y + c_7 xy^2 \tag{6.81}$$

where the coefficients, c_i, can be evaluated by a curve-fitting process on the estimated ULS curvature of the damaged structure on a gapped grid of measuring points as shown in Figure 6.1. Particularly, to obtain the smoothed ULS curvature at point (x_i, y_j), curvature data at all the adjacent points, but not the point (x_i, y_j) itself, are used to evaluate the coefficients, c_i. This process is repeated for each measuring point to give a smooth ULS curvature to model the undamaged plate structure. The presence of the peak in the ULS curvature due to local damage can then be detected by subtracting the smoothed curvature from the estimated curvature of the damaged structure. The damage index map is given similar to Equation (6.80) as

$$d(x_i, y_j) = [\alpha_{xx}|u_{xx}^D - \tilde{u}_{xx}| + \alpha_{yy}|u_{yy}^D - \tilde{u}_{yy}| + \alpha_{xy}|u_{xy}^D - \tilde{u}_{xy}|]^2 \tag{6.82}$$

6.7.5.1 The uniform load surface curvature sensitivity

Consider a damaged plate divided into n_e rectangular elements. The characteristic equation for this undamped structural dynamic system can be expressed as

$$K\{\phi\}_r - \omega_r^2 M\{\phi\}_r = 0 \tag{6.83}$$

It is assumed that any occurrence of structural damage only causes a reduction in local stiffness, while the mass matrix remains unchanged. The stiffness and mass matrices of the damaged plate are expressed as

$$K = K_0 + \Delta K = \sum_{e=1}^{n_e} p_e k_e, \quad M = M_0 \tag{6.84}$$

in which k_e is the element stiffness matrix of the intact plate and p_e indicates the health status of the corresponding element, where $p_e = 1$ stands for the intact and $p_e < 1$ for the damaged state. M_0 and M are the mass matrices of the damaged and intact states.

Taking the derivative of Equation (6.83) with respect to the stiffness parameter, p_e, and noting the relations in Equation (6.84), gives

$$(K - \omega_r^2 M)\frac{\partial\{\phi\}_r}{\partial p_e} = \left(2\omega_r \frac{\partial\omega}{\partial p_i} M - k_e\right)\{\phi\}_r \tag{6.85}$$

The sensitivity of the natural frequency and mode shapes with respect to the parameter, p_e, can be calculated using Nelson's method (Nelson, 1976).

To find the sensitivity of the ULS curvature $\partial uc(x, y)/\partial p_e$, the sensitivity of the ULS at the measuring points needs to be calculated first. Taking the derivative of Equation (6.65) with respect to the parameter, p_e, gives

$$\frac{\partial u(x_i, y_j)}{\partial p_e} = \frac{\partial u_k}{\partial p_e} = \sum_{r=1}^{m} \sum_{l=1}^{n} \left[\frac{1}{\omega_r^2} \left(\frac{\partial \phi_{kr}}{\partial p_e} \phi_{lr} + \frac{\partial \phi_{lr}}{\partial p_e} \phi_{kr} \right) - \frac{2}{\omega_r^3} \frac{\partial \omega_r}{\partial p_k} \phi_{kr} \phi_{lr} \right] \qquad (6.86)$$

Substituting the sensitivities of the mode shape and the modal frequency in Equation (6.15) into Equation (6.86), the ULS sensitivities on the sensor grids are obtained. Further, taking the derivative of Equation (6.76) with respect to the parameter, p_e, and substituting Equation (6.86), the sensitivity of the Chebyshev coefficients is obtained as

$$\left\{ \frac{\partial c_{ij}}{\partial p_e} \right\}_{P \times 1} = \left([T(x_i)T(y_j)]_{Q \times P}^T [T(x_i)T(y_j)]_{Q \times P} \right)^{-1} [T(x_i)T(y_j)]_{Q \times P}^T \left\{ \frac{\partial u(x_i, y_j)}{\partial p_e} \right\}_{Q \times 1}$$

$$(6.87)$$

If the measuring points are located at the Chebyshev zeros $(\hat{x}_i, \hat{y}_j)$, substituting Equation (6.86) into Equation (6.78) also gives the sensitivity of Chebyshev coefficients as

$$\frac{\partial c_{ij}}{\partial p_k} = \frac{\lambda}{P} \sum_{r=1}^{N} \sum_{s=1}^{M} \frac{\partial u(\hat{x}_i, \hat{y}_j)}{\partial p_k} \cos \left(\frac{i(r - 0.5)\pi}{N} \right) \cos \left(\frac{j(s - 0.5)\pi}{M} \right), \quad \begin{cases} i = 1, \cdots, N \\ j = 1, \cdots, M \end{cases}$$

$$(6.88)$$

Finally the sensitivity of the ULS curvature with respect to the stiffness parameter, p_e, can be derived from the Equation (6.79) and calculated as

$$\frac{\partial uc_{xx}(x, y)}{\partial p_e} = \sum_{i=1}^{N} \sum_{j=1}^{M} \frac{\partial c_{ij}}{\partial p_e} \frac{\partial T_i^2(x)}{\partial x^2} T_j(y), \qquad \frac{\partial uc_{yy}(x, y)}{\partial p_e} = \sum_{i=1}^{N} \sum_{j=1}^{M} \frac{\partial c_{ij}}{\partial p_e} T_i(x) \frac{\partial T_j^2(y)}{\partial y^2}$$

$$(6.89a)$$

and

$$\frac{\partial uc_{xy}(x, y)}{\partial p_e} = \sum_{i=1}^{N} \sum_{j=1}^{M} \frac{\partial c_{ij}}{\partial p_e} \frac{\partial T_i(x)}{\partial x} \frac{\partial T_j(y)}{\partial y} \qquad (6.89b)$$

Since the structural damage is assumed as a reduction of local stiffness in Equation (6.84), the changes in the ULS curvature caused by damage can be expanded with the first-order Taylor series approximation.

$$\Delta uc_{xx}(x_i, y_j) = uc_{xx}^D - uc_{xx} = \frac{\partial uc_{xx}}{\partial p_1} \Delta p_1 + \frac{\partial uc_{xx}}{\partial p_2} \Delta p_2 + \cdots + \frac{\partial uc_{xx}}{\partial p_{ne}} \Delta p_{ne} \qquad (6.90)$$

in which n_e indicates the number of suspected damaged elements. Similar relations can be obtained for the curvature changes $\Delta uc_{yy}(x_i, y_j)$ and $\Delta uc_{xy}(x_i, y_j)$.

Applying Equation (6.90) to all the measuring points and rearranging in matrix form, gives

$$
\begin{bmatrix}
\dfrac{\partial uc_{xx}(x_1,y_1)}{\partial p_1} & \dfrac{\partial uc_{xx}(x_1,y_1)}{\partial p_2} & \cdots & \dfrac{\partial uc_{xx}(x_1,y_1)}{\partial p_{ne}} \\[2ex]
\vdots & \vdots & \vdots & \vdots \\[1ex]
\dfrac{\partial uc_{xx}(x_N,y_M)}{\partial p_1} & \dfrac{\partial uc_{xx}(x_N,y_M)}{\partial p_2} & \cdots & \dfrac{\partial uc_{xx}(x_N,y_M)}{\partial p_{ne}} \\[2ex]
\vdots & \vdots & \vdots & \vdots \\[1ex]
\dfrac{\partial uc_{yy}(x_N,y_M)}{\partial p_1} & \dfrac{\partial uc_{yy}(x_N,y_M)}{\partial p_2} & \cdots & \dfrac{\partial uc_{yy}(x_N,y_M)}{\partial p_{ne}} \\[2ex]
\vdots & \vdots & \vdots & \vdots \\[1ex]
\dfrac{\partial uc_{xy}(x_N,y_M)}{\partial p_1} & \dfrac{\partial uc_{xy}(x_N,y_M)}{\partial p_2} & \cdots & \dfrac{\partial uc_{xy}(x_N,y_M)}{\partial p_{ne}}
\end{bmatrix}
\begin{Bmatrix}
\Delta p_1 \\ \Delta p_2 \\ \vdots \\ \Delta p_{ne}
\end{Bmatrix}
$$

$$
= \begin{Bmatrix}
uc_{xx}^{D}(x_1,y_1) - uc_{xx}(x_1,y_1) \\[1ex]
\vdots \\[1ex]
uc_{xx}^{D}(x_N,y_M) - uc_{xx}(x_N,y_M) \\[1ex]
\vdots \\[1ex]
uc_{yy}^{D}(x_N,y_M) - uc_{yy}(x_N,y_M) \\[1ex]
\vdots \\[1ex]
uc_{xy}^{D}(x_N,y_M) - uc_{xy}(x_N,y_M)
\end{Bmatrix}
\tag{6.91}
$$

or

$$
S \cdot \{\Delta p_e\} = \{\Delta uc\}
\tag{6.92}
$$

It should be noted that Equation (6.91) comprises n_e unknown variables and a total of $n_{eq} = 3 \times Q$ linear equations. To ensure that a unique over-determined solution can be found, the condition $n_{eq} \geq n_e$ is required. The Singular Value Decomposition (SVD) with error-truncation technique (Wu and Law, 2005a) is adopted to solve the linearized equations. To reflect the nonlinear effects from large magnitude damage or interaction between the variables, the equations are solved by an iterative algorithm as follows.

Step 1. Initialize the vector of stiffness parameter as $\{p_e|_{e=1}^{n_e}\}_0 = 1$.

Step 2. At the beginning of each iteration, construct the ULS curvature sensitivity matrix, S_i, for the ith iteration, and calculate the curvature change, $\{\Delta uc\}_i$, due to damage.

Step 3. Solve the increment of parameter vector as, $\{\Delta p_e\}_i = S_i^*\{\Delta uc\}_i$, where S_i^* is the generalized inverse of S_i from the SVD technique.

Step 4. Normalize the increment vector, $\{\Delta p_e\}_i = \{\Delta p_e\}_i \times \alpha$, so that the condition $\beta_{min} \leq \Delta p_e|_{e=1}^{n_e} \leq \beta_{pos}$ is satisfied. α is a scale factor and β_{min} and β_{pos} are the convergence limits as shown below.

Step 5. Evaluate the new stiffness parameter vector as $\left\{p_e\big|_{e=1}^{n_e}\right\}_i = \left\{p_e\big|_{e=1}^{n_e}\right\}_{i-1} \cdot$ $(1 + \{\Delta p_e\}_i)$. If $\varepsilon_1 = |\{\Delta p_e\}| < \bar{\varepsilon}_1$ or $\varepsilon_2 = \frac{|\{\Delta uc\}_{i+1}|}{|\{\Delta uc\}_i|} < \bar{\varepsilon}_2$, the solution is considered converged at $\left\{p_e\big|_{e=1}^{n_e}\right\}_i$. Otherwise, go to Step 2 and repeat the process until the condition is satisfied.

The scale factor, α, in Step 4 is calculated as $\alpha = \min\left(\frac{\beta_{\min}}{\min\left(\Delta p_e\big|_{e=1}^{n_e}\right)}, \frac{\beta_{pos}}{\max\left(\Delta p_e\big|_{e=1}^{n_e}\right)}\right)$, in which the range $[\beta_{\min}, \beta_{pos}]$ is set to compromise the convergence speed and solution stability. A wider range could speed up the solution convergence, but for some cases, it may lead to a divergent solution, while a narrower range would have the opposite effect. The application of this technique in the damage assessment of a plate structure is referred to in Wu and Law (2005a).

6.7.6 Numerical examples of damage localization

Two plates with different boundary conditions, namely a four-side simply supported plate and a cantilever plate, are used as examples to demonstrate how the change in the ULS curvature can be used as an index to locate damage in a plate. These examples were chosen because each of them exhibits a different behaviour with the load distribution. For example, in a uniformly loaded four-side simply supported plate, both the maximum bending moment and flexural displacement occur at the geometrical centre of the plate, where flexural damage would most likely occur. In the cantilever plate, the maximum bending moment and shear force occur at the clamped edge where the flexural displacement is a minimum, so that the damage is usually in the form of a crack along the fixed edge.

The configuration of the cantilever plate is shown in Figure 6.2. The dimensions of the plate are 600 mm × 480 mm with a thickness of 20 mm. The finite element model of the plate consists of $15 \times 12 = 180$ equal size square Reissner–Mindlin plate elements. Three DOFs, which are translational along the Z-axis and rotational along the X- and Y-axes, are used at each node. The simply supported plate has the same dimensions and finite element mesh as the cantilever plate, but with different boundary conditions.

It is assumed that damage will affect only the stiffness matrix of a structure. The change in the stiffness matrix due to damage is modelled by a reduction in the elastic modulus of the corresponding element. The extent of damage is then linearly related to the degree of reduction in the elastic modulus, E.

For each damage case, the natural frequencies and corresponding mode shapes are obtained from the finite element analysis. The ULS curvatures are calculated separately using Equations (6.69) and (6.79). Since the three DOFs at each node in the Reissner–Mindlin plate elements are uncoupled, one can choose whether to calculate the translational DOF-based ULS curvature or rotational DOF-based ULS curvature. Results from the simulation study show that the latter is more sensitive to local damage and less sensitive to random noise than the former. However, since there it is difficult to measure the rotational DOFs with current dynamic testing techniques, only the translational DOFs along the Z-axis are used in the study. Equations (6.70), (6.80) and (6.82) are then used separately in the calculation of the damage index based on changes in the ULS curvature and the gapped-smoothing technique. The weights α_{xx}, α_{yy} and α_{xy} are all taken equal to unity. The effectiveness of the methods from the

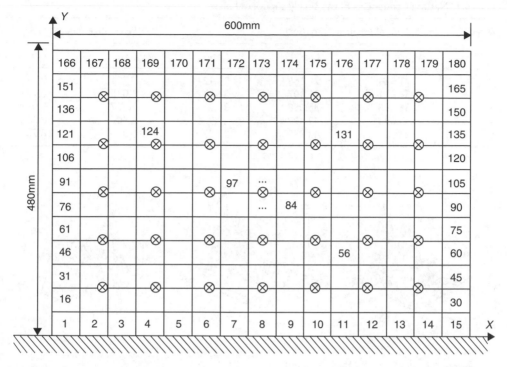

Figure 6.2 Finite element model of the cantilever plate (⊗ gapped grid of measuring points)

central difference and the Chebyshev polynomial are compared. The effects of measurement noise, mode truncation, and sensor sparsity on the ULS curvature changes are also studied with particular damage cases.

6.7.6.1 Simply supported plate

For the four-side simply supported plate, two different damage patterns are simulated to study the capability of the ULS curvature method for sparsely distributed and closely distributed damage, respectively. Case 1 has 75% damage in element 56, 50% damage in element 131 and 25% damage in element 124. Case 2 has 50% damage in both elements 84 and 97. A comparison of the first five natural frequencies for the intact case and the two damaged cases are shown in Table 6.1.

6.7.6.2 Study on truncation effect

As mentioned in Section 6.6.2, the ULS, as well as the modal flexibility, can be approximately obtained from the few lower modes. However, if too few modes are identified experimentally, the flexibility or ULS from modal parameters will generally appear stiffer than it really is, and this affects the results of damage detection. The study of how many modes and what frequency band is sufficient for the modal-based ULS from Equation (6.52) leading to reliable damage detection is called truncation effect analysis. Figure 6.3 compares the changes in the exact modal flexibility and the changes in

Table 6.1 Natural frequencies of the simply supported plate

Mode	Natural frequency (Rad/sec)			Percentage reduction (%)	
	Intact	Case 1	Case 2	Case 1	Case 2
1	21.324	21.165	21.121	0.751	0.961
2	31.631	31.307	31.572	1.035	0.187
3	36.298	35.858	36.266	1.227	0.088
4	43.094	42.499	42.936	1.400	0.368
5	44.011	43.875	43.670	0.310	0.781

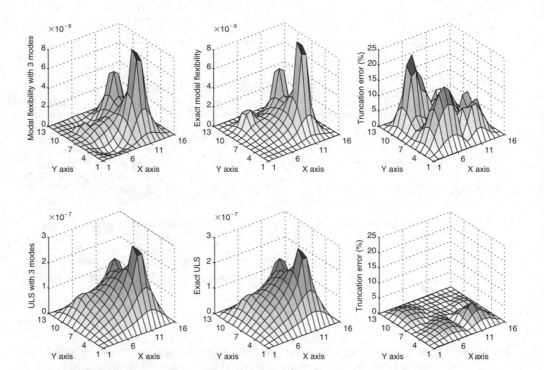

Figure 6.3 Comparison of truncation effect on ULS with modal flexibility

the exact ULS due to damage Case 1 is with the modal parameters from the first three modes. The exact ULS is calculated using all the modes available in the finite element model. The percentage truncation errors were evaluated as

$$\{e_{ij}\}_{du} = \frac{\{du_{ij}\}_R - \{du_{ij}\}_T}{\max \{du_{ij}\}_R} \times 100\% \qquad (6.93)$$

where $\{du_{ij}\}_R$, $\{du_{ij}\}_T$ are the changes of the exact ULS and the changes of modal truncated ULS respectively. The truncation error on the modal flexibility was computed similarly. It can be seen that the ULS converges more rapidly than the modal flexibility with the first three lower modes. The truncation errors in ULS are less than

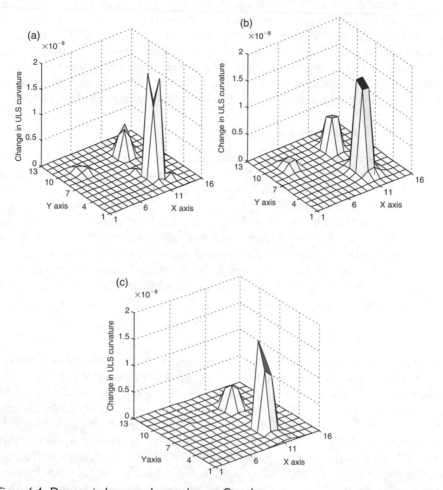

Figure 6.4 Damage index map due to damage Case 1
(a) curvature from central difference; (b) curvature from Chebyshev polynomial;
(c) curvature from Chebyshev polynomial with gapped-smoothing technique

6%, whereas the maximum truncation error in modal flexibility approaches 25%. In the following studies, all the ULS curvatures for the intact and damaged plates are estimated from the first three modes.

6.7.6.3 Comparison of curvature methods

The changes in the curvature of the uniform load surface for the plate with damage Case 1 are plotted in Figure 6.4. Figures 6.4(a), 6.4(b) and 6.4(c) show the results computed from Equations (6.70), (6.70) and (6.82), respectively, when there is no information on the intact state. It can be clearly seen that there is a peak located at each damage element. The more severely the element is damaged, the more sharp and tall the peak looks. According to Equations (6.70), (6.70) and (6.82), the absolute value of the damage indices, or visually the height of the peak, increases exponentially with the change in ULS curvature, and the 25% damage in element 124 shows only a very tiny peak compared with the 75% damage in element 56. Nevertheless, it does not mean the damage with 25% stiffness reduction is the limit the methods can detect. For this case, both the central difference method (Figure 6.4(a)) and the Chebyshev polynomial method (Figure 6.4(b)) can successfully locate all three damaged elements, whereas, when prior knowledge of the intact structure is not available, the Chebyshev polynomial method with gapped-smoothing technique failed to locate the damage in element 124.

6.7.6.4   Resolution of damage localization

It is well known that most damage-index methods can localize the spatially distributed damage quite accurately, but have difficulty detecting contiguous multiple damages. Damage Case 2 is specially simulated to study the effectiveness of the ULS curvature method for closely distributed damages. Figure 6.5 shows the results of damage detection for this case. It can be seen that the two damaged elements 84 and 97 are located close together at the centre of the plate, and they can be separately detected by inspecting the change in the ULS curvature, computed either from central difference or by the Chebyshev polynomial. However, when given the modal data for the damaged state only, although the damaged region can be localized from Figure 6.5(c), it is hard to tell exactly which element is damaged.

6.7.6.5   Cantilever plate

It is always a difficult problem for damage index methods to reliably identify damage near the supported boundary. For a one-side clamped slab, it is intuitive that the damage is more likely to develop near and along the fixed edge where the maximum bending moment occurs. Two typical damage cases for the cantilevered slab are simulated. Case 3 has 75% damage in element 15, 50% damage in elements 13 and 14 and 25% damage in element 12. Case 4 has 75% damage in element 8, 50% damage in elements 7 and 9 and 25% damage in elements 6 and 10. The damage in Case 3 starts from one end of the fixed edge and continues along the edge across four elements. The damage in Case 4 models a band of damage symmetrically located in the middle and along the fixed edge across five elements. The natural frequencies for the intact and damaged structures are listed in Table 6.2.

The results of damage detection for damage Cases 3 and 4 are shown in Figures 6.6 and 6.7, respectively. It can be seen from Figure 6.6 that the damage band near the boundary can be detected by inspecting the ULS change calculated by both methods, even without the initial curvature of intact structure. The ULS curvature by Chebyshev polynomial method gives relatively better localization of the damage than those from central difference with a parabolic curve surface compared with a sharp change from central difference method with more than one peak, while the damage location cannot be exactly validated when the intact structure curvature is absent. A similar observation can be obtained from Figure 6.7.

6.7.6.6   Effect of sensor sparsity

In a real experiment, it is not practical to have a very fine sensor mesh to measure the dynamic response of all the nodes in the finite element model. To study the effect

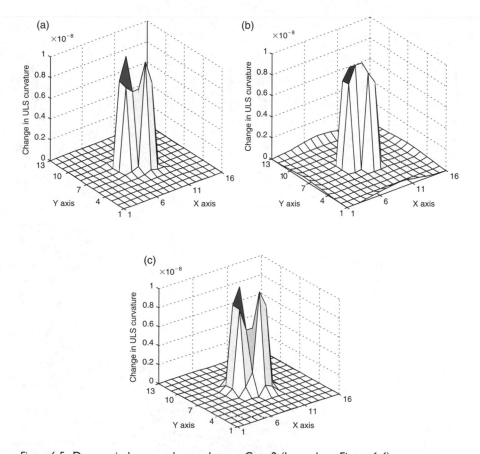

Figure 6.5 Damage index map due to damage Case 2 (Legends as Figure 6.4)

Table 6.2 Natural frequencies of the cantilever plate

Mode	Natural frequency (Rad/sec)			Percentage reduction (%)	
	Intact	Case 3	Case 4	Case 3	Case 4
1	8.051	7.932	7.879	1.478	2.183
2	11.305	10.987	11.296	2.813	0.080
3	16.561	16.32	16.541	1.455	0.121
4	22.109	21.667	22.105	1.999	0.018
5	24.765	24.442	24.310	1.304	1.872

of sensor sparsity on the ULS curvature method, the sensor mesh is reduced to 7×5 and the locations are shown in Figure 6.2, while the nodes grid of the finite element model is 16×13. The damage Case 1 is studied again with this new sensor grid on the four-side plate. First of all, the sensors are placed in an equal spatial grid as shown in Figure 6.2, and the central difference method is applied to estimate the ULS curvature

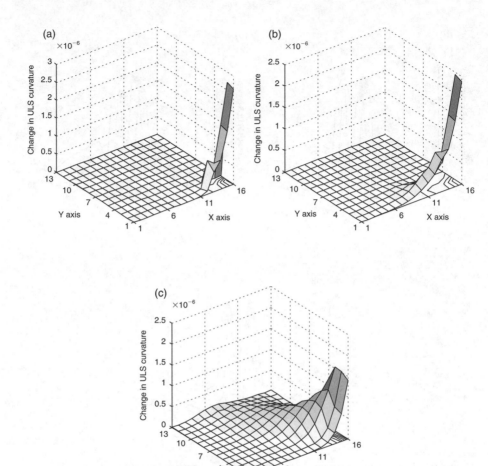

Figure 6.6 Damage index map due to damage Case 3 (Legends as Figure 6.4)

changes due to damage. Then the same number of sensors is placed on the grid points corresponding to the Chebyshev zeros, and the Chebyshev polynomial approximation method is used to calculate the ULS curvature changes in the finite element grid with $M = 7$ and $N = 5$. The Chebyshev polynomial method is applied again when the modal data of the intact structure is absent, and the ULS curvature for the intact structure is approximated by a cubic smooth polynomial function with the gapped-smoothing technique. Corresponding results of damage detection are plotted in Figure 6.8.

It can be seen from Figure 6.8(b) that the Chebyshev polynomial method can still localize, with confidence, all three damaged elements with different extents of stiffness reduction from 25% to 75%. The central difference method fails to detect the damage with 25% stiffness reduction. The absolute values of curvature change in the damaged region degrade dramatically when compared with Figure 6.4(b), and the detected area of suspected damaged region is much larger than before. When information on the intact structure is not available, it is fortuitous to find in Figure 6.8(c) that the damaged

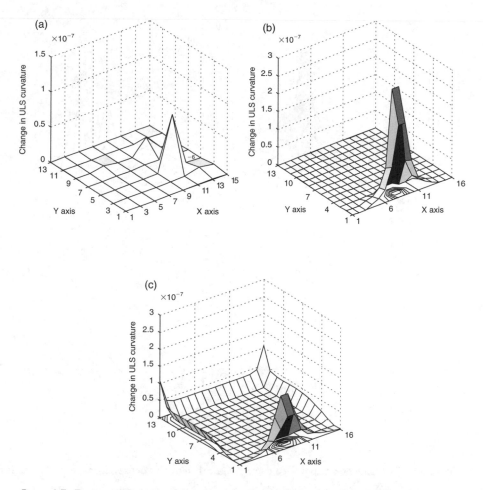

Figure 6.7 Damage index map due to damage Case 4 (Legends as Figure 6.4)

element 124, which is missing in Figure 6.4(c), can just be detected even with the polynomial interpolation from the coarse sensor mesh. However, the reason for this observation is unknown.

6.7.6.7 *Effect of measurement noise*

According to Equation (6.52), the ULS is estimated from experimentally measured natural frequencies and mode shapes, which, in practice, are liable to be contaminated by the measurement noise. Thus, to take account of the noise in experimentally measured modal parameters, 1% random noise is added to the natural frequencies and 5% noise is added to the mode shapes. It is assumed that the random noise is uniformly distributed with zero mean and unit variance.

Damage Case 3 in the cantilever plate is studied. Firstly, the ULS curvature change was estimated by the central difference method and plotted in Figure 6.9(a). It can

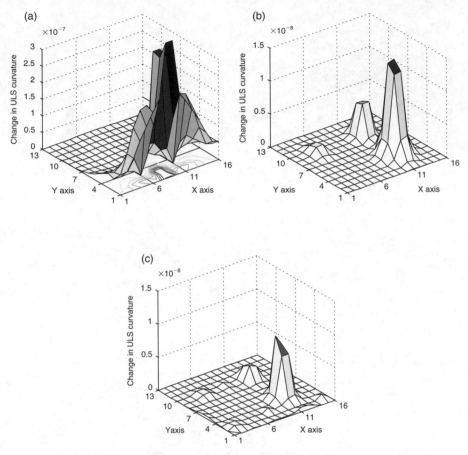

Figure 6.8 Damage index map due to damage Case 1 with less dense sensor mesh (Legends as Figure 6.4)

be seen that the damage index is highly influenced by the measurement noise, and the damaged elements cannot be located from this noisy map of curvature change. To remove the noise effect on the ULS curvature, especially the high peaks near the free edge of the plate, a low order ($M = 6$, $N = 6$) Chebyshev polynomial function of two variables is used to smooth the oscillatory ULS. The coefficients of this approximation are obtained from Equation (6.76). Figures 6.9(b) and 6.9(c) show the estimated ULS curvature changes as the damage index map, with and without the prior knowledge of intact structure, respectively. It is clear that a low order Chebyshev polynomial approximation on the noisy ULS could dramatically suppress the random noise effect in comparison with Figure 6.9(a).

6.7.6.8 When the damage changes the boundary condition of the structure

An alternative approach to model damages similar to Cases 3 and 4 is adopted by free-ing all the boundary connections within the damage region of the plate. The resulting

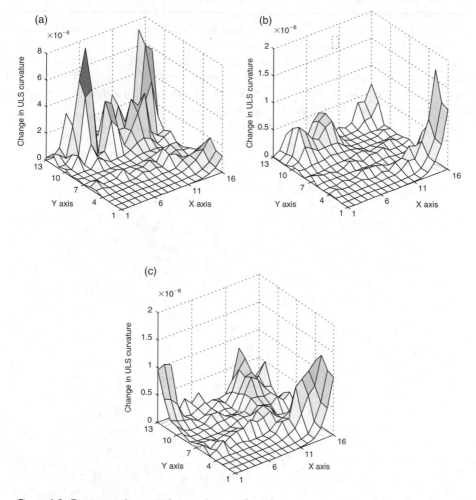

Figure 6.9 Damage index map due to damage Case 3 with random noise (Legends as Figure 6.4)

damage index map looks similar to what the one in the last study. This indicates that the ULS curvature method is applicable even though the damage changes the boundary condition of the structure.

6.8 Conclusions

This chapter presents the recent developments in the modal-based approach for damage detection of structures. The modal information of mode shape and modal frequencies has been integrated into modal flexibilities and uniform load surface for a higher sensitivity to local changes. Their curvatures are subsequently used to model the spatial information of the local damage. Further, the sensitivities of these parameters with respect to local damage have been analytically presented for the quantification of the local changes. All the above have helped to answer the questions: Does any damage occur? Where does it occur? and what is the amplitude? Other questions such as: what type of damage it is? will need the inclusion of different types of damage models described in Chapter 3 to provide the answer.

These modal-based methods have the disadvantage of limited measured information and may not be applicable to large civil structures with thousands of DOFs. Other types of approaches are sought to provide more measured information economically and they are discussed in Chapters 7 and 8.

Chapter 7

System identification based on response sensitivity

7.1 Time-domain methods

System identification can also be conducted with information directly from the time response. The time-domain approach has an advantage over the frequency-domain method in that there is an unlimited supply of measured information. Moreover, the damping properties of structures cannot be readily estimated in the frequency-domain method. Another advantage is that there is no requirement for complete measurements in space and state. However, the detrimental effect from measurement noise (Hjelmstad and Banan, 1995) remains significant with time-domain methods (Ge and Soong, 1998). This chapter addresses the system identification approach based on the sensitivity of the measured response with respect to a system parameter. Features of the approach with different applications are presented.

7.2 The response sensitivity

For a general finite element model of a linear elastic time-invariant structure, the equation of motion is given by

$$[M]\{\ddot{x}\} + [C]\{\dot{x}\} + [K]\{x\} = [B]\{F\} \tag{7.1}$$

where $[M]$, $[C]$ and $[K]$ are the system mass, damping and stiffness matrices, respectively. Rayleigh damping is adopted, which is of the form $[C] = a_1[M] + a_2[K]$, where a_1 and a_2 are constants to be determined from two modal damping ratios; $\{\ddot{x}\}$, $\{\dot{x}\}$ and $\{x\}$ are the acceleration, velocity and displacement response vectors of the structure; $\{F\}$ is a vector of applied forces with matrix $[B]$ mapping these forces to the associated degrees-of-freedom (DOFs) of the structure. The dynamic responses of the structures can be obtained by direct numerical integration using the Newmark method.

7.2.1 The computational approach

For the perturbation of a system parameter, $\Delta\alpha$, the perturbed equation of motion is obtained by differentiating both sides of Equation (7.1) with respect to the system parameter. Assuming the parameter is related only to the stiffness of the dynamic system, the following is obtained

$$[M]\left\{\frac{\partial \ddot{x}}{\partial \alpha_i}\right\}+[C]\left\{\frac{\partial \dot{x}}{\partial \alpha_i}\right\}+[K]\left\{\frac{\partial x}{\partial \alpha_i}\right\} = -\frac{\partial [K]}{\partial \alpha_i}\{x\}-a_2\frac{\partial [K]}{\partial \alpha_i}\{\dot{x}\}(i = 1, 2, \ldots N) \quad (7.2a)$$

where $\{\partial x/\partial \alpha_i\}$, $\{\partial \dot{x}/\partial \alpha_i\}$ and $\{\partial \ddot{x}/\partial \alpha_i\}$ are vectors of the displacement, velocity and acceleration sensitivities with respect to the unknown parameter; and α^i is the parameter in the ith element or other stiffness parameter of the system. A similar equation can be written for the perturbation of parameters which are related to the system mass and damping matrices.

Let $\qquad \ddot{Y} = \dfrac{\partial \ddot{x}}{\partial \alpha_i}, \dot{Y} = \dfrac{\partial \dot{x}}{\partial \alpha_i}\text{and} Y = \dfrac{\partial x}{\partial \alpha_i},$

thus,

$$[M]\{\ddot{Y}\} + [C]\{\dot{Y}\} + [K]\{Y\} = -\frac{\partial [K]}{\partial \alpha_i}\{x\} - a_2\frac{\partial [K]}{\partial \alpha_i}\{\dot{x}\} \quad (7.2b)$$

Since $\{x\}$ and $\{\dot{x}\}$ have been obtained from Equation (7.1), the right-hand-side of Equation (7.2b) can be considered as an equivalent forcing function, and the equation is of the same form as Equation (7.1). Therefore, the sensitivities $\ddot{Y}, \dot{Y}$ and Y can also be obtained using the Newmark method.

7.2.2 The analytical formulation

Equation (7.1) may be rewritten in the state-space formulation as

$$\dot{X} = K^*X + \overline{F} \quad (7.3)$$

where $X = \begin{bmatrix} x \\ \dot{x} \end{bmatrix}_{2n \times 1}$, $K^* = \begin{bmatrix} 0 & I \\ -M^{-1}K & -M^{-1}C \end{bmatrix}_{2n \times 2n}$, $\overline{F} = \begin{bmatrix} 0 \\ M^{-1}[B]F \end{bmatrix}_{2n \times 1}$

and X represents a vector of state variables with a length $2n$ containing the displacements and velocities of the nodes. These differential equations are then converted to discrete equations using exponential matrix representation

$$X_{k+1} = AX_k + \overline{DF}_k \quad (7.4)$$

$$A = e^{K^*h}, \quad \overline{D} = K^{*-1}(A - I)$$

where A is the exponential matrix and $(k + 1)$ denotes the value at the $(k + 1)$th time step of computation. The time step h represents the time increment between the variable states X_k and X_{k+1} in the computation. I is a unit matrix. The dynamic response of the system can be obtained from Equation (7.4) with zero initial conditions. Once the displacement and velocity responses are obtained, the acceleration response can be obtained by directly differentiating the velocity response.

The first differential of the dynamic response with respect to a physical parameter of the system, α_i, can be obtained by differentiating both sides of Equation (7.3) with respect to parameter α_i of an element as

$$\frac{\partial \dot{X}}{\partial \alpha_i} = K^* \frac{\partial X}{\partial \alpha_i} + \frac{\partial K^*}{\partial \alpha_i} X$$

$$= K^* \frac{\partial X}{\partial \alpha_i} + \begin{bmatrix} 0 \\ -M^{-1} \frac{\partial \overline{K}}{\partial \alpha_i} x \end{bmatrix} + \begin{bmatrix} 0 \\ -a_2 M^{-1} \frac{\partial \overline{K}}{\partial \alpha_i} \dot{x} \end{bmatrix} (i = 1, 2, \dots N) \qquad (7.5)$$

It is noted that the system mass matrix is not dependent on the physical parameter, α_i, and thus the partial derivative $\partial M / \partial \alpha_i$ in Equation (7.5) vanishes. Let $Y = \partial X / \partial \alpha_i$, where Y is the vector of displacement and velocity differentials with respect to parameter α_i in the time domain.

$$\text{Put } \overline{P} = \begin{bmatrix} 0 \\ -M^{-1} \frac{\partial \overline{K}}{\partial \alpha_i} x \end{bmatrix} \text{ and } \overline{G} = \begin{bmatrix} 0 \\ -a_2 M^{-1} \frac{\partial \overline{K}}{\partial \alpha_i} \dot{x} \end{bmatrix}$$

Equation (7.5) can then be rewritten as

$$\dot{Y} = K^* Y + \left[\overline{P} + \overline{G} \right] \qquad (7.6)$$

It is noted that Equation (7.6) has the same form as Equation (7.3). The displacement and velocity response sensitivities can then be obtained in a discretized form similar to Equation (7.4) as,

$$Y_{k+1} = A Y_k + \overline{D} [\overline{P}_k + \overline{G}_k] \qquad (7.7)$$

and the acceleration response sensitivity can be obtained by direct differentiating the velocity sensitivity.

7.2.3 Main features of the response sensitivity

The main features of the response sensitivity have been studied (Lu and Law, 2007a) with an example of a plane frame structure to study the sensitivities generated for different types of excitation. The frame structure consists of eleven Euler–Bernoulli beam elements with twelve nodes each with three DOFs as shown in Figure 7.1. The frame is fixed at nodes 1 and 12, modelled with large translational and rotational stiffnesses of 1.5×10^{10} kN/m and 1.5×10^9 kN-m/rad, respectively. The mass density of the material is 2.7×10^3 kg/m^3 and the elastic modulus of the material is 69×10^9 N/m^2. The height and width of the frame are 1.2 m and 0.6 m, respectively, and the cross-sectional dimensions of the member are $b = 0.01$ m and $h_0 = 0.02$ m with the second moment of inertia in the plane of bending equal to 6.67×10^{-9} m^4. The first five undamped natural frequencies of the intact frame are 13.095, 57.308, 76.697, 152.410 and 196.485 Hz. The Rayleigh damping model is adopted with the damping ratios of the first two modes taken equal to 0.01. The equivalent Rayleigh coefficients a_1 and a_2 are 1.3395 and 4.52×10^{-5}, respectively.

Sinusoidal, impulsive and random excitations are applied separately at node 2 in the x-direction to generate the dynamic responses and their sensitivities with respect to a

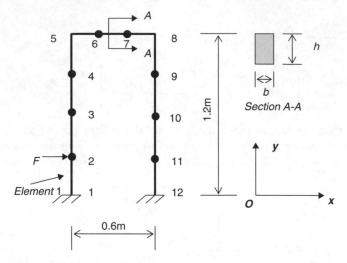

Figure 7.1 The plane frame structure

system parameter of the frame. The elastic modulus of material in element 1 is selected as the perturbed parameter. The response is measured along the x-direction at node 9 with a sampling rate of 500 Hz including the first five modes of the structure. The sinusoidal force is taken as $F(t) = 10 \sin (2\pi ft)$N, where f is the excitation frequency taken equal to the fundamental modal frequency of the frame and at 25 Hz, which is between the first and second modal frequencies. The impulsive force lasting for 0.1 second is expressed in the following form with a magnitude of 10 N.

$$F(t) = \begin{cases} 200(t - 0.05) & 0.05 \leq t \leq 0.1 \\ 200(0.15 - t) & 0.1 \leq t \leq 0.15 \end{cases}$$

The normally distributed random force is between 0 N and 10 N.

The time histories of these excitation forces, the displacement response and its sensitivity, plus the acceleration response sensitivity with respect to the perturbed elastic modulus of material are shown in the following figures. Since the magnitudes of all the excitation forces are equal, a direct comparison of the responses and their sensitivities is possible.

Figure 7.2 gives the sensitivities from sinusoidal excitation at the undamped fundamental frequency of the structure. The amplitude of the displacement response increases gradually until the energy input is balanced by the energy dissipated from damping where the amplitude becomes relatively stable. In the computation of the sensitivities from Equation (7.2), the forcing function consists of both the displacement response and the velocity response. The velocity is approximately one thousand times larger than the displacement. However, the second term on the right-hand-side of the equations is one hundred times smaller than the first term because of the small damping coefficient, a_2. Therefore, the forcing term in Equation (7.2) is dominated by the displacement response, which increases in the first stage and becomes relatively

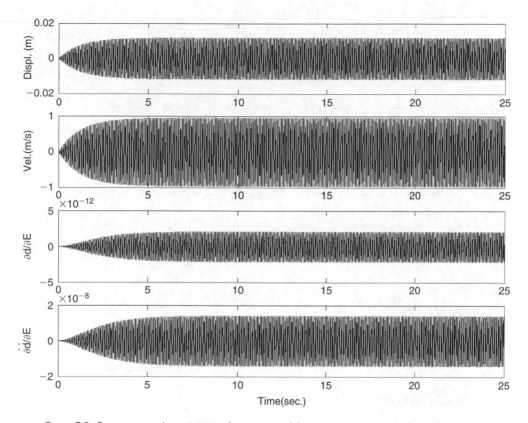

Figure 7.2 Response and sensitivities from sinusoidal excitation at a modal frequency

constant later. Therefore, the amplitude of the sensitivities obtained also increases with time and it becomes stable when the energy input and energy dissipation are balanced.

Figure 7.3 gives the sensitivities from sinusoidal excitation at 25 Hz, which is between the first and the second natural frequencies of the structure. The displacement response consists of a combination of responses mainly at the first and second natural frequencies of the structure. This can be explained by the modal superposition principle. This response becomes relatively stable with time under the damping effect. The sensitivities obtained under the forcing function are dominated by the displacement response and they consist of components at both frequencies, but with the component at the first natural frequency dominating. The sensitivities diminish with time but maintain a small amplitude of vibration because of the relatively stable input contributed by the displacement response.

Figure 7.4 gives the response sensitivities from the impulsive excitation. The displacement response is dominated by the first natural frequency of the structure with the higher frequency components diminish rapidly with damping. The response reduces to zero with time. The sensitivities obtained from Equation (7.2) increase when the

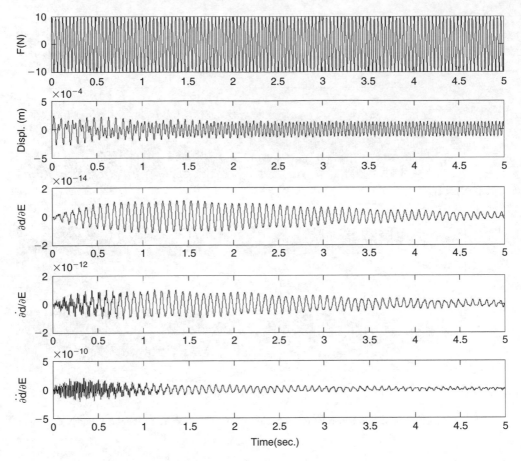

Figure 7.3 Response and sensitivities from sinusoidal excitation at 25 Hz

input energy is larger than the dissipated energy and they reach a maximum at around 1.2 s. All the sensitivities diminish to zero with time under the damping effect.

Figure 7.5 gives the response sensitivities from normally distributed random excitation. The first few modes of the structure are excited with a strong contribution to the displacement response. The sensitivities obtained from Equation (7.2) are under the force excitation dominated by the displacement response with vibration at the first five natural frequencies of the structure. The sensitivities consist of a combination of components at the first few natural frequencies of the structure with both increasing and decreasing amplitude under the damping effect. The sensitivities will not diminish to zero with time as the displacement responses always exist under the random excitation.

The shapes of the three types of response sensitivities in time are similar but they are different for different excitation. A comparison of the sensitivities in Figures 7.2 to 7.5 shows that sinusoidal excitation gives higher sensitivities than random force excitations, while those from impulsive excitation exhibit the smallest sensitivity. This may be because there is only one impact acting on the structural frame in the duration

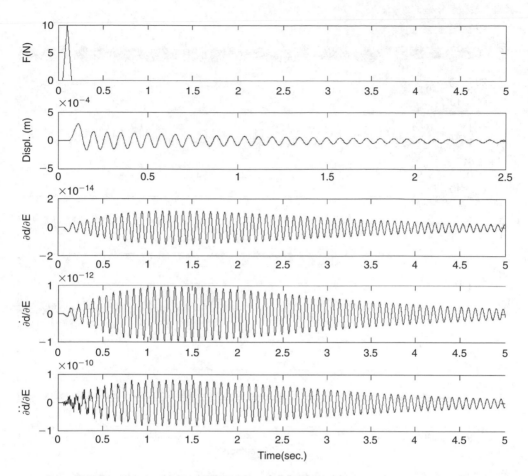

Figure 7.4 Response and sensitivities from an impulsive excitation

studied, while the other excitations act on the structure continuously in the same period giving a higher energy input. The sensitivities from excitation at a modal frequency of the structure is large compared to those from excitation at a frequency which is not a modal frequency. The sensitivities are dependent on the displacement response of the structure as seen in Equation (7.2). These observations can further be explained from the viewpoint of energy input: the largest energy input on the structure is from the sinusoidal excitation at a natural frequency, such that the sensitivities are largest under this excitation. The smallest energy input is from the impulsive force and the sensitivities are the smallest. For further discussion on this, see Lu and Law (2007a).

7.3 Applications in system identification

7.3.1 *Excitation force identification*

An important type of inverse problem in structural dynamics is the force identification or force reconstruction from measured dynamic responses. There are three main classes

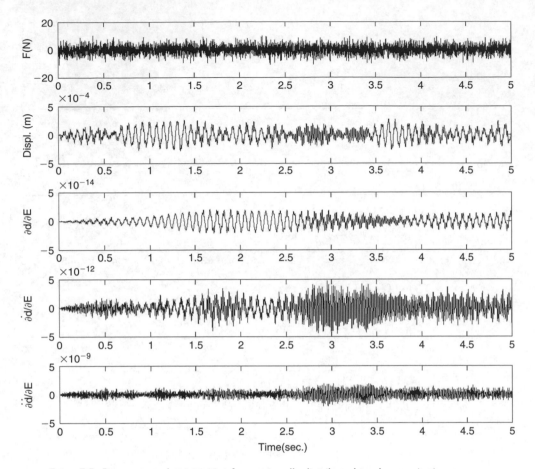

Figure 7.5 Response and sensitivities from normally distributed random excitation

of this problem. One major class is where the initiation site is known (Williams and Jones, 1948). Another major class of the problem is where both the force history and its location are unknown. Examples include using the modal response data to determine the location of impact forces on the read/write head of computer disks (Briggs and Tse, 1992) and using wave propagation responses to determine the location of structural impacts (Doyle, 1994). The third class of the problem is the identification of moving forces on structures. Examples include vehicle–bridge interaction forces, which are very important for bridge engineering. Since the forces are moving, it is difficult to measure them directly, while direct measurement of the forces using instrumented vehicles are expensive and are subjected to bias (Yang and Yau, 1997) and results from computation simulations are subject to modelling errors (Peters, 1986).

In this section, the sensitivities of the dynamic response with respect to the parameters of the force (frequency and amplitude) are calculated in the time domain. The force identification problem thus becomes a parameter identification problem.

7.3.1.1 *The response sensitivity*

For a general finite element model of a linear elastic time-invariant structure, the dynamic governing equation is given by Equation (7.1). The *j*th input force can be modelled in the following generalized form of a sine series

$$F^j(t) = F_0^j + \sum_{i=1}^{n} F_i^j \sin \omega_i^j t \tag{7.8}$$

where F_0^j, F_i^j and ω_i^j are the parameters of the *jth* force. Substituting Equation (7.8) into Equation (7.1), and performing differentiations on both sides of the equation with respect to the parameters of each force, gives

$$[M] \left\{ \frac{\partial \ddot{x}}{\partial F_0^j} \right\} + [C] \left\{ \frac{\partial \dot{x}}{\partial F_0^j} \right\} + [K] \left\{ \frac{\partial x}{\partial F_0^j} \right\} = [B] \qquad (j = 1, 2, \ldots N_f)$$

$$[M] \left\{ \frac{\partial \ddot{x}}{\partial F_i^j} \right\} + [C] \left\{ \frac{\partial \dot{x}}{\partial F_i^j} \right\} + [K] \left\{ \frac{\partial x}{\partial F_i^j} \right\} = [B] \sin \omega_i^j t \qquad (i = 1, 2, \ldots n;$$
$$j = 1, 2, \ldots N_f) \tag{7.9}$$

$$[M] \left\{ \frac{\partial \ddot{x}}{\partial \omega_i^j} \right\} + [C] \left\{ \frac{\partial \dot{x}}{\partial \omega_i^j} \right\} + [K] \left\{ \frac{\partial x}{\partial \omega_i^j} \right\} = [B] F_i^j t \cos \omega_i^j t \quad (i = 1, 2, \ldots n;$$
$$j = 1, 2, \ldots N_f)$$

where $\{\partial x / \partial(\bullet)\}$, $\{\partial \dot{x} / \partial(\bullet)\}$ and $\{\partial \ddot{x} / \partial(\bullet)\}$ are the displacement, velocity and acceleration sensitivities with respect to the unknown parameter, and N_f is the number of external forces. The dynamic responses of the structure and its sensitivities can be obtained from these equations by direct numerical integration using the Newmark method.

The inverse problem is solved using the penalty function method (Friswell and Mottershead, 1995) with

$$\{\delta z\} = [S]\{\delta P\} \tag{7.10}$$

where $\{\delta z\}$ is the error in the measured output; $\{\delta P\}$ is the perturbation in the parameters; and $[S]$ is the two-dimensional sensitivity matrix, which is the change in acceleration response with respect to the force parameters in the time domain. Equation (7.10) is an ill-conditioned problem. In order to provide bounds to the solution, the damped least-squares method (DLS) (Tikhonov, 1963) discussed in Chapter 2 is adopted and singular-value decomposition is used in the pseudo-inverse calculation.

7.3.1.2 *Experimental verification*

The above method is verified experimentally in the laboratory with a steel beam. The length, width and height of the beam are 2.1 m, 0.025 m and 0.019 m, respectively, and the elastic modulus and mass density of material are 2.07×10^{11} N/m^2 and 7.832×10^3 kg/m^3 respectively. The beam is suspended at its two ends as shown in Figure 7.6. It is discretized into twenty Euler–Bernoulli beam elements with two

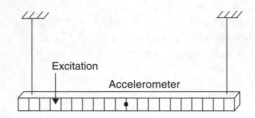

Figure 7.6 The steel beam

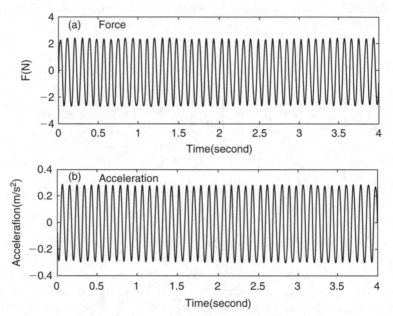

Figure 7.7 Time histories of the true and identified force and acceleration
(___ experiment; ---- identified)

DOFs at each node. The first five measured natural frequencies of the beam are 22.87, 62.76, 123.05, 203.24 and 303.45 Hz, respectively, from hammer test and the first five calculated natural frequencies of the beam are 22.83, 62.74, 123.05, 203.00 and 302.87 Hz, respectively. The sinusoidal force is applied at the nodal point of the first vibration mode of the beam 480 mm from the left free end with an exciter model LDS V450. A 10-second horizontal acceleration obtained from the mid-span of the beam is used to identify the input excitation force. The sampling frequency is 2000 Hz.

A sinusoidal excitation with an amplitude of 2.5 N and a frequency equal to half of the first natural frequency of the beam is applied. The force is modelled with five sinusoidal terms plus a constant term. The identified force converges after 19 iterations with the optimal regularization parameter equals to 0.15. Figure 7.7 shows the first four seconds of the true and identified input forces and the measured and reconstructed accelerations from the experiment, respectively. The identified input force is seen as

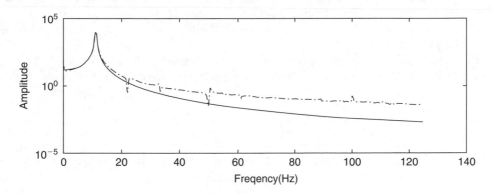

Figure 7.8 Spectrum of true and identified force (___ true; ---- identified)

matching the true input force very well except at the peaks, and the reconstructed acceleration almost overlaps the measured acceleration.

The frequency spectrum of the true and the identified force time histories are shown in Figure 7.8. There is a general reduction in the higher frequency components of the identified force compared with the true force. This is because the initial model of the force does not cover a wide spectrum with the high frequency components.

7.3.2   Condition assessment from output only

Recent development with system identification is more related to the identification of structural parameters from output-only measurements (Chen and Li, 2004; Shi et al., 2000a). The sensitivities of the dynamic response with respect to the structural physical parameters and parameters of the input excitation force are calculated from Equations (7.2) and (7.9). Perturbations in the structural parameters are identified together with the input excitation forces using an iterative algorithm (Lu and Law, 2007b) as shown below. The location of the input force is assumed known in the identification.

7.3.2.1   Algorithm of iteration

Measurements from two states of the structure are required. The first set of measurements from the undamaged structure serves to update the system parameters with a known set of force inputs. While in the measurements on the second stage with damage, both the excitation force and the damaged structure are unknown, and the following iterative algorithm is used in the identification. The updated finite element model of the structure serves as the reference model in the subsequent comparison.

(A) Iteration to update the excitation force parameters

Starting with an initial guess of the unknown force parameter vector $\{(P_F)_0\}$ and the set of physical parameter $\{P_S\}_0$ from the updated finite element model of the structure, the procedure of iteration is given as:

Step 1: With the initial force vector and vector of the undamaged system, Equation (7.1) is solved at the $j = k + 1$ iteration step for the displacement vector $\{x\}$ using the Newmark method and subsequently for the acceleration vector $\{\ddot{x}\}$.

Step 2: The sensitivity matrix $[S_F]$ of the response with respect to the different force parameters is obtained from Equations (7.1) and (7.9) using again the Newmark method at the $j = k + 1$ iteration step with the force vector $\{(P_F)_k\}$ obtained from the previous step.

Step 3: Find $\{(P_F)_{k+1}\}$ from the damped least-squares solution of Equation (7.10).

Step 4: Repeat Steps 1 to 3 until the following convergence criteria is satisfied.

$$\left\| \frac{\{(P_F)_{k+1}\} - \{(P_F)_k\}}{\{(P_F)_{k+1}\}} \right\| \leq Tol_F.$$

Step 5: The final vector $\{(P_F)_{k+1}\}$ obtained is taken as the modified set of force parameters $\{P_F\}$ for the second stage of iteration.

(B) Iteration to update for the physical parameters of the structure

With the modified excitation-force parameter vector $\{P_F\}$ obtained from (A) above, the set of physical parameters is then obtained as below:

Step 6: The vector of the physical parameter $\{P_S\}_0$ from the updated finite element model of the structure is taken as the set of initial values. Equation (7.1) is solved at the $j = k + 1$ iteration step for the displacement vector $\{x\}$ using the Newmark method and subsequently for the acceleration vector, $\{\ddot{x}\}$.

Step 7: The sensitivity matrix $[S_S]$ of the response with respect to the different physical parameters of the structure is obtained from Equations (7.1) and (7.2) again using the Newmark method at the $j = k + 1$ iteration step with the initial physical parameter vector $\{(P_S)_k\}$ obtained from a previous step.

Step 8: Find $\{(P_S)_{k+1}\}$ from the damped least-squares solution of Equation (7.10).

Step 9: Repeat Steps 6 to 8 until the following convergence criteria is reached.

$$\left\| \frac{\{(P_S)_{k+1}\} - \{(P_S)_k\}}{\{(P_S)_{k+1}\}} \right\| \leq Tol_S.$$

Step 10: The final vector $\{(P_S)_{k+1}\}$ obtained is taken as the modified set of physical parameters $\{P_S\}$ for the next cycle of iteration on the force parameters.

The identified excitation force obtained in (A) can be further improved using the updated physical parameters obtained in (B) and repeating Steps 1 to 5. The vector of physical parameters can also be further improved using the modified excitation force and repeating Steps 6 to 10.

The convergence of this computation strategy has been proved by Li and Chen (1999) in the estimation of the wind load and system parameters at the same time. While the uniqueness of the solution is not checked in this work. This and other algorithms for solving both the unknown forces and system parameters, such as Ling and Haldar (2004) and Shi et al. (2000a), do not guarantee a unique solution. They all depend on the effectiveness of the minimization of the objective function not falling into the local minimum. The uniqueness of the algorithms remains an unsolved problem for further research.

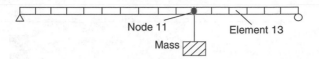

Figure 7.9 Experimental setup

7.3.2.2 Experimental verification

The algorithm is demonstrated with an experiment on a simply supported steel beam as shown in Figure 7.9. The parameters of the beam are: length 2.0 m, width 25 mm and height 19 mm, the elastic modulus and mass density of material are 2.065×10^{11} N/m^2 and 7.832×10^3 kg/m^3, respectively. It is discretized into sixteen Euler beam elements with three DOFs at each node. A mass of 2.61 kg is hanging by a fine nylon rope at node 11 of the beam, and the excitation generated by cutting the rope will serve as the input force. The true value of the force is 25.58 N and is a 'step force' acting at the initial time $t = 0$. Mathematically, it is expressed as

$$f(t) = \begin{cases} Mg & t = 0 \\ 0 & t > 0 \end{cases}$$

The flexural rigidities of all the elements and the assumed impulsive force are taken as the unknowns in the inverse analysis. The initial values of the damage parameters for all the finite elements are all zero. The initial vector of the force parameters is $\{(P_F)_0\} = [0, 0, 0, 0, 0, 0, 2\pi, 4\pi, 6\pi, 8\pi, 10\pi]^T$.

The sampling frequency is 2000 Hz. The acceleration responses collected by B&K 4370 accelerometers at nodes 7 and 9 within the duration 0.5 to 1.5 seconds were used for the identification. The first 0.5-second data is skipped because of the many high frequency components in the response caused by the impulsive force generated by the falling weight. A commercial data logging system, INV303, and the associated signal analysis package, DASP2003, are used in the data acquisition. Damage is introduced by removing an equal thickness of 0.5 mm of material from both sides of the beam over a length of 9 mm in element 13, with one edge of the damage zone starting at node 13. The equivalent reduction in the second moment of inertia of element 13 is found to be 11.3% after condensing the middle DOFs to the two end nodes 13 and 14 by Guyan reduction. The first five natural frequencies of the undamaged and the damaged beam are found to be very close to the measured values, indicating a model which is accurate enough for the subsequent damage identification.

The algorithm is applied for the system identification. Both the convergence limits in the force and parameter identification are 1.0×10^{-6}. The iteration stops after two cycles. All the force parameters are identified simultaneously. The required number of iterations for convergence in the second cycle of iteration is 19 and 154 for the force and the damages, and the corresponding optimal regularization parameters are 6.02 and 13.74, respectively. Figure 7.10 shows the identified damage where element 13 is noted to have 13.5% reduction, which is close to the true value. But there is a large false identification in element 12. This observation can be explained since element 12 is

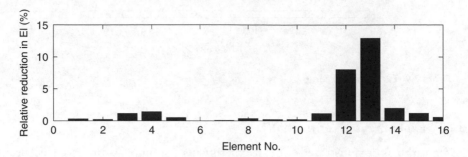

Figure 7.10 Identified results after the second round of iterations

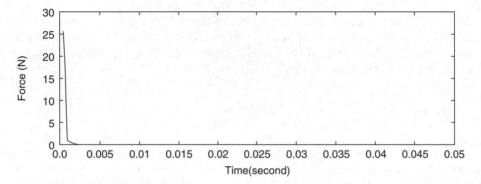

Figure 7.11 Identified force time history

immediately adjacent to the damage and the vibration energy in the element would be much more disturbed than those in other elements, as discussed by Shi et al. (2000b). Figure 7.11 shows the identified time history of the force with a peak of 25.6 N at $t = 0$, which is very close to the true value. Figure 7.12 shows the time histories of the reconstructed acceleration using the identified input force and the corresponding measured acceleration smoothed with a twenty terms orthogonal polynomial function to remove the measurement noise (Zhu and Law, 2001). The time series match each other very well except for the high frequency components at the beginning of the time histories.

7.3.3 Removal of the temperature effect

A plane truss element is taken to illustrate this approach. The elemental stiffness matrix of the element is

$$[k^e]_0 = \frac{EA}{l_0} \begin{bmatrix} 1 & 0 & -1 & 0 \\ 0 & 0 & 0 & 0 \\ -1 & 0 & 1 & 0 \\ 0 & 0 & 0 & 0 \end{bmatrix} \tag{7.11}$$

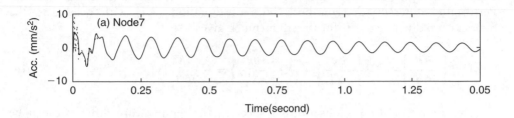

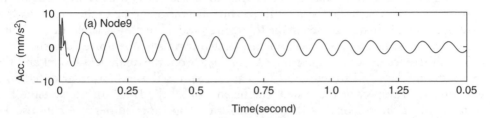

Figure 7.12 Reconstructed and experimental accelerations
(____reconstructed; ─── experiment)

where l_0 is the original element length. When taking into account a temperature difference ΔT, the elemental stiffness matrix is given as

$$[k^e]_{\Delta T} = \frac{EA}{l} \begin{bmatrix} 1 & 0 & -1 & 0 \\ 0 & 0 & 0 & 0 \\ -1 & 0 & 1 & 0 \\ 0 & 0 & 0 & 0 \end{bmatrix} \tag{7.12}$$

$$l = l_0 + \alpha \Delta T l_0 = l_0(1 + \alpha \Delta T) \tag{7.13}$$

where α is the thermal expansion coefficient which equals $12.5 \times 10^{-6}/°C$ for steel. Since $\alpha \Delta T \ll 1$,

$$\frac{1}{l} = \frac{1}{l_0(1 + \alpha \Delta T)} \approx \frac{1}{l_0}(1 - \alpha \Delta T) \tag{7.14}$$

Equation (7.12) can be rewritten as

$$[k^e]_{\Delta T} = \frac{EA(1 - \alpha \Delta T)}{l_0} \begin{bmatrix} 1 & 0 & -1 & 0 \\ 0 & 0 & 0 & 0 \\ -1 & 0 & 1 & 0 \\ 0 & 0 & 0 & 0 \end{bmatrix} \tag{7.15}$$

with the temperature effect on the member cross-sectional dimensions ignored. Performing differentiation on both sides of Equation (7.1) with respect to the temperature difference ΔT_j for the jth member, gives

$$[M]\left\{\frac{\partial \ddot{x}}{\partial \Delta T_j}\right\} + [C]\left\{\frac{\partial \dot{x}}{\partial \Delta T_j}\right\} + [K]\left\{\frac{\partial x}{\partial \Delta T_j}\right\} = -\frac{\partial [K]}{\partial \Delta T_j}\{x\} \tag{7.16}$$

The sensitivity of the response with respect to the temperature difference can be obtained from Equation (7.16) by direct integration. It is common in large-scale structure to find viscoelastic materials, the properties of which can drastically change with temperature affecting not only the stiffness, but also the damping of the structure. It is assumed that the materials used in the following studies are non-viscoelastic, and thus the system mass and damping will not be affected by this temperature difference.

The difference of responses at time t_i obtained from the analytical model and the experimental damaged structures, ΔR_{t_i}, can be expressed as a first-order differential equation with respect to the system coefficients of all the DOFs of the system. The differential of the response with respect to the temperature difference can also be calculated for each finite element. When writing in the form of Taylor's first-order approximation,

$$\Delta R_{t_i} = \sum_{k=1}^{ne}\sum_{i=1}^{6}\sum_{j=1}^{6}\frac{\partial R_{t_i}}{\partial m_{ijk}}\Delta m_{ijk} + \sum_{k=1}^{ne}\sum_{i=1}^{6}\sum_{j=1}^{6}\frac{\partial R_{t_i}}{\partial c_{ijk}}\Delta c_{ijk} + \sum_{k=1}^{ne}\sum_{i=1}^{6}\sum_{j=1}^{6}\frac{\partial R_{t_i}}{\partial k_{ijk}}$$

$$\times \Delta k_{ijk} + \sum_{j=1}^{ne}\frac{\partial R_{t_i}}{\partial \Delta T_j}\Delta T_j \tag{7.17}$$

The pattern of temperature distribution in a structure can be obtained from temperature sensors or from a theoretical model on the temperature distribution. However, the temperature differences in all members are assumed equal for simplicity in this study, i.e. $\Delta T_1 = \Delta T_2 = \ldots = \Delta T_N$. The terms of the temperature difference can then be moved to the left-hand-side of Equation (7.29) as

$$\Delta R_{t_i} - \sum_{j=1}^{ne}\frac{\partial R_{t_i}}{\partial \Delta T_j}\Delta T_j = \sum_{k=1}^{ne}\sum_{i=1}^{6}\sum_{j=1}^{6}\frac{\partial R_{t_i}}{\partial m_{ijk}}\Delta m_{ijk} + \sum_{k=1}^{ne}\sum_{i=1}^{6}\sum_{j=1}^{6}\frac{\partial R_{t_i}}{\partial c_{ijk}}$$

$$\times \Delta c_{ijk} + \sum_{k=1}^{ne}\sum_{i=1}^{6}\sum_{j=1}^{6}\frac{\partial R_{t_i}}{\partial k_{ijk}}\Delta k_{ijk}$$

$$\text{or} \quad \Delta R'_{t_i} = \sum_{k=1}^{ne}\sum_{i=1}^{6}\sum_{j=1}^{6}\frac{\partial R_{t_i}}{\partial m_{ijk}}\Delta m_{ijk} + \sum_{k=1}^{ne}\sum_{i=1}^{6}\sum_{j=1}^{6}\frac{\partial R_{t_i}}{\partial c_{ijk}}\Delta c_{ijk} + \sum_{k=1}^{ne}\sum_{i=1}^{6}\sum_{j=1}^{6}\frac{\partial R_{t_i}}{\partial k_{ijk}}\Delta k_{ijk}$$

$$\tag{7.18}$$

When a lump mass matrix is adopted, Equation (7.18) becomes

$$\Delta R'_{t_i} = \sum_{k=1}^{n} \frac{\partial R_{t_i}}{\partial m_k} \Delta m_k + \sum_{k=1}^{ne}\sum_{i=1}^{6}\sum_{j=1}^{6} \frac{\partial R_{t_i}}{\partial c_{ijk}} \Delta c_{ijk} + \sum_{k=1}^{ne}\sum_{i=1}^{6}\sum_{j=1}^{6} \frac{\partial R_{t_i}}{\partial k_{ijk}} \Delta k_{ijk}$$

where Δm_k is the kth lump mass and $\Delta R'_{t_i}$ is the difference of dynamic response from the two states of the structure with the temperature effect removed. This response with the temperature effect removed can then be used in Equation (7.18) for the system identification of the structure.

7.3.4 Identification with coupled system parameters

For a structural component made of isotropic elastic material, the elemental stiffness matrix is proportional to the elastic modulus of material and the geometric parameters of the element. The stiffness matrix of the structure is expressed as the summation of the elemental stiffness matrices as,

$$[K] = \sum_{i=1}^{N} [k]_e^i \tag{7.19}$$

where N is the number of elements in the finite element model and $[k]_e^i$ is the stiffness matrix of the ith element.

Take a planar beam-column element as an example. Let E^i, I^i and A^i be the elastic modulus of material, the second moment of inertia of cross-section and the sectional area of the ith element, respectively. A physical damage affects all these parameters differently, and they become

$$
\begin{aligned}
E_d^i &= E_0^i(1 + \alpha_E^i) & (-1 \le \alpha_E^i \le 0)\\
I_d^i &= I_0^i(1 + \alpha_I^i) & (-1 \le \alpha_I^i \le 0)\\
A_d^i &= A_0^i(1 + \alpha_A^i) & (-1 \le \alpha_A^i \le 0)
\end{aligned}
\tag{7.20}
$$

where E_0^i, I_0^i and A_0^i are the parameters of the intact structure; E_d^i, I_d^i and A_d^i are the parameters of the damaged structure; and α_E^i, α_I^i and α_A^i represent the damage indices for the above parameters denoted by their subscripts. Equation (7.20) gives

$$E_d^i I_d^i = E_0^i I_0^i(1 + \alpha_E^i + \alpha_I^i + \alpha_E^i \alpha_I^i); \quad E_d^i A_d^i = E_0^i A_0^i(1 + \alpha_E^i + \alpha_A^i + \alpha_E^i \alpha_A^i) \tag{7.21}$$

Performing differentiations on both sides of Equation (7.1) with respect to the different damage indices, e.g. α_E^i of the ith element, gives,

$$[M]\left\{\frac{\partial \ddot{x}}{\partial \alpha_E^i}\right\} + [C]\left\{\frac{\partial \dot{x}}{\partial \alpha_E^i}\right\} + [K]\left\{\frac{\partial x}{\partial \alpha_E^i}\right\} = -\frac{\partial [k]_i^e}{\partial \alpha_E^i}\{x\} - a_2 \frac{\partial [k]_i^e}{\partial \alpha_E^i}\{\dot{x}\}(i = 1, 2, \ldots N)$$

$$\tag{7.22}$$

where $\{\partial x/\partial \alpha_E^i\}$, $\{\partial \dot{x}/\partial \alpha_E^i\}$ and $\{\partial \ddot{x}/\partial \alpha_E^i\}$ are the displacement, velocity and acceleration sensitivities with respect to the unknown index α_E^i, and they can then be obtained by direct integration of Equation (7.22) using the Newmark method. The sensitivities with respect to other parameters can similarly be obtained.

It should be noted that the indices α_E^i, α_I^i and α_A^i are used for the system identification instead of the original physical parameters E^i, I^i and A^i.

7.3.5 Condition assessment of structural parameters having a wide range of sensitivities

When two sets of parameters with distinctly different sensitivities with respect to local damage exist, the model updating approach updates those parameters with the larger sensitivity while those which are less sensitive remain relatively constant in the updating process. This scenario often exists with system identification, and the following strategy should be applied for a good updated result.

The set of parameters discussed in the last section is adopted in an example with a frame structure. The initial set of damage indices $\{\alpha_E\}$, $\{\alpha_I\}$ and $\{\alpha_A\}$ is set equal to zero. The sensitivity with respect to these three indices have been checked and it is found that the one for $\{\alpha_A\}$ is several orders smaller than those for the other two indices (Lu, 2005). Therefore, the following two-stage iterative algorithm is adopted here for the solution of Equation (7.10):

Stage 1: Fix α_A and update the indices α_E and α_I.

Step 1: Solve the dynamic response vector $\{R\}$ from Equation (7.1) at time step $j = k$ with known $(\alpha_E, \alpha_I)_k^T$ and compute the error vector $\{\delta R_k\}$.

Step 2: Solve Equation (7.22) at time step $j = k$ with known $(\alpha_E, \alpha_I)_k^T$ for the sensitivities $\{\partial R / \partial \alpha_E^i\}$ and $\{\partial R / \partial \alpha_I^i\}$ to form the sensitivity matrix.

Step 3: Find $(\alpha_E, \alpha_I)_{k+1}^T$ from Equation (7.10).

Step 4: Repeat Steps 1 to 3 until $\left\| (\alpha_E, \alpha_I)_{k+1}^T - (\alpha_E, \alpha_I)_k^T \right\| \leq toler1$, where $toler1$ is a small prescribed value.

Stage 2: Fix the updated values of α_E and α_I obtained from Stage 1 and update α_A.

Step 1: Solve the response vector $\{R\}$ from Equation (7.1) at time step $j = k$ with known $(\alpha_A)_k^T$ and compute the error vector $\{\delta R_k\}$.

Step 2: Solve Equation (7.22) at time step $j = k$ with known $(\alpha_A)_k^T$ for the sensitivity $\{\partial R / \partial \alpha_A^i\}$ to form the sensitivity matrix.

Step 3: Find $(\alpha_A)_{k+1}^T$ from Equation (7.1).

Step 4: Repeat Steps 1 to 3 until $\left\| (\alpha_A)_{k+1}^T - (\alpha_A)_k^T \right\| \leq toler2$, where $toler2$ is a small prescribed value.

Repeat Stages 1 and 2 in the next iteration until $\left\| R_{k+1} - R_k \right\| \leq$ a small prescribed value.

7.4 Condition assessment of load resistance of isotropic structural components

The last few sections have presented algorithms and considerations in applying the response-sensitivity based approach for system identification. The following gives an experimental verification of this approach with a three-dimensional frame structure in

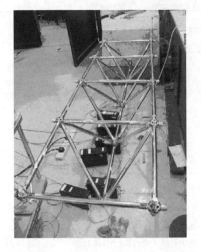

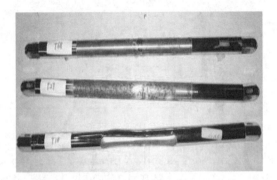

(a) The five-bay cantilever truss structure   (c) Damage members

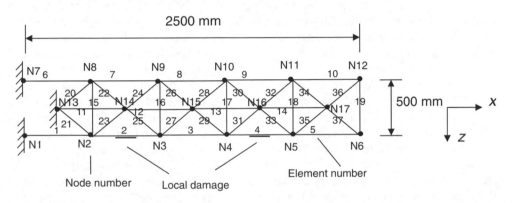

(b) Finite element model

Figure 7.13 The truss structure and its finite element model

the laboratory. It is similar to the frame structure discussed in Chapter 3 but with a triangular-shaped cross-section.

The frame was assembled using the Meroform M12 construction system as shown in Figure 7.13(a). The structure consists of thirty-seven 22 mm diameter alloy steel tubes jointed together with seventeen standard Meroform ball nodes. The main material and geometric properties of a frame member are shown in Table 7.1. Each tube is fitted with a screwed end connector which, when tightened into the node, also clamps the tube by means of an internal compression fitting. All the connection bolts are tightened with the same torsional moment to avoid asymmetry or nonlinear effects caused by man-made assembly errors. The length of all the horizontal, vertical and diagonal tube members between the centres of two adjacent balls is exactly 0.5m after assembly. The structure orients horizontally and is fixed into a rigid concrete support at three nodes at one end.

Table 7.1 Material and geometrical properties of frame member

Properties	Member
Young modulus [N/m^2]	2.10E−11
Area [m^2]	6.597E−5
Density [kg/m^3]	1.2126E−4
Mass [kg]	0.32
Poisson ratio	0.3
Moment of area I_y[m^4]	3.645E−9
Moment of area I_z[m^4]	3.645E−9
Torsional rigidity J[m^4]	7.290E−9

The finite element model consists of thirty-seven Euler–Bernoulli beam elements and seventeen nodes as shown in Figure 7.13(b) and the dimensions of the structure are also shown in this figure. Each node has six DOFs, and altogether there are 102 DOFs for the whole structure.

The total weight of the ball and half weight of the bolt, which connects the ball to the beam elements, are placed on each node as the lumped mass. Another half of the weight of the bolt is included as part of the beam element. Each ball node and bolt weighs 230 grams and 90 grams respectively. An additional mass of 72 grams is added to each joint to balance the mass of the moving accelerometers.

Model errors may come from two sources. The first one is the lumped mass at each node of the structure. Half the weight of each bolt is assigned to the related node as a lumped mass, and this may be incorrect. Another possible error is the elastic modulus of the material. Therefore, the proportion of the weight of each bolt at the related node and the elastic modulus of the material are taken as the design parameters for model updating. It is noted that there may be other sources of modelling errors, for example, the uncertainty of the boundary conditions and the flexibility at the ball joints. Since the structure was fixed into a rigid concrete support at three ball joints at one end and it was assembled using the Meroform M12 construction system with all the connection bolts tightened using the same torsional moment, these possible errors are considered very small and are not updated in the calculation.

7.4.1 Dynamic test for model updating

Excitation by impact hammer was applied at node N6 in the y and z directions using a B&K model 8202 force hammer. There were only 14 free nodes with 28 measurable DOFs in the vertical and horizontal directions, while those in the axial direction of the truss were not measured. All the sensors were aligned in the same direction as the impact excitation. A commercial data logging system, INV303e, and the associated signal analysis package, DASP2003, were used in the data acquisition and modal analysis. Four seconds of responses were recorded for each group of sensors with five hammer impacts during this period. The sampling rate was 2020 Hz. A convergence criteria of 1×10^{-6} was adopted.

The first two seconds of the vertical dynamic responses obtained at nodes 3 and 4 were used. Due to the limitation of the computer, these measured responses were re-sampled to have a sampling rate of 505 Hz. There are two unknown parameters in the first stage of updating, which is much smaller than the $505 \times 2 \times 2$ equations formed from the measured responses. Results show that 62.5% of the weight of each bolt should be placed at the related node as a lumped mass and the elastic modulus of material is 99.3% of the original value. The first eleven updated natural frequencies agree well with the measured natural frequencies with a maximum relative error of 1.86%. The calculated and the experimental mode shapes were also checked with the Modal Assurance Criteria (Moller and Friberg, 1998) and the two sets of mode shapes match each other very well. These show that the updated finite element model can accurately represent the intact structure for the next stage of damage detection.

7.4.2 Damage scenarios

Local faults are introduced by replacing intact members with damaged ones. The artificial damage is of two types. Type I is a perforated slot cut in the central length of the member. The slot is 134.77 mm long, and the remaining depth of the tube in the cut-cross-section is 14.75 mm. Type II is the removal of a layer of material from the surface of the member. The external diameter of the tube is reduced from 22.02 mm to 21.47 mm, and the weakened section is 202 mm long, located in the middle of the beam leaving 99 mm and 75 mm length of original tube cross-section on both sides. Figure 7.13(c) gives a close up view of the damaged frame members.

Damage scenario E1 has Type II damage in the fourth member. Damage scenario E2 includes an additional Type I damage in the second element. The slot opens vertically (global z-direction). Damage scenario E3 is similar to the last scenario but with the slot in element two opened horizontally (global y-direction). The equivalent damages computed by the Guyon method are listed in Table 7.2. Each of the above scenarios involves four types of damage in the beam element, i.e. a reduction in cross-sectional area A, the polar moment of inertia J and the moment of inertia I_z and I_y.

7.4.3 Dynamic test for damage detection

A falling weight test was then conducted on the structure for each damage scenario. A 5.15 kg mass was hung at node N17. Free vibration was introduced by a sudden release of the mass. The time histories from selected accelerometers were recorded. These responses were intentionally selected to be not in close proximity to the damaged elements. The sampling rate was 2020 Hz and the time duration was 8 seconds for each test covering the whole duration of vibration caused by the falling weight excitation. The orthogonal polynomial function (Zhu and Law, 2001) was used to remove the measurement noise in the acceleration data. The required iteration number for convergence and the corresponding optimal regularization parameter are shown in Table 7.3.

Damage scenario E1

Only data in the first two seconds of the response from two accelerometers (node 3, z-direction and node 11, y-direction) were used for the damage identification. The measured acceleration was re-sampled at 505 Hz.

Four sets of damage indices, namely, α_J, α_{I_y}, α_{I_z} and α_A were updated with a total of $37 \times 4 = 148$ unknowns, which is smaller than $505 \times 2 \times 2$ equations. Figure 7.14

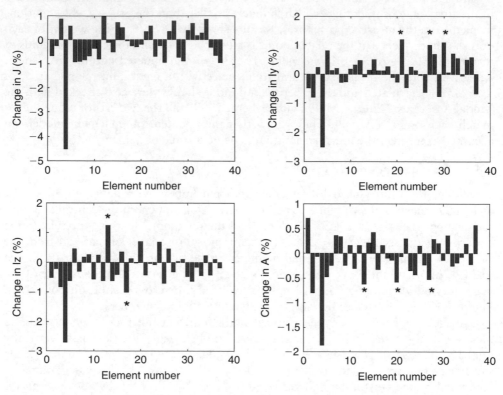

Figure 7.14 Identified results for scenario E1 (* denotes false positive)

Table 7.2 Equivalent and identified damage extent in the damaged elements.

Damage parameter	Scenario E1		Scenario E2		Scenario E3	
	Element 2	Element 4	Element 2	Element 4	Element 2	Element 4
α_J	—	6.8%/4.5%	26.7%/24.5%	6.8%/4.4%	26.7%/24.6%	6.8%/4.2%
α_{Iy}	—	3.4%/2.5%	6.7%/1.7%	3.4%/2.3%	23.3%/21.5%	3.4%/2.6%
α_{Iz}	—	3.4%/2.7%	23.3%/22.0%	3.4%/2.8%	6.7%/1.4%	3.4%/2.9%
α_A	—	2.3%/1.8%	8.8%/8.0%	2.3%/1.6%	8.8%/7.5%	2.3%/1.5%

Note: •/• denotes the equivalent and identified damage extent respectively

shows the identified changes in the four sets of physical parameters. The damaged elements are correctly identified, while there are alarms in elements 2, 13, 17, 20 and 27 with prominent identified values in the group of parameters. It is noted that both the location and the pattern of damage are identified successfully. The identified values at the damaged element in Table 7.2 are very close to the equivalent damage values (Table 7.2).

Table 7.3 Iteration number and regularization parameter required for convergence

	Damage indices	Scenario E1	Scenario E2	Scenario E3
Iteration number	α_E and α_I	19	21	22
	α_A	15	18	18
λ_{opt}	α_E and α_I	0.0031	0.0047	0.0053
	α_A	0.0024	0.0026	0.0026

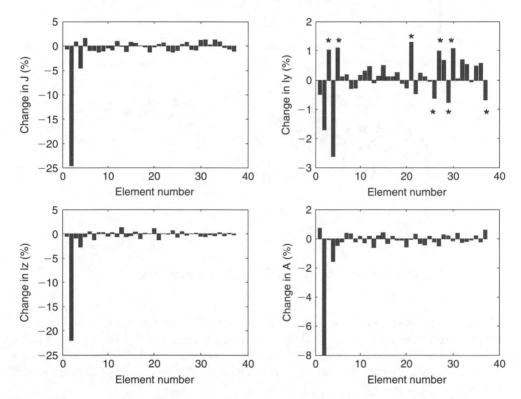

Figure 7.15 Identified results for scenario E2 (* denotes false positive)

Damage scenario E2

Data in the first two seconds of the response from accelerometers at node 9 (z-direction) and node 11 (y-direction) were used for the damage identification. Results in Figure 7.15 show that the location of damage and the damage patterns can be identified correctly, apart from some alarms in I_y, where elements 26, 29 and 37 are incorrectly identified as having a change larger than 1%. The identified values at the damaged element in Table 7.2 are also very close to the equivalent damage values (Table 7.2).

Damage scenario E3

Data in the first two seconds of the response from accelerometers at node 9 (z-direction) and node 11 (y-direction) were used for the damage identification. Results

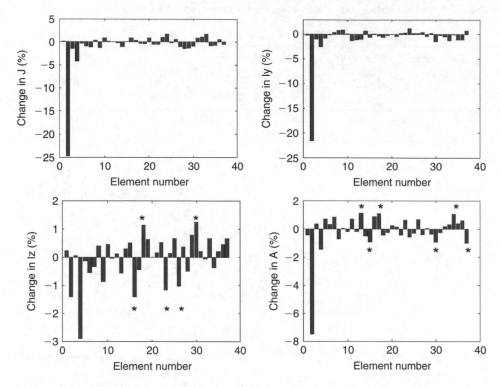

Figure 7.16 Identified results for scenario E3 (* denotes false positive)

in Figure 7.16 show that the location of the damage and the damage patterns are identified correctly, apart from some alarms in I_z, where elements 15, 16, 23, 30, 26 and 37 are incorrectly identified as having a change larger than 1.0%. The identified values at the damaged element in Table 7.2 are very close to the equivalent damage values (Table 7.2).

The false positives in the identified results

There are cases where an element is identified with a relatively large change in the parameter as shown in the results for the damage scenarios studied. They are classified as false positives at the first glance. But, after a more detailed inspection of the four sets of parameter changes, all these alarms correspond to only a change in one of the parameters, not all of them. It is known that local damage in a member will have changes in all the physical properties of the member and the presence of all these changes are found only in the damaged elements. But the elements with the alarm correspond to only one occurrence among the four parameters. This does not meet the required conditions for a true damage. Therefore, these occurrences with a large change in one single parameter are indeed false positives and are errors arising from the inverse computation.

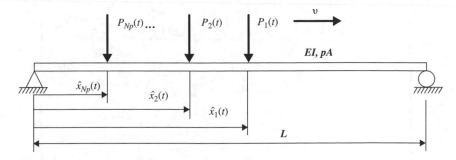

Figure 7.17 A vehicle-bridge system

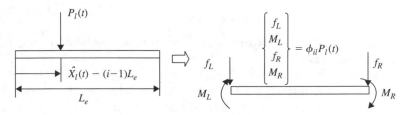

Figure 7.18 Equivalent nodal loads for a beam element

7.5 System identification under operational loads

7.5.1 Existing approaches

System identification from output only requires separate identification of the excitation forces and the system parameters iteratively as shown in Section 7.3.2. The following example of identification of the structural parameters under a group of moving loads on top of a continuous beam is presented for illustration.

7.5.1.1 The equation of motion

A continuous uniform Euler–Bernoulli beam subject to a set of moving loads $P_l(t), (l = 1, 2, \ldots, Np)$ is shown in Figure 7.17. The loads are assumed to be moving as a group at a prescribed velocity v along the axial direction of the beam from left to right. The governing equation can be written as (Lin and Trethewey, 1990)

$$\mathbf{M}_b \ddot{\mathbf{u}} + \mathbf{C}_b \dot{\mathbf{u}} + \mathbf{K}_b \mathbf{u} = \mathbf{\Phi P} \tag{7.23}$$

where $\mathbf{M}_b$, $\mathbf{C}_b$ and $\mathbf{K}_b$ are the structural mass, damping and stiffness matrices of the beam; $u, \dot{u}$ and $\ddot{u}$ denote the nodal displacement, velocity and acceleration vectors, respectively; $\mathbf{P} = \{P_1(t), P_2(t), \ldots, P_{Np}(t)\}^T$ are the moving loads; $\Phi = \{\Phi_1 \Phi_2 \ldots \Phi_l \ldots \Phi_{Np}\}$ is a $2(N+1) \times N_p$ matrix; and N is the number of finite element in the beam. Figure 7.18 shows the equivalent nodal loads for the ith beam element loaded by a moving force $P_l(t)$. $\Phi_l = \{00 \ldots \psi_i \ldots 0\}^T$ and ψ_i is the vector of the shape functions evaluated for the lth force written as:

$$
\psi_i = \left\{ \begin{array}{c} 1 - 3\left(\dfrac{\hat{x}_l(t) - (i-1)l_e}{l_e}\right)^2 + 2\left(\dfrac{\hat{x}_l(t) - (i-1)l_e}{l_e}\right)^3 \\[2ex] (\hat{x}_l(t) - (i-1)l_e)\left(\dfrac{\hat{x}_l(t) - (i-1)l_e}{l_e} - 1\right)^2 \\[2ex] 3\left(\dfrac{\hat{x}_l(t) - (i-1)l_e}{l_e}\right)^2 - 2\left(\dfrac{\hat{x}_l(t) - (i-1)l_e}{l_e}\right)^3 \\[2ex] (\hat{x}_l(t) - (i-1)l_e)\left(\left(\dfrac{\hat{x}_l(t) - (i-1)l_e}{l_e}\right)^2 - \left(\dfrac{\hat{x}_l(t) - (i-1)l_e}{l_e}\right)\right) \end{array} \right\}^T ,
$$

$$((i-1)^*l_e \le \hat{x}_l(t) < (i)^*l_e) \quad (7.24)$$

where $\{\hat{x}_l(t), l = 1, 2, \ldots, Np\}$ are the locations of the moving loads at time t and l_e is the length of the element.

The stiffness matrix of a finite element is assumed to decrease uniformly with damage, and the flexural rigidity, EI_i, of the ith finite element of the beam becomes $\alpha_i EI_i$ when there is damage. The fractional change in stiffness of an element can be expressed as

$$\Delta K_{bi} = (K_{bi} - \tilde{K}_{bi}) = (1 - \alpha_i)K_{bi} \qquad (7.25)$$

where $\mathbf{K}_{bi}$ and $\tilde{\mathbf{K}}_{bi}$ are the ith element stiffness matrices of the undamaged and damaged beam, respectively and $\Delta \mathbf{K}_{bi}$ is the stiffness reduction of the element.

7.5.1.2 Damage detection from displacement measurement

The displacement $w(x, t)$ at measurement location x_s can be obtained from the shape functions and nodal displacements of the beam as

$$w(x_s, t) = \boldsymbol{\varphi}_s \mathbf{u} \qquad (j-1)l_e \le x_s < jl_e \qquad (7.26)$$

where

$$\boldsymbol{\varphi}_s = \{0, 0, \ldots, 0, \varphi_j, 0, \ldots, 0\},$$

$$\varphi_j = \left\{ 1 - 3\left(\frac{x}{l_e}\right)^2 + 2\left(\frac{x}{l_e}\right)^3, x\left(\frac{x}{l_e} - 1\right)^2, 3\left(\frac{x}{l_e}\right)^2 - 2\left(\frac{x}{l_e}\right)^3, x\left(\frac{x}{l_e}\right)^2 - \frac{x^2}{l_e} \right\},$$

$$x = x_s - (j-1)l_e.$$

and for N_s measuring points at $\{x_s, s = 1, 2, \ldots, N_s\}$, Equation (7.26) can be written as

$$\mathbf{w} = \boldsymbol{\varphi}\mathbf{u} \qquad (7.27)$$

where $\mathbf{w} = \{w(x_1, t), w(x_2, t), \ldots, w(x_{N_s}, t)\}^T$ and $\boldsymbol{\varphi} = \{\boldsymbol{\varphi}_1, \boldsymbol{\varphi}_2, \ldots, \boldsymbol{\varphi}_{N_s}\}^T$.

The nodal responses in Equation (7.27) can be written as

$$u = (\boldsymbol{\varphi}^T \boldsymbol{\varphi})^{-1} \boldsymbol{\varphi}^T \mathbf{w} \tag{7.28}$$

and the required $\ddot{u}$ and $\dot{u}$ can be obtained using the orthogonal function expansion (Zhu and Law, 2001). The strain at point x_s and time t can be written as

$$\varepsilon(x_s, t) = -z \frac{\partial^2 w(x_s, t)}{\partial x^2} \tag{7.29}$$

where z represents the distance from the neutral axis of the beam to the bottom surface. Similar to Equation (7.27), Equation (7.29) can be written in matrix form as

$$\boldsymbol{\varepsilon} = \boldsymbol{\varphi}'' u \tag{7.30}$$

where $\boldsymbol{\varepsilon} = \{\varepsilon(x_1, t), \varepsilon(x_2, t), \ldots, \varepsilon(x_{N_s}, t)\}^T$.

$$\boldsymbol{\varphi}_s'' = \{0, 0, \ldots, \varphi_j'', 0, \ldots, 0\};$$

$$\varphi_j'' = \left\{ -\frac{6}{l_e^2} + \frac{12x}{l_e^3}, \frac{6x}{l_e^2} - \frac{4}{l_e}, \frac{6}{l_e^2} - \frac{12x}{l_e^3}, \frac{6x}{l_e^2} - \frac{2}{l_e} \right\};$$

$$x = x_s - (j - 1)l_e$$

Similar to Equation (7.25), the equation of motion of the damaged beam under moving loads can be written as

$$\mathbf{M}_b \ddot{\mathbf{u}} + \mathbf{C}_b \dot{\mathbf{u}} + \mathbf{f}(\mathbf{u}) = \boldsymbol{\Phi} \mathbf{P}(t) \tag{7.31}$$

where $\mathbf{f}(\mathbf{u})$ is the elastic restoring force of the system, and

$$\mathbf{f}(\mathbf{u}) = \tilde{K}_b \mathbf{u} \tag{7.32}$$

From Equation (7.31), the moving loads can be obtained as follow if the restoring forces are known

$$\mathbf{P}(t) = (\boldsymbol{\Phi}^T \boldsymbol{\Phi})^{-1} \boldsymbol{\Phi}^T [\mathbf{M}\ddot{\mathbf{u}} + \mathbf{C}\dot{\mathbf{u}} + \mathbf{f}(\mathbf{u})] \tag{7.33}$$

where I is the unity matrix. The moving loads in Equation (7.31) can be obtained from Equation (7.10) using the damped least-squares method. And the element damage index matrix is obtained from Equation (7.25) after the moving loads are identified from minimizing the following function

$$J(\boldsymbol{\alpha}) = \|F(\mathbf{u}) - f(\mathbf{u})\|^2 \tag{7.34}$$

where $F(\mathbf{u}) = \boldsymbol{\Phi} \mathbf{P}_{identify}(t) - \mathbf{M}\ddot{\mathbf{u}} - \mathbf{C}\dot{\mathbf{u}}$; and $\mathbf{P}_{identify}(t)$ is the set of identified moving loads from the last iteration of the moving load identification.

Since the restoring forces and moving loads are all unknown, the iterative algorithm shown below is adopted to solve the problem (Zhu and Law, 2007a).

1. Calculate the nodal displacements or strains from measurements by Equation (7.27) or (7.30).
2. Use the orthogonal function expansion to calculate the nodal velocity and accelerations. Twenty terms in the expansion is recommended for modelling the moving loads.
3. Initially assume there is no damage in the beam: $\alpha_0 = \{1, 1, \ldots, 1\}^T$.
4. Identify the moving loads $P_{identify}(t)$ from the measured responses using Equation (7.10) with regularization (Zhu and Law, 2002).
5. Calculate the elastic restoring forces from $F(\mathbf{u}) = \mathbf{\Phi}P_{identify}(t) - \mathbf{M}\ddot{\mathbf{u}} - \mathbf{C}\dot{\mathbf{u}}$.
6. Identify the damage index using Equations (7.25) and (7.34).
7. Calculate the error for convergence:

$$Error1 = \frac{\|\mathbf{P}_{i+1} - \mathbf{P}_i\|}{\|\mathbf{P}_i\|} \times 100\%, \quad Error2 = \frac{\|\alpha_{i+1} - \alpha_i\|}{\|\alpha_i\|} \times 100\%$$

Convergence is achieved when the sum of these two errors is a minimum.
8. When the computed errors do not converge, repeat Steps 4 to 7.

7.5.2 The generalized orthogonal function expansion

The excitation was modelled in the generalized form of a sine series in Section 7.3.1, and the identification was successful using five terms in the series. But in the case of a moving force, which is a function of both space and time, the modelling with a sine series would be cumbersome. A more general form of representation is presented in this section, which is in the form of a generalized orthogonal function expansion.

The excitation force $f(t)$ in the time period $[0, T]$ can be expressed in terms of a Chebyshev series, $f(t) = \sum_{i=1}^{N_f} C_i T_i(t)\, t \in [0, T]$, where $T_i(t)$ is the first kind of orthogonal Chebyshev polynomial, and

$$\begin{cases} T_1(t) = \dfrac{1}{\sqrt{\pi}}, \quad T_2(t) = \sqrt{\dfrac{2}{\pi}}\left(\dfrac{2}{T}t - 1\right) \\[3mm] T_3(t) = \sqrt{\dfrac{2}{\pi}}\left[2\left(\dfrac{2}{T}t - 1\right)^2 - 1\right], \; \ldots, \quad T_{i+1}(t) = 2\left(\dfrac{2}{T}t - 1\right)T_i(t) - T_{i-1}(t) \end{cases}$$

$$(7.35)$$

where N_f is the degree of the polynomial and $\{C_i, i = 1, 2, \ldots N_f\}$ is the vector of coefficients in the expansion. The frequency component of the forcing function can be modified by adopting different numbers of terms in the orthogonal function. The sensitivities of each of these coefficients, C_i, with respect to the different system parameters can be found from equations similar to Equation (7.9) for the damage detection.

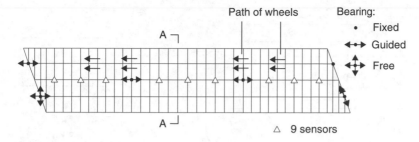

Figure 7.19 Sensor arrangements on bridge deck.

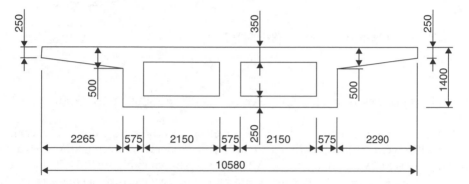

Figure 7.20 Section A-A of bridge deck.

7.5.3 Application to a bridge-vehicle system

7.5.3.1 The vehicle and bridge system

This section gives a numerical example of residual pre-stress force identification in a three-span concrete bridge deck. A full-scale bridge deck, as shown in Figure 7.19, is adopted for this example. The bridge is a three-span continuous structure with a total length of 73 m and a skew angle of 27°. It has two lanes in a single carriageway with a total width of 10.58 m. This bridge has a longitudinal slope of 6.75% and a two cells box-girder cross-section. The mesh of the finite element model and the typical cross-section of the bridge deck are shown in Figures 7.19 and 7.20 respectively. The elastic modulus and mass density of concrete are $E = 2.6 \times 10^{10}$ Pa and $\rho = 2450$ kg/m^3, respectively, and the damping ratio for the first two modes are assumed as $\xi = 0.012$. The pre-stress force along each of the three vertical webs of the box-section has a distribution calculated from the design drawing with $P = 2.0648 \times 10^7$ N at the two ends of the tendon.

The Mindlin–Reissner plate, including the effect of the transverse shear deformation, is adopted for this study with the shear deformable plate element and the membrane element combined. The finite element model of the bridge deck consists of 296 nodes and 330 flat shell elements, and there are $3 \times 36 = 108$ elements with pre-stress in the webs. The model of the pre-stress effect described in Section 3.2.6 is adopted. The

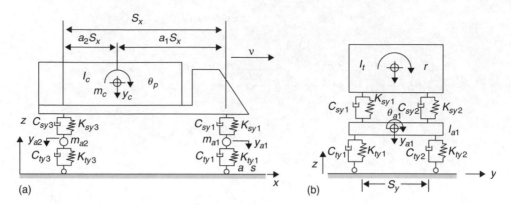

Figure 7.21 The seven degrees-of-freedom vehicle.

natural frequencies from the finite element model are 4.60, 6.27, 7.60, 11.13 and 14.27 Hz.

A two-axle three-dimensional vehicle travels across the bridge deck. The vehicle is simulated as a two-axle three-dimensional vehicle model with seven DOFs according to the H20-44 vehicle design loading in AASHTO, as shown in Figure 7.21. The vehicular body is assigned three DOFs, corresponding to the vertical displacement (y_{t1}), rotation about the transverse axis (pitch or θ_{t1}), and rotation about the longitudinal axis (roll or ϕ_{t1}). The four wheels are provided with four DOFs in the vertical displacements (y_1, y_2, y_3, y_4). Therefore, the total number of independent DOFs is seven. The coupled vehicle-bridge system equations of motion are presented as follows:

$$
\begin{bmatrix} M_b & 0 & H_b M_{v1} \\ 0 & M_{v1} & 0 \\ 0 & 0 & M_{v2} \end{bmatrix} \begin{Bmatrix} \ddot{u} \\ \ddot{Z} \end{Bmatrix} + \begin{bmatrix} C_b & H_b C_{v21} & H_b C_{v22} \\ 0 & C_{v11} & C_{v12} \\ -C_t H_b^T & C_{v21} & C_{v22} + C_t \end{bmatrix} \begin{Bmatrix} \dot{u} \\ \dot{Z} \end{Bmatrix}
$$
$$
+ \begin{bmatrix} K_b & H_b K_{v21} & H_b K_{v22} \\ 0 & K_{v11} & K_{v12} \\ -K_t H_b^T & K_{v21} & K_{v22} + K_t \end{bmatrix} \begin{Bmatrix} u \\ Z \end{Bmatrix} = \begin{Bmatrix} H_b M_s \\ 0 \\ f(t) \end{Bmatrix}
\tag{7.36}
$$

where M_b, C_b and K_b are the mass, damping and stiffness matrices of the bridge model, respectively; $Z = \{y_{t1} \phi_{t1} \theta_{t1} y_1 y_2 y_3 y_4\}^T$; M_{v1}, M_{v2}, C_{v11}, C_{v12}, C_{v21}, C_{v22}, K_{v11}, K_{v12}, K_{v21} and K_{v22} are the mass, damping and stiffness sub-matrices of the vehicle, respectively; M_s is the static load vector of the vehicle; and $\ddot{u}$, $\dot{u}$ and u are the nodal acceleration, velocity and displacement vectors, respectively; $f(t)$ is the matrix of moving interaction forces. The force vectors acting at an arbitrary location on a shell element are transformed as nodal loads using the Hermite interpolation function (Wu, 2007).

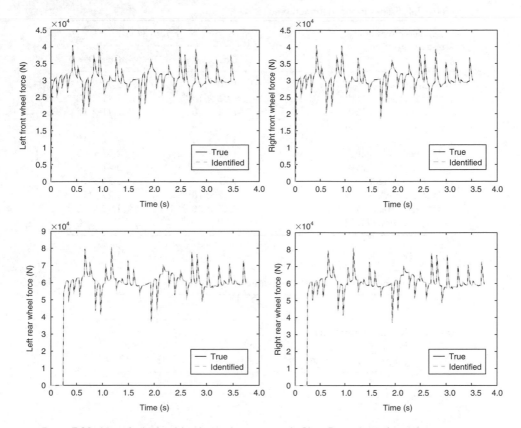

Figure 7.22 Identified wheel-load time histories with Class D road roughness.

7.5.3.2 The residual pre-stress identification

The three masses of the vehicle are: $m_c = 17000\,\text{kg}$, $m_{a1} = 600\,\text{kg}$ and $m_{a2} = 1000\,\text{kg}$, while other parameters are referred to in Zhu and Law (2002). The velocity of the vehicle is $20\,\text{m/s}$ and the travelling path is as shown in Figure 7.19. The sampling rate is $100\,\text{Hz}$ and road roughness (ISO, 1995) Classes C and D are considered. The response generated by the passage of the vehicle is used for the identification.

The changes in local pre-stress are simulated as 5% pre-stress loss in the 9th element of the left web, in the 18th element of the middle web and in the 18th, 19th, 27th and 29th elements of the right web and 10% in the 28th element of the right web. This simulates the pre-stress changes in a single element or a group of elements in the structure.

The location of the vehicle is assumed to be known in the identification. The inter-action forces are modelled with 600 terms in the orthogonal expansion of the forces as shown in Equation (7.35). The initial set of coefficients of the orthogonal function is set equal to unity. The initial set of pre-stress forces in the webs is taken equal to those of the original structure. Nine vertical acceleration responses, as shown in Figure 7.19, are used in the identification. The convergence criterion is 1.0×10^{-5}.

Table 7.4 Errors in the identified results.

Identified forces or residual pre-stress		Error (%)	
		Class C	Class D
Left web	9th	1.20×10^{-7}	9.95×10^{-7}
Middle web	18th	3.86×10^{-8}	2.49×10^{-7}
	18th	1.74×10^{-7}	2.41×10^{-7}
	19th	4.42×10^{-7}	5.50×10^{-7}
	27th	6.15×10^{-7}	1.27×10^{-6}
Right web	28th	1.26×10^{-7}	2.07×10^{-6}
	29th	5.61×10^{-7}	3.71×10^{-6}
Left front wheel load		9.77×10^{-9}	3.34×10^{-8}
Right front wheel load		7.27×10^{-9}	3.29×10^{-8}
Left rear wheel load		9.43×10^{-9}	1.99×10^{-8}
Right rear wheel load		5.50×10^{-9}	1.45×10^{-8}

Table 7.5 Information at convergence of results.

	Class C	Class D
Required iterations	5	4
Error of convergence	8.48×10^{-8}	3.39×10^{-6}
Regularization parameter λ	1.9103×10^{-14}	3.0366×10^{-15}

There are 375 time instances with the vehicle on the bridge deck, and hence there are $9 \times 375 = 3375$ simultaneous equations in Equation (7.36), which is larger than $4 \times 600 + 108 = 2508$ unknowns in the inverse problem.

Figure 7.22 gives the identified force time histories with Class D road roughness compared with the true values. The identified forces are overlapping with the true values with very small errors as shown in Table 7.4. The errors in the identified pre-stress of the selected elements are also shown in Table 7.4 while those in the unselected elements are close to zero. Table 7.5 gives the associated information for convergence. The method is shown to be able to effectively identify the interaction forces and local pre-stress change simultaneously in a large complex bridge structure.

7.6 Conclusions

The response sensitivity approach leads to analytical relationships between the dynamic response and the system parameters, including the excitation force. Different features of system identification using this approach could be performed with these analytical relationships as illustrated in this chapter. Developments described in this chapter lead to further development of another time response approach based on wavelets as discussed in the next chapter.

Chapter 8

System identification with wavelet

8.1 Introduction

In the last decade, wavelet theory has been one of the emerging and fast-evolving mathematical and signal processing tools for vibration analysis. Staszewski (1998) presented a summary of recent advances and applications of wavelet analysis for damage detection. Kijewski and Kareem (2003) studied wavelet transforms for system identification in civil engineering. The main advantage of the continuous wavelet transform is its ability to provide information simultaneously in time and scale with adaptive windows. Wang and Deng (1999) proposed that the wavelet transform be directly applied to spatially distributed structural response signals, such as surface profile, displacement, strain and acceleration measurements. The continuous wavelet transform of the fundamental mode shape and its Lipschitz exponent was used to detect the damage location and extent in a beam (Hong et al., 2002; Chang and Chen, 2003; Douka et al., 2003; Gentile and Messina, 2003). The main purpose of this application is to check the spatially distributed response signals that can pick up damage information.

8.1.1 *The wavelets*

Many existing vibration-based approaches for damage detection require the modal properties with the aid of the traditional Fourier transform. There are a few inherent characteristics of the Fourier transform that might affect the accuracy of the damage identification. Firstly, the Fourier transform is a data reduction process and information on the structural condition might be lost during the process (Sun and Chang, 2002a). Secondly, the Fourier transform is a global analysis technique, and its basis functions are global functions. Any perturbation of the function at any point in the time domain influences every point in the frequency domain. This means that the Fourier transform does not exhibit the time dependency of signals and it cannot capture the evolutionary characteristics that are commonly observed in the measured signals from structures under random excitation (Gurley and Kareem, 1999). Damage is typically a local phenomenon which tends to be captured in high frequency modes. These high frequencies are normally closely spaced but poorly excited. All these factors add difficulty to the implementation of Fourier-transform-based damage detection techniques (Sun and Chang, 2002a).

There has been increasing interest in the wavelet-based approach in recent years due to its success in several applications. The wavelet transform decomposes a signal using short duration waves, allowing a refined decomposition, rather than decomposition with infinite duration sinusoids as with Fourier transforms. The short duration wave, known as the wavelet basis, has a higher energy density than the sinusoid in Fourier transforms in general. Therefore, in many cases, the wavelets require fewer coefficients than Fourier transforms to describe a signal. The wavelet transform is a two-parameter transform. For the time signal, the two domains of the wavelet transform are time, t, and scale, a. The scale, a, can be approximately related to the frequency, ω. The main advantage gained by using wavelets in signal analysis is the ability to perform local analysis of a signal, i.e. to zoom in on any interval of time or space. Wavelet analysis is thus capable of revealing some hidden aspects of the data that other signal analysis techniques fail to detect. This property is particularly important for damage detection applications. The wavelet transform is becoming a promising technique for the damage identification of structures (Staszewski, 1998). A brief background to wavelet analysis is given below.

The continuous wavelet transform of a signal $f(t)$ is defined as

$$Wf(a,b) = \frac{1}{\sqrt{a}} \int_{-\infty}^{+\infty} f(t)\psi^* \left(\frac{t-b}{a}\right) dt \qquad (8.1)$$

where a and b are the dilation and translation parameters, respectively, and $\psi(t)$ is the mother wavelet. Both a and b are real numbers and a must be positive; $\psi^*(t)$ indicates its complex conjugate. It should be noted that a wavelet family associated with a mother wavelet, $\psi(t)$, is generated by two operations: dilation and translation. The translation parameter, b, indicates the location of the moving wavelet window in the wavelet transform. Shifting the wavelet window along the time axis implies examining the signal in the neighbourhood of the current window location. Therefore, information in the time domain remains, in contrast with the Fourier transform where the time domain information becomes invisible after the integration over the entire time domain. The dilation parameter, a, indicates the width of the wavelet window. A small value of a implies a higher-resolution filter, i.e. the signal is examined through a narrow wavelet window in a smaller scale.

The signal $f(t)$ may be recovered or reconstructed by an inverse wavelet transform of $Wf(a,b)$ defined as

$$f(t) = \frac{1}{C_\psi} \int_{-\infty}^{+\infty} \int_{-\infty}^{+\infty} (Wf)(a,b)\psi \left(\frac{t-b}{a}\right) \frac{1}{a^2} da\, db \qquad (8.2)$$

where C_ψ is defined as following with the mother wavelet satisfying the admissibility condition to ensure the existence of the inverse wavelet transform.

$$C_\psi = \int_{-\infty}^{+\infty} \frac{|\psi(\omega)|^2}{|\omega|} d\omega < +\infty \qquad (8.3)$$

The existence of the integral in Equation (8.3) requires that

$$\Psi(0) = 0, \; i.e. \int_{-\infty}^{+\infty} \psi(x)dx = 0 \qquad (8.4)$$

In practical signal processing, a discrete version of the wavelet transform is often employed by discretizing the dilation parameter a and the translation parameter b. In general, the procedure becomes much more efficient if dyadic values of a and b are used, i.e.

$$a = 2^j \quad b = 2^j k \qquad j, k \in Z \qquad (8.5)$$

where Z is a set of positive integers. For a special choice of $\psi(t)$, the corresponding discretized wavelets $\{\psi_{j,k}\}$ are defined as

$$\psi_{j,k}(t) = 2^{-j/2}\psi(2^{-j}t - k) \qquad j, k \in Z \qquad (8.6)$$

constitute an orthonormal basis for $L^2(R)$. Using the orthogonal basis, the wavelet expansion of a function $f(t)$ and the coefficients of the wavelet expansion are defined as

$$f(t) = \sum_{j,k} d_{j,k}\psi_{j,k}(t) \qquad (8.7)$$

and

$$d_{j,k} = \int_{-\infty}^{+\infty} f(t)\psi_{j,k}^*(t)dt \qquad (8.8)$$

In the discrete wavelet analysis, a signal can be represented by its approximations and details. The detail D_j and the approximation A_j at level j are defined as

$$D_j = \sum_{k \in Z} \alpha_{j,k}\psi_{j,k}(t), \quad A_J = \sum_{j>J} D_j \qquad (8.9)$$

8.1.2 The wavelet packets

A possible drawback of the wavelet transform is that the frequency resolution is quite poor in the high frequency region. Hence, there are difficulties when discriminating signals containing high frequency components. The wavelet packet transform is an extension of the wavelet transform that provides complete level-by-level decomposition. The wavelet packets are alternative bases formed by linear combinations of the usual wavelet functions (Coifman and Wickerhauser, 1992). As a result, the wavelet packet transform enables the extraction of features from signals containing stationary and non-stationary components with arbitrary time-frequency resolution. See Sun and Chang, 2002a, on the use of wavelet packet signature to detect damage.

Wavelet packets inherit properties such as orthonormality and time-frequency localization from their corresponding wavelet functions. A wavelet packet $\psi_{j,k}^i(t)$ is a function with three indices where integers i, j and k are the modulation, scale and translation parameters, respectively,

$$\psi^i_{j,k}(t) = 2^{j/2}\psi^i(2^j t - k), \quad i = 1, 2, 3, \ldots \tag{8.10}$$

The wavelet packet decomposition (WPD) of a time-domain signal $f(t)$ can be calculated using a recursive filter-decimation operation (Coifman and Wickerhauser, 1992).

The wavelet functions $\psi^i_{j,k}(t)$ can be obtained from the following recursive relationships:

$$\psi^{2i}(t) = \sqrt{2} \sum_{k=-\infty}^{\infty} h(k)\psi^i(2t - k) \tag{8.11}$$

$$\psi^{2i+1}(t) = \sqrt{2} \sum_{k=-\infty}^{\infty} g(k)\psi^i(2t - k) \tag{8.12}$$

The first two wavelets are the usual scaling function $\phi(t)$ and the mother wavelet function $\psi(t)$, where

$$\psi^0(t) = \phi(t), \quad \psi^1(t) = \psi(t) \tag{8.13}$$

The discrete filters $h(k)$ and $g(k)$ are the quadrature mirror filters associated with the scaling function and the mother wavelet function. Most of the existing mother wavelets are developed to satisfy essential properties such as the invertibility and orthogonality.

The Wavelet Transform (WT) consists of one high frequency term from each level and one low-frequency residual from the last level of decomposition. The Wavelet Packet Transform (WPT), however, contains a complete decomposition at every level and hence can achieve a higher resolution in the high frequency region. The recursive relations between the jth and the $(j+1)$th level components are

$$f^i_j(t) = f^{2i-1}_{j+1}(t) + f^{2i}_{j+1}(t), \tag{8.14}$$

$$f^{2i-1}_{j+1}(t) = Hf^i_j(t), \tag{8.15}$$

$$f^{2i}_{j+1}(t) = Gf^i_j(t), \tag{8.16}$$

where H and G are the filtering–decimation operators related to the discrete filters $h(k)$ and $g(k)$ in such a way that

$$H\{\cdot\} = \sum_{k=-\infty}^{\infty} h(k - 2t), \tag{8.17}$$

$$G\{\cdot\} = \sum_{k=-\infty}^{\infty} g(k - 2t), \tag{8.18}$$

After the j level of decomposition, the original signal $f(t)$ can be expressed as

$$f(t) = \sum_{i=1}^{2^j} f_j^i(t) \tag{8.19}$$

The wavelet packet component signal $f_j^i(t)$ can be represented by a linear combination of wavelet packet functions, $\psi_{j,k}^i(t)$, as

$$f_j^i(t) = \sum_{k=-\infty}^{\infty} c_{j,k}^i(t)\psi_{j,k}^i(t), \tag{8.20}$$

where the wavelet packet coefficients $c_{j,k}^i(t)$ can be obtained from

$$c_{j,k}^i(t) = \int_{-\infty}^{\infty} f(t)\psi_{j,k}^i(t)dt \tag{8.21}$$

provided that the wavelet packet functions are orthogonal, i.e.

$$\psi_{j,k}^m(t)\psi_{j,k}^n = 0 \qquad \text{if } m \neq n$$

Each component in the wavelet packet decomposition can be viewed as the output of a filter tuned to a particular basis function. At the lower decomposition level, the WPD yields a good resolution in the time domain but a poor resolution in the frequency domain. While, at the higher decomposition level, the WPD results in a good resolution in the frequency domain but a poor resolution in the time domain. For the purpose of structural health monitoring, discrete information in the frequency domain is more important and thus a high level of the WPD is often required to detect the minute changes in the signals.

8.2   Identification of crack in beam under operating load

An early application of wavelet theory in the spatial domain crack identification of structures was proposed by Liew and Wang (1998). The wavelet in the spatial domain is calculated based on finite difference solutions of a mathematical representation of the structure in question. The crack location is indicated by a peak in the variations of the wavelets along the length of the beam. A classical measurement system such as the impulse hammer technique, is only able to measure mode shapes at a few discrete points of a transversely vibrating beam. Therefore, new sensors or measuring techniques are needed to pick up the perturbations caused by the presence of a crack. Recently the possibility of measuring displacements on denser grids (a few hundred of points) by using a laser scanning vibrometer was reported (Pai and Young, 2001).

One of the questions that is attracting significant research attention is related to the use of the structural response from operational dynamic loads in a damage detection procedure. The operational loads for bridges are moving vehicular loads, and the operational deflection shapes are the deflections of the bridge deck subject to moving vehicular loads. Mazurek and Dewolf (1990) conducted the laboratory studies on simple two-span girders under moving loads with structural deterioration by vibration analysis. Structural damages were artificially introduced by a release of supports and insertion of cracks. Piombo et al. (2000) modelled the vehicle-bridge interaction system as a three-span orthotropic plate subject to a seven degrees-of-freedom (DOF) multi-body system with linear suspensions and tyre flexibility. The wavelet technique was used to extract the modal parameters. Lee et al. (2002) studied the identification of the operational modal properties of a bridge structure using vibration data caused by the traffic loadings, with damage assessment based on the estimated modal parameters using the neural networks technique. Majumder and Manohar (2003) developed a time-domain formulation to detect damages in a beam using data originating from the linear beam-oscillator dynamic interaction and extended the capabilities of this formulation to include the damaged beam structure that undergoes nonlinear vibrations. The study combines finite element modelling for the vehicle-bridge system with a time-domain formulation to detect changes in structural parameters. The structural properties and motion characteristics of the moving vehicle are assumed to be available, and the elemental stiffness loss is used to simulate the different damage scenarios.

In this section, the operational deflection time history of a bridge subject to a moving vehicular load is analyzed using the continuous wavelet transform. The identification of a crack in a beam is based on the spatial wavelet analysis of response measurements at one point of the bridge deck. The damage index based on the wavelet coefficient is used as an indicator of the damage extent. The effect of the parameters of the vehicle-bridge interaction system and noise in the measurements on the damage detection is illustrated with an experimental study on a reinforced concrete bridge deck modelled with a Tee-section subject to vehicular loadings.

8.2.1 Dynamic behaviour of the cracked beam subject to moving load

The bridge-vehicle system is modelled as a continuous beam subject to a moving load, $P(t)$, as shown in Figure 8.1. The load is assumed to be moving at a prescribed velocity, $v(t)$, along the axial direction of the beam from left to right. The beam is assumed to be an Euler–Bernoulli beam. The equation of motion can be written as

$$\rho A \frac{\partial^2 w(x,t)}{\partial t^2} + C \frac{\partial w(x,t)}{\partial t} + \frac{\partial^2}{\partial x^2}\left(EI(x)\frac{\partial^2 w(x,t)}{\partial x^2}\right) = P(t)\delta(x - \hat{x}(t)) \qquad (8.22)$$

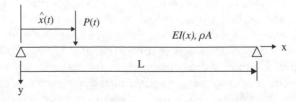

Figure 8.1 A continuous beam subject to moving loads

where ρA and C are the mass per unit length and the damping of the beam; $EI(x)$ is the flexural stiffness of the Euler–Bernoulli beam; $w(x, t)$ is the displacement function of the beam; $\hat{x}(t)$ is the location of moving load $P(t)$ at time t; and $\delta(t)$ is the Dirac delta function. Expressing the transverse displacement $w(x, t)$ in modal coordinates, gives

$$w(x, t) = \sum_{i=1}^{\infty} \phi_i(x) q_i(t) \tag{8.23}$$

where $\phi_i(x)$ is the mode shape function of the ith mode and $q_i(t)$ is the ith modal amplitude. Substituting Equation (8.23) into Equation (8.22), and multiplying by $\phi_i(x)$, then integrating with respect to x between 0 and L, and applying the orthogonality conditions, gives

$$\frac{d^2 q_i(t)}{dt^2} + 2\xi_i\omega_i \frac{dq_i(t)}{dt} + \omega_i^2 q_i(t) = \frac{1}{M_i} P(t) \phi_i(\hat{x}(t)) \tag{8.24}$$

where ω_i, ξ_i and M_i are the modal frequency, damping ratio and the modal mass of the ith mode, and

$$M_i = \int_0^L \rho A \phi_i^2(x) dx \tag{8.25}$$

The displacement of the beam at point x and time t can be found from Equations (8.23) and (8.24) as

$$w(x, t) = \sum_{i=1}^{\infty} \frac{\phi_i(x)}{M_i} \int_0^t h_i(t - \tau) P(\tau) \phi_i(\hat{x}(\tau)) d\tau \tag{8.26}$$

where

$$h_i(t) = \frac{1}{\omega_i'} e^{-\xi_i\omega_i t} \sin \omega_i' t; \quad \omega_i' = \omega_i\sqrt{1 - \xi_i^2} \tag{8.27}$$

8.2.2 The crack model

Figure 8.2 shows a uniform bridge beam structure with N cracks. The damaged continuous beam is discretized into $N + 1$ segments of constant linear density, ρA; bending stiffness, EI (undamaged beam stiffness); and length l_i, $(i = 1, 2, \ldots, N + 1)$. The segments are connected together through rotational springs (damage section) whose stiffnesses are denoted by k_i, $(i = 1, 2, \ldots, N)$.

The eigenfunction of an Euler–Bernoulli beam segment can be written as

$$r_i(x_i) = A_i \sin \beta x_i + B_i \cos \beta x_i + C_i \sinh \beta x_i + D_i \cosh \beta x_i, \quad (i = 1, 2, \ldots, N + 1) \tag{8.28}$$

where $r_i(x_i)$ is the eigenfunction for the ith segment, and β is the eigenvalue of the beam. There are $N + 1$ segments connected by rotational springs here. The boundary

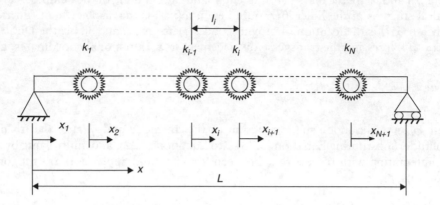

Figure 8.2 Beam with rotational springs representing damaged section

conditions for the damaged beam are:

$$r_1(x_1)|_{x_1=0} = r_{N+1}(x_{N+1})|_{x_{N+1}=l_{N+1}} = 0$$

$$\left.\frac{\partial^2 r_1(x_1)}{\partial x_1^2}\right|_{x_1=0} = \left.\frac{\partial^2 r_{N+1}(x_{N+1})}{\partial x_{N+1}^2}\right|_{x_{N+1}=l_{N+1}} = 0$$

$$
\begin{cases}
\qquad\quad r_i(x_i)|_{x_i=l_i} = r_{i+1}(x_{i+1})|_{x_{i+1}=0} \\[2mm]
\left.\dfrac{\partial r_i(x_i)}{\partial x_i}\right|_{x_i=l_i} + \left.\dfrac{EI}{k_i}\dfrac{\partial^2 r_i(x_i)}{\partial x_i^2}\right|_{x_i=l_i} = \left.\dfrac{\partial r_{i+1}(x_{i+1})}{\partial x_{i+1}}\right|_{x_{i+1}=0} \\[3mm]
\left.\dfrac{\partial^2 r_i(x_i)}{\partial x_i^2}\right|_{x_i=l_i} = \left.\dfrac{\partial^2 r_{i+1}(x_{i+1})}{\partial x_{i+1}^2}\right|_{x_{i+1}=0} \qquad , \quad (i=1,2,\ldots,N) \;\;(8.29) \\[3mm]
\left.\dfrac{\partial^3 r_i(x_i)}{\partial x_i^3}\right|_{x_i=l_i} = \left.\dfrac{\partial^3 r_{i+1}(x_{i+1})}{\partial x_{i+1}^3}\right|_{x_{i+1}=0}
\end{cases}
$$

Substituting Equation (8.28) into the boundary conditions in Equation (8.29), the mode shape of the continuous beam with N damage locations can be written as

$$\phi(x) = r_1(x)(1 - H(x - l_1)) + \sum_{i=2}^{N+1} r_i\Big(x - \sum_{j=1}^{i-1} l_j\Big)\Big(H\Big(x - \sum_{j=1}^{i-1} l_j\Big) - H\Big(x - \sum_{j=1}^{i} l_j\Big)\Big)$$

$$(8.30)$$

where $H(x)$ is the unit step function.

$$
\begin{cases}
r_1(x) = A_1 \sin \beta x + C_1 \sinh \beta x, & (0 \le x < l_1) \\
r_i(x) = A_i(x) \sin \beta x + B_i \cos \beta x \\
\qquad\quad + C_i \sinh \beta x + D_i \cosh \beta x, & (0 \le x < l_i, i = 2,3,\ldots,N+1)
\end{cases}
\qquad (8.31)
$$

and parameters $\{A\} = \{\beta, A_1, C_1, A_i, B_i, C_i, D_i\}$ $(i = 2, 3, \ldots, N + 1)$ are determined from the following equation.

$$[S]\{A\} = 0 \tag{8.32}$$

and the elements of matrix S are referred to (Zhu and Law, 2006).

8.2.3 Crack identification using continuous wavelet transform

Equation (8.30) shows that there are discontinuities at the damage points, particularly the slope discontinuities at the cracks. Mode shape curvature is widely used to find these discontinuous points (Pandey et al., 1991). However, the first problem for damage detection using curvature directly is to calculate the curvature by derivation which is very difficult in practice, and the differentiation of the mode shape will further amplify the measurement error. The wavelet transform is used here to measure the local regularity of a signal.

The continuous wavelet transform of a square-integrable signal, $f(x)$, where x is time or space, is defined similar to Equation (8.1) (Mallat and Hwang, 1992)

$$Wf(u, s) = f(x) \otimes \psi_s(x) = \frac{1}{\sqrt{s}} \int_{-\infty}^{+\infty} f(x) \psi^* \left(\frac{x - u}{s} \right) dx \tag{8.33}$$

where $\otimes$ denotes the convolution of two functions; $\psi_s(x)$ is the dilation of $\psi(x)$ by the scale factor s; u is the translation indicating the locality; and $\psi^*(x)$ is the complex conjugate of $\psi(x)$, which is a mother wavelet satisfying the following admissibility condition in Equation (8.3). From Equation (8.26), the displacement at x_m can be written as

$$w(x_m, t) = \sum_{i=1}^{\infty} \frac{\phi_i(x_m)}{M_i} \int_0^t h_i(t - \tau) P(\tau) \phi_i(\hat{x}(\tau)) d\tau \tag{8.34}$$

The second derivation of the displacement with respect to the position of the moving load can be obtained as

$$\frac{\partial^2 w(x_m, t)}{\partial^2 \hat{x}_l(t)} = \sum_{i=1}^{\infty} \frac{\phi_i(x_m)}{M_i} \int_0^t h_i(t - \tau) P(\tau) \frac{\partial^2 \phi_i(\hat{x}(\tau))}{\partial \hat{x}(\tau)^2} d\tau \tag{8.35}$$

where $\partial^2 \phi_i(\hat{x})/\partial \hat{x}^2$ is the second-order derivation of the ith mode, which is the curvature of the displacement mode shape. The second derivative of the displacement with respect to the load position is shown to include the curvature information of the mode.

Let us take the Gaussian function, $\theta(x)$, the wavelet of which can be defined as the second derivative of the function

$$\psi(x) = \frac{d^2 \theta(x)}{dx^2} \tag{8.36}$$

The wavelet, $\psi(x)$, in Equation (8.36) is continuous differentiable and is usually referred to as the Mexican Hat wavelet that has the following explicit expression:

$$\psi(x) = \frac{2}{\sqrt{3\sigma}} \pi^{-1/4} \left(\frac{x^2}{\sigma^2} - 1\right) exp\left(\frac{-x^2}{2\sigma^2}\right) \tag{8.37}$$

where σ is the standard deviation.

The wavelet transform for the displacement $w(x_m, t)$ is then expressed by the following relation (Mallat and Hwang, 1992) when the Mexican Hat wavelet is adopted, as

$$Ww(\widehat{x}(t), s) = w(x_m, t) \otimes \psi_s(\widehat{x}(t)) = s^2 \frac{d^2}{d\widehat{x}(t)^2}(w(x_m, t) \otimes \theta_s)(\widehat{x}(t)) \tag{8.38}$$

Equation (8.38) is the multi-scale differential operator of the second order, and the relation between the second differentiability of $w(x_m, t)$ and its wavelet transform decays at fine scales. The wavelet transform $Ww(\widehat{x}(t), s)$ is proportional to the second derivative of $w(x_m, t)$ smoothed by the Gaussian function $\theta_s(x)$. So the wavelet transform can be used to replace the direct differentiation of the displacement to obtain the curvature properties. The damage can be determined using the wavelet transform of the operational displacement time history at one point when the beam structure is subject to the action of the moving load. A similar formulation can be obtained for accelerations.

8.2.4 Numerical study

A simply supported beam, 50 m long, 1.0 m high and 0.5 m wide (Mahmoud, 2001) is used. The elastic modulus and density of the beam are $E = 2.1 \times 10^{11}$ Pa and $\rho = 7860$ kg/m^3, and the moving load is a constant $F_0 = 10$ kN. The first six natural frequencies are 0.94, 3.75, 8.44, 15.00, 23.44 and 33.75 Hz. The crack compliance, C_C, of a rectangular beam with the crack depth ratio δ is (Sun and Chang, 2002b)

$$C_C = \frac{1}{k} = \frac{2h}{EI} \left(\frac{\delta}{1-\delta}\right)^2 [5.93 - 16.69\delta + 37.14\delta^2 - 35.84\delta^3 + 13.12\delta^4] \tag{8.39}$$

The above beam is used in the simulation, and the damping ratio for all modes is taken equal to 0.02. The crack is at $1/3L$ from the left-hand-side with a depth ratio of 0.5. White noise is added to the calculated responses of the beam to simulate the polluted measurements, and 1%, 3% and 5% noise levels are studied separately.

$$w = w_{calculated} + E_p \cdot N_{noise} \cdot \sigma(w_{calculated}) \tag{8.40}$$

where w is the polluted displacement; E_p is the noise level; N_{noise} is a standard normal distribution vector with zero mean value and unit standard deviation; and $w_{calculated}$ is the calculated displacement, and $\sigma(w_{calculated})$ are their standard deviations. The continuous wavelet transform on the displacement time history at the mid-span are calculated with dilation s equal to 1 to 512 in unit increments.

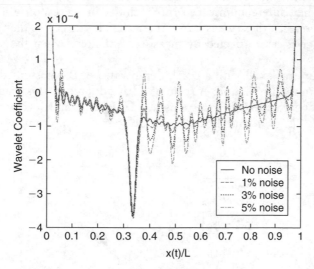

Figure 8.3 Wavelet coefficients of the displacement at the mid-span when a moving load is moving on the beam at 1 m/s.

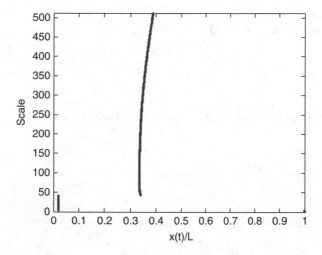

Figure 8.4 The dip peak position versus scale plot

Figure 8.3 shows the wavelet coefficients of the displacement at the mid-span with scale 64 when the moving load is on top of the beam. There is a dip in the curve at $1/3L$ indicating the location of the damage.

Figure 8.4 shows the location of the peak from using different scales. The position of the dip in the wavelet coefficient curve is close to $1/3L$ when the scale is not less than 42 and they are close to the two ends when the scale is less than 42. The latter is associated with the impacts on the entry and exit of the moving load. When the scale is larger than 42, the position of the dip indicates the location of the damage.

Figure 8.5 shows the dip value of the minimum wavelet coefficient versus the scale in a log-log plot when different noise levels are included in the response. The results are close to each other when the scale is larger than 28. This indicates that the identified location of damage is least affected by measurement noise when the scale is larger than 28.

The case of four cracks all with a crack depth ratio of 0.5 is studied further; 3% noise is included in the simulation. A scale of 64 is adopted in this study. Figure 8.6 shows the wavelet coefficients of the responses at 1/4L, 1/2L and 3/4L with four cracks located at 1/5L, 2/5L, 3/5L and 4/5L. There are four dips in the wavelet coefficient

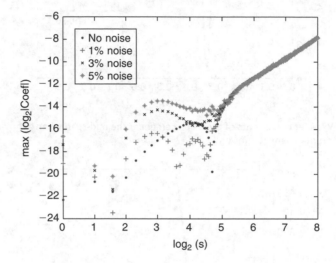

Figure 8.5 Wavelet coefficients of the response with different noise levels

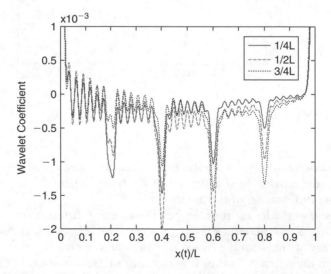

Figure 8.6 Wavelet coefficients of the responses with four damages (3% noise)

curves which are close to the damage locations at $1/5L$, $2/5L$, $3/5L$ and $4/5L$, and the dip value varies with the measuring location. The multiple damage locations can be determined accurately from the wavelet coefficient of the response obtained from a single measuring point.

8.2.5 *Experimental verification*

The experimental set-up shown in Figure 8.7 includes three Tee-section concrete beams, i.e. the leading beam, the main beam and the tailing beam. The length of the leading and tailing beams are 4.5 m each, and the main beam is 5.0 m long. The gaps between the beams are 10 mm. A vehicle is pulled along the beam by an electric motor at an approximate speed of 0.5 m/s. The axle spacing of the vehicle is 0.8 m, and the wheel spacing is 0.39 m. The vehicle weighs 10.60 kN, with the front axle load weighing 5.58 kN and the rear axle load weighing 5.02 kN. As the total mass of the concrete beam is 1050 kg, the weight ratio between the vehicle and bridge is 1.01.

Seven displacement transducers (sensors 1 to 7) are evenly distributed at the bottom and along the beam to measure the responses as marked. Thirteen photoelectric sensors are installed on the leading beam and the main beam at 0.56 m spacing to monitor the speed of the vehicle. The third and thirteenth photoelectric sensors are located at the entry and exit points of the main beam separately. An INV300 data acquisition system is used to collect the data from all eight channels. The sampling frequency is 2024.292 Hz, and the sampling period is 30 s for each test.

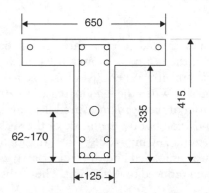

(a) Cross-section layout of the reinforced concrete beam

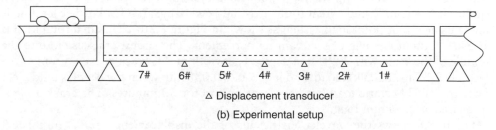

△ Displacement transducer

(b) Experimental setup

Figure 8.7 Experimental setup

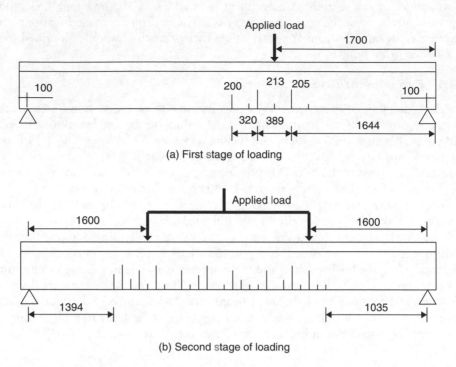

(a) First stage of loading

(b) Second stage of loading

Figure 8.8 Damage loading and the crack zone

Damage in the beam is created using a three-point load system applied at $1/3L$ from the right support of the beam, as shown in Figure 8.8(a). The load is gradually increased with 2 kN increments. When 36 kN is reached, several tensile cracks are clearly seen on the beam rib. When the load increases to 50 kN, the crack width of the largest crack at the bottom of the beam measured 0.10 mm. The location of this crack is close to the loading position but on the inside of the span with a visual crack depth of 213 mm and a crack zone 760 mm wide. When the load had been on the beam for 30 minutes, the beam was unloaded and the crack partly closed with the crack width at the bottom of the beam reduced to 0.025 mm. These observations are referred to as the small-damage case.

For the large-damage case, the beam is first loaded at $2/3L$ of the beam from the right support up to 50 kN using the three-point load system. This creates a crack pattern similar in magnitude and extent to the existing crack zone at $1/3L$. Further loading is made using a four-point load system as shown in Figure 8.8(b). The final total load is 105 kN without yielding of the main reinforcement. The largest crack is close to the middle of the beam with a 281 mm depth. The width of this crack at the bottom of the beam is 0.1 mm at 105 kN load, and it becomes 0.038 mm when the beam is unloaded, once the 105 kN static load has been kept on top for 30 minutes. The crack zone is measured as 2371 mm long.

Figure 8.9 shows the wavelet coefficients of the displacement at $3/8L$ (number 3 transducer) when the model car is moving on the concrete beam. There are six main

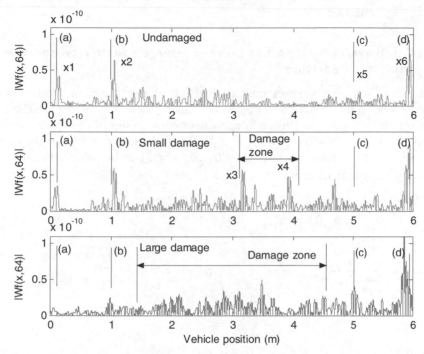

Figure 8.9 Spatial wavelet coefficient at scale 64 (a) denotes entry of first axle, (b) denotes entry of second axle, (c) denotes exit of first axle, (d) denotes exit of second axle.

peaks for the small-damage case. The first and second peaks are associated with impacts on the entry of the front and rear axles. The fifth and sixth peaks are associated with impacts on the exit of the front and rear axles. The third and fourth peaks are related to the locations of the damage in Figure 8.9(a). The results show that the damage location can be determined using the peaks in the wavelet coefficient of the response from a single measuring point. For the large-damage case, there are many cracks created in the reinforced concrete beam. There are also many peaks in the curve of the wavelet coefficient besides those associated with the entry and exit of the vehicle. While the damage zone can be clearly estimated, the crack location cannot be determined separately. This can be explained by the fact that the large static load of 105kN causes bond slippage between the steel bar and concrete, and the damage cannot be simply modelled as an open crack.

The above studies show that the operating load can be used effectively to determine the damage location accurately even though there are multiple damages in the bridge beam. The location is determined as the position of the dip in the wavelet coefficient curve and there is no baseline requirement in determining the damage location.

8.3 The sensitivity approach

The response sensitivity approach for damage detection has been discussed in detail in Chapter 7. The response is expressed in terms of wavelets and the wavelet packets, and their sensitivities are expressed as functions of the response sensitivity for a finer scale damage detection.

8.3.1 The wavelet packet component energy sensitivity and the solution algorithm

The measured response is represented by the Haar wavelet basis function through the dyadic wavelet transformation. The bandwidths of each level of the dyadic wavelet transform are octaves, and this enables a direct comparison of the energy content of the wavelet packets as shown below. The WPT component function of the measured acceleration, $s(t)$, and strain, $\varepsilon(t)$, i.e. $s_j^i(t)$ and $\varepsilon_j^i(t)$, can be reconstructed from the wavelet packet coefficients as,

$$s_j^i(t) = \sum_{k=-\infty}^{+\infty} c_{j,k}^i \psi_{j,k}^i(t) = R_j^i c_j^i = R_j^i D_j^i s(t),$$

$$\varepsilon_j^i(t) = \sum_{k=-\infty}^{+\infty} d_{j,k}^i \psi_{j,k}^i(t) = R_j^i d_j^i = R_j^i D_j^i \varepsilon(t)$$

where

$$R_j^i = [\psi_{j,0}^i \quad \psi_{j,2}^i \quad \cdots \quad \psi_{j,l}^i], \quad (l = 0, 1, \ldots, N/2^j - 1)$$

and $D_{j+1}^{2i} = H^{j+1} D_j^i$, $D_{j+1}^{2i+1} = G^{j+1} D_j^i$, $D_1^0 = H^1$ and $D_1^1 = G^1$; H^{j+1} and G^{j+1} are matrices formed by the low-pass filter function and high-pass filter function, respectively (Sun and Chang, 2002b); and c_j^i and d_j^i are the wavelet packet coefficients for the acceleration and strain, respectively, with $c_j^i = D_j^i s(t)$ and $d_j^i = D_j^i \varepsilon(t)$. The ith wavelet packet transform component energy of the acceleration, $s(t)$, and strain, $\varepsilon(t)$, at the jth level of decomposition, E_{sj}^i and $E_{\varepsilon j}^i$, are energy terms defined as

$$
\begin{aligned}
E_{sj}^i &= (s_j^i)^T(s_j^i) & E_{\varepsilon j}^i &= (\varepsilon_j^i)^T(\varepsilon_j^i) \\
&= s^T(R_j^i D_j^i)^T(R_j^i D_j^i)s, & &= \varepsilon^T(R_j^i D_j^i)^T(R_j^i D_j^i)\varepsilon \\
&= s^T T_j^i s & &= \varepsilon^T T_j^i \varepsilon
\end{aligned}
\qquad (8.41)
$$

The sensitivity of these energy terms with respect to the elemental damage is then computed as

$$\frac{\partial E_{sj}^i}{\partial \alpha_i} = \frac{\partial s^T}{\partial \alpha_i} T_j^i s + s^T T_j^i \frac{\partial s}{\partial \alpha_i}, \quad \frac{\partial E_{\varepsilon j}^i}{\partial \alpha_i} = \frac{\partial \varepsilon^T}{\partial \alpha_i} T_j^i \varepsilon + \varepsilon^T T_j^i \frac{\partial \varepsilon}{\partial \alpha_i} \qquad (8.42)$$

where α_i is the damage index for the ith element as described in Equation (7.25); and $T_j^i = (R_j^i D_j^i)^T(R_j^i D_j^i)$, which is not a function of the signal, is determined only by the wavelet type, and therefore $\partial T_j^i / \partial \alpha_i = 0$. The strain and acceleration sensitivities are referred to Law et al., 2005.

The solution algorithm

The damage identification equation can be written as

$$E_a - E_e = \frac{\partial E_a}{\partial \alpha_i} \cdot \Delta\alpha \tag{8.43}$$

where vectors E_a and E_e are the wavelet packet transform component energies of the analytical model and the experimental model, respectively; $\partial E_a/\partial \alpha_i$ and $\partial E_e/\partial \alpha_i$ are their sensitivity with respect to a change in the system parameter; and $\Delta\alpha$ is the vector of parameter changes of the system. Equation (8.43) is ill conditioned, like many inverse problems. To provide bounds to the solution, the damped least-squares method (Tikhonov, 1963) discussed in Chapter 2 is adopted, and singular value decomposition is used to find the pseudo-inverse.

The principle of the solution algorithm is to compare the structural behaviour in the damaged and intact states. To determine the location and the extent of the damage, the analyst requires a satisfactory model of the structure in its intact state. This reliable model is often achieved by comparing experimentally obtained data of the structure in its initial state with corresponding predictions from an initial mathematical model.

When measurement from the first state of the structure is obtained, the wavelet-packet component energy and its sensitivity are first computed based on the analytical model of the structure and the input force obtained in the experiment. The vector of parameter increments is then obtained from Equation (8.43) using the experimentally obtained WPT component energy. The analytical model is then updated and the corresponding component energy and its sensitivity are again computed for the next iteration. Convergence is considered achieved when both the following two criteria are met:

$$\left\| \frac{\{E_{k+1}\} - \{E_k\}}{\{E_{k+1}\}} \right\| \leq toler\ 1, \quad \left\| \frac{\{\Delta\alpha_{k+1}\} - \{\Delta\alpha_k\}}{\{\Delta\alpha_{k+1}\}} \right\| \leq toler\ 2$$

where k denotes the kth time instance. The tolerance limits for both convergence criteria have been set equal to 1.0×10^{-6} in the present study.

When measurement from the second state is obtained, the updated analytical model is used in the iteration in the same way as that using the measurement from the first state. The final vector of identified parameter increments corresponds to the changes that occurred between the two states of the structure.

Equation (8.43) is usually solved only once for small deviations from the finite element model (FEM). But with the present iterative approach to update the deviations by small increments at a time, any large deviation from the initial model can be updated with the improved finite element model closer to the real structure after each iteration of model improvement.

8.3.2 The wavelet sensitivity and the solution algorithm

The formulation on the wavelet coefficient sensitivity can be derived for any physical parameter of the structural system. In the following derivation, α_h represents a physical parameter of the hth finite element, e.g. the elastic modulus of material, a dimension or the second moment of inertia of a cross-section. Both analytical and computational approaches to obtain the sensitivity of the wavelet coefficient are given below.

Analytical approach

Expressing the responses z, $\dot{z}$ and $\ddot{z}$ in terms of the wavelet transforms and substituting into the equation of motion of the structural system, gives

$$
M \left\{ \begin{array}{c} \sum\limits_{j,k} d_{j,k}^1 \ddot{\psi}_{j,k}(t) \\ \sum\limits_{j,k} d_{j,k}^2 \ddot{\psi}_{j,k}(t) \\ \cdots \\ \sum\limits_{j,k} d_{j,k}^N \ddot{\psi}_{j,k}(t) \end{array} \right\} + C \left\{ \begin{array}{c} \sum\limits_{j,k} d_{j,k}^1 \dot{\psi}_{j,k}(t) \\ \sum\limits_{j,k} d_{j,k}^2 \dot{\psi}_{j,k}(t) \\ \cdots \\ \sum\limits_{j,k} d_{j,k}^N \dot{\psi}_{j,k}(t) \end{array} \right\} + K \left\{ \begin{array}{c} \sum\limits_{j,k} d_{j,k}^1 \psi_{j,k}(t) \\ \sum\limits_{j,k} d_{j,k}^2 \psi_{j,k}(t) \\ \cdots \\ \sum\limits_{j,k} d_{j,k}^N \psi_{j,k}(t) \end{array} \right\} = D \sum\limits_{j,k} d_{j,k}^F \psi_{j,k}(t)
$$

(8.44)

where M, C and K are the mass, damping, stiffness matrices, respectively; z, $\dot{z}$ and $\ddot{z}$ are the displacement, velocity and acceleration vectors, respectively; D is the mapping matrix relating the force vector $F(t)$ to the corresponding DOFs of the system; and $d_{j,k}^F = \int_R F(t)\psi_{j,k}(t)dt$. Computing the inner product with $\psi_{j,k}(t)$ on both sides of Equation (8.44), and noting the orthogonal property of the wavelets, gives

$$
\left(M \int \ddot{\psi}_{j,k}(t)\psi_{j,k}(t)dt + C \int \dot{\psi}_{j,k}(t)\psi_{j,k}(t)dt + K \right) \left\{ \begin{array}{c} d_{j,k}^1 \\ d_{j,k}^2 \\ \cdots \\ d_{j,k}^N \end{array} \right\} = D d_{j,k}^F
$$

(8.45)

Since $\int \ddot{\psi}_{j,k}(t)\psi_{j,k}(t)dt$ and $\int \dot{\psi}_{j,k}(t)\psi_{j,k}(t)dt$ are functions of the wavelets only, they can be expressed as

$$
a_{j,k} = \int \ddot{\psi}_{j,k}(t)\psi_{j,k}(t)dt, \quad b_{j,k} = \int \dot{\psi}_{j,k}(t)\psi_{j,k}(t)dt
$$

and Equation (8.45) becomes

$$
(Ma_{j,k} + Cb_{j,k} + K)\left\{ \begin{array}{cccc} d_{j,k}^1 & d_{j,k}^2 & \cdots & d_{j,k}^N \end{array} \right\}^T = Dd_{j,k}^F
$$

(8.46)

rewriting this gives

$$
\left\{ \begin{array}{cccc} d_{j,k}^1 & d_{j,k}^2 & \cdots & d_{j,k}^N \end{array} \right\}^T = (Ma_{j,k} + Cb_{j,k} + K)^{-1}Dd_{j,k}^F
$$

(8.47)

Differentiating both sides of Equation (8.46) with respect to the damage index, α_h, gives

$$\left(\frac{\partial M}{\partial \alpha_h}a_{j,k} + \frac{\partial C}{\partial \alpha_h}b_{j,k} + \frac{\partial K}{\partial \alpha_h}\right)\left\{\begin{array}{c} d_{j,k}^1 \\ d_{j,k}^2 \\ \cdots \\ d_{j,k}^N \end{array}\right\} + (Ma_{j,k} + Cb_{j,k} + K)\left\{\begin{array}{c} \dfrac{\partial d_{j,k}^1}{\partial \alpha_h} \\[2mm] \dfrac{\partial d_{j,k}^2}{\partial \alpha_h} \\[2mm] \cdots \\[2mm] \dfrac{\partial d_{j,k}^N}{\partial \alpha_h} \end{array}\right\} = 0 \qquad (8.48)$$

Substituting Equation (8.47) into (8.48), finally gives the wavelet coefficient sensitivity in terms of the system parameter as

$$\left\{\begin{array}{cccc} \dfrac{\partial d_{j,k}^1}{\partial \alpha_h} & \dfrac{\partial d_{j,k}^2}{\partial \alpha_h} & \cdots & \dfrac{\partial d_{j,k}^N}{\partial \alpha_h} \end{array}\right\}^T = -(Ma_{j,k} + Cb_{j,k} + K)^{-1}$$

$$\times \left(\frac{\partial M}{\partial \alpha_h}a_{j,k} + \frac{\partial C}{\partial \alpha_h}b_{j,k} + \frac{\partial K}{\partial \alpha_h}\right)(Ma_{j,k} + Cb_{j,k} + K)^{-1}Dd_{j,k}^F \qquad (8.49)$$

Computational approach

Differentiating the formulation of the wavelet coefficient in Equation (8.8) with respect to α_h, gives

$$\frac{\partial d_{j,k}}{\partial \alpha_h} = \frac{\partial}{\partial \alpha_h}\left(\int_R f(t)\psi_{j,k}(t)dt\right) \qquad (8.50a)$$

Since $\psi_{j,k}(t)$ is not related to α_h, we have

$$\frac{\partial d_{j,k}}{\partial \alpha_h} = \int_R \frac{\partial f(t)}{\partial \alpha_h}\psi_{j,k}(t)dt \qquad (8.50b)$$

where $\partial f(t)/\partial \alpha_h$ is the sensitivity of response to a local change in α_h.

Equation (8.50) can also be obtained in an alternative formulation. Express the response sensitivity $\partial f(t)/\partial \alpha_h$ in terms of wavelets, and the wavelet coefficient, $c_{j,k}$, is obtained as

$$c_{j,k} = \int_R \frac{\partial f(t)}{\partial \alpha_h}\psi_{j,k}(t)dt \qquad (8.51)$$

Comparing Equations (8.50) and (8.51), gives

$$\frac{\partial d_{j,k}}{\partial \alpha_h} = c_{j,k} \qquad (8.52)$$

Therefore, the wavelet coefficient sensitivity of function $f(t)$ can be computed from the wavelet transform of the sensitivity of response $f(t)$. Note that the response sensitivity was discussed in Chapter 7.

The response sensitivity can also be obtained through the analytical state-space formulation (Law et al., 2005) for a general structural system. In the measurement state, since

$$f = L\ddot{z} \tag{8.53}$$

where f is the response vector at an arbitrary set of measured locations, and L is the mapping vector relating the measured DOFs to the total DOFs of the system. Thus,

$$\frac{\partial f(t)}{\partial \alpha_b} = L \frac{\partial \ddot{z}}{\partial \alpha_b} \tag{8.54}$$

Substituting Equation (8.54) into Equation (8.50), the wavelet coefficient sensitivity from measurement can be given as

$$\frac{\partial d_{j,k}}{\partial \alpha_b} = \int_R L \frac{\partial \ddot{z}}{\partial \alpha_b} \psi_{j,k}(t) dt \tag{8.55}$$

The solution algorithm

The structure is assumed to behave linearly before and after the occurrence of damage. D_0 and D_d are vectors of the wavelet coefficient of the two states of the structure, e.g. the intact and damaged states, respectively. $\partial D_0/\partial \alpha$ and $\partial D_d/\partial \alpha$ are the sensitivity matrices of the wavelet coefficient of the two corresponding states of the system. $\Delta \alpha$ is the vector of parameter changes of the system. Thus,

$$
\begin{aligned}
D_d - D_0 &= \frac{\partial D_0}{\partial \alpha} \cdot \Delta \alpha \\
&= S \cdot \Delta \alpha
\end{aligned}
\tag{8.56}
$$

The solution algorithm is similar to that for using the wavelet-packet component energy sensitivity in Section 8.3.1 with the following two convergence criteria.

$$\left\| \frac{\{d_{j,k}\}_{i+1} - \{d_{j,k}\}_i}{\{d_{j,k}\}_{i+1}} \right\| \leq toler\ 1, \quad \left\| \frac{\{\Delta\alpha_{i+1}\} - \{\Delta\alpha_i\}}{\{\Delta\alpha_{i+1}\}} \right\| \leq toler\ 2$$

where i refers to the ith iteration. The tolerance limits for both convergence criteria have been set equal to 1.0×10^{-6}.

8.3.3 *The wavelet packet transform sensitivity*

The wavelet packet transform was discussed in Section 8.1.2. Assuming that $f(t)$ is the signal to be transformed, the discrete wavelet packet transform (WPT) coefficients (Mallat and Hwang, 1992) of the signal can be obtained as

$$c_{j,k}^i = \int_{-\infty}^{+\infty} f(t)\psi_{j,k}^i(t)dt \tag{8.21}$$

where i, j and k are the wavelet packet number, the decomposition level and the translation respectively (Daubechies, 1988; 1992), and $\psi_{j,k}^i(t)$ satisfies the following relation

$$\psi_{j,k}^i(t) = 2^{-\frac{j}{2}}\psi^i(2^{-j}t - k) \tag{8.10}$$

where $\psi^i(t)$ is the ith wavelet packet with the following properties

$$\psi^{2i}(t) = \sqrt{2}\sum_{k=-\infty}^{+\infty} h(k)\psi^i(2t - k) \tag{8.11}$$

$$\psi^{2i+1}(t) = \sqrt{2}\sum_{k=-\infty}^{+\infty} g(k)\psi^i(2t - k), \quad i = 0, 1, \ldots, 2^j - 1 \tag{8.12}$$

where $h(k)$ and $g(k)$ are the low-pass and high-pass analysis filters. It is noted that $\psi_{0,k}^0$ can be defined as

$$\psi_{0,k}^0 = \psi^0(t - k) = \varphi(t - k) \tag{8.57}$$

where $\varphi(t)$ is the scale function of the wavelet.

The original signal can also be reconstructed from the WPT coefficients as

$$f(t) = \sum_{i=0}^{2^j-1} f_j^i(t), \quad \text{and} \quad f_j^i(t) = \sum_{k=-\infty}^{+\infty} c_{j,k}^i \psi_{j,k}^i(t) \tag{8.58}$$

where $f_j^i(t)$ denotes the signal component in the ith frequency band.

The WPT is applied to $\partial\ddot{x}_l/\partial\alpha_i$ and $\ddot{x}_l$ separately, where l denotes the lth DOF of the structure. The relation between $WPT(\partial\ddot{x}_l/\partial\alpha_i)$ and $WPT(\ddot{x}_l)$ was given in Section 8.3.2 as

$$WPT\left(\frac{\partial\ddot{x}_l}{\partial\alpha_i}\right) = \frac{\partial}{\partial\alpha_i}WPT(\ddot{x}_l) \tag{8.59}$$

where $WPT(\partial \ddot{x}_l/\partial \alpha_i)$ and $WPT(\ddot{x}_l)$ are the WPT coefficients of $\partial \ddot{x}_l/\partial \alpha_i$ and $\ddot{x}_l$, respectively. We can then form the sensitivity matrix of the WPT coefficients as

$$ S = \left[\frac{\partial}{\partial \alpha_1} WPT(\ddot{x}_l) \quad \frac{\partial}{\partial \alpha_2} WPT(\ddot{x}_l) \quad \cdots \quad \frac{\partial}{\partial \alpha_m} WPT(\ddot{x}_l) \right] \tag{8.60} $$

when the response sensitivity $\partial \ddot{x}_l/\partial \alpha_i$ is available.

8.3.4 Damage information from different wavelet bandwidths

Information from local damages may be carried in the signal within a particular bandwidth of the response and it is usually masked in the conventional approach of damage detection using modal or response data. The wavelet packet decomposition separates the response signal into packets of different bandwidths, and thus, enables the following study on the damage information from different wavelet packets.

A four-metre long simply supported concrete beam, 0.2 m wide and 0.3 m high, with a uniform rectangular cross-section subject to different types of excitations is considered. The elastic modulus and Poisson ratio are respectively 35.8 GPa and 0.197, and the density of the material is 2376.21 kg/m³. The beam is divided into twelve equal Euler–Bernoulli beam elements as shown in Figure 8.10. The vertical stiffnesses of the left and right supports are simulated with springs of 1.93×10^8 kN/m and 0.47×10^8 kN/m, respectively. The damping ratios for the first six modes are 3.1, 14.3, 7.6, 10.4, 1.6 and 1.4 %, respectively.

The intact beam is first excited with a triangular impulsive force applied at one-third point of the beam from the left support with 2197N peak value and it lasts for 0.005 seconds. The sampling rate is 2000 Hz. The first five seconds of acceleration response collected from node 4 of the beam is decomposed into four levels of wavelets and the associated wavelet packets. The wavelet packets are Fourier transformed with an FFT size of 8192 for an inspection of the frequency content, and the frequencies of the spectral peaks observed are shown in Table 8.1. Mode 4 cannot be detected from the spectrum because of the location of the sensor. Modes 5 and 6 are shifted greatly relative to the analytical values under the effect of the excitation. The first two thousand data points of the wavelet packets are used in the identification, and the first 0.25 second of acceleration response is shown in Figure 8.11.

Damage is then created in element 4 of the beam by reducing its bending rigidity by 20%, and the beam is subject again to the same excitation. The spectral frequencies obtained from the wavelet packets are also shown in Table 8.1. A downward shift in

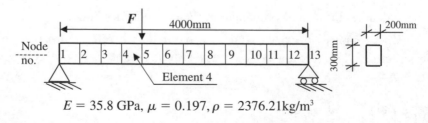

$$ E = 35.8 \text{ GPa}, \mu = 0.197, \rho = 2376.21 \text{kg/m}^3 $$

Figure 8.10 Simply supported concrete beam for the numerical study

the natural frequency of the two different states occurs in all the modes. The largest shift in the natural frequency between the two states of the beam is in the first mode, which is 1.61%. The frequency bandwidth of the decomposed wavelet packets are also shown in Table 8.1. All the packets have the same bandwidth of 62.5 Hz and it is noted that each of the 1st, 2nd, 3rd, 7th and 13th wavelet packets encompass one natural frequency of the structure.

The flexural rigidity in element 4 is varied with a 0% to 50% reduction in 2% decrements, and the WPT component energy is computed from the response collected at node 5 of the beam under the same impulsive excitation as for last study. The WPT

Table 8.1 Natural frequencies of the response from the two states of the structure

Natural Frequency (Hz)				Wavelet packet number	Frequency Bandwidth (Hz)
Mode no.	FEM	FFT on the WPT			
		Intact	Damaged		
1	30.94 ⟶	31.01	30.52	1st	0~62.5
2	100.78 ⟶	104.25	103.03	2nd	62.5~125
3	184.14 ⟶	185.06	184.57	3rd	125~187.5
—	—	—	—	4th	187.5~250
4	305.29	—	—	5th	250~312.5
—	—	—	—	6th	312.5~375
—	—	404.30	400.39	7th	375~437.5
5	468.48	—	—	8th	437.5~500
—	—	—	—	9th	500~562.5
—	—	—	—	10th	562.5~625
—	—	—	—	11th	625~687.5
6	706.31 ↘	—	—	12th	687.5~750
—	—	⟶ 767.09	766.11	13th	750~812.5
—	—	—	—	14th	812.5~875
—	—	—	—	15th	875~937.5
—	—	—	—	16th	937.5~1000

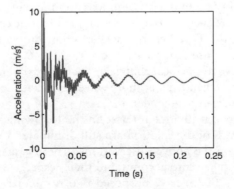

Figure 8.11 Acceleration response from node 4

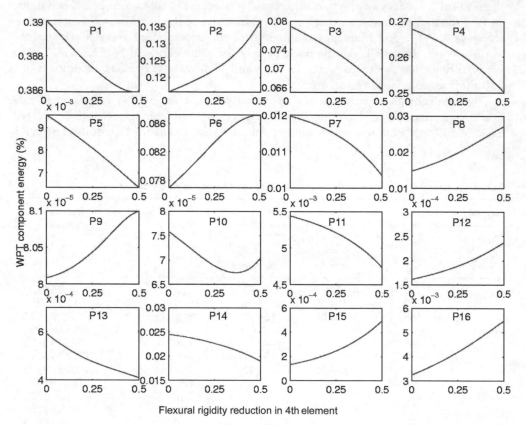

Figure 8.12 Variation of the wavelet-packet transform component energy under impulsive excitation

component energy is plotted against the damage extent in Figure 8.12 as a percentage of the total energy in the response. The wavelet-packet transform component energy can be categorized into two types: Type I energy varies monotonically with the damage extent and Type II energy exhibits a maximum or minimum in the variation. It is known that if the Type II packet energy alone is used in the identification, it gives non-unique solution. Fortunately, the Type I energy packets contribute a large proportion of the total energy of the response (packet numbers 1, 4, 2, 6, 3, 14, 8 and 7) and they are much more sensitive to the local change in the physical parameter than the Type II packet energy. It is recommended that those wavelet packets dominated by a structural vibration mode are selected for the identification.

All 16 packet-energy sensitivities are used in the following simulation studies. The accuracy and uniqueness of the solution are still dominated by the Type I WPT component energy based on the above observations. There is also no need to select the wavelet packets to be taken into account and their respective weightings in the analysis. Sun and Chang (2002b) have proposed two normalized packet-energy change parameters for the identification, which enhances the sensitivity of Type I packets, and

Table 8.2 Damage scenarios

Damage Scenario	Damage extent	Damage locations	Response type	Excitation type
1	10%	11th element	acceleration	Impulsive excitation at node 5
2	10%	11th element	strain	
3	10%	11th element	acceleration and strain	
4	5%,10%	6th, 7th element	acceleration	
5	5%,10%	6th, 7th element	strain	
6	5%,10%	6th, 7th element	acceleration and strain	
7	5%,10%,15%	3rd, 4th, 8th element	acceleration	
8	5%,10%,15%	3rd, 4th, 8th element	strain	
9	5%,10%,15%	3rd, 4th, 8th element	acceleration and strain	
10	5%,10%	6th, 7th element	acceleration	Sinusoidal excitation at node 5
11	5%,10%	6th, 7th element	strain	
12	5%,10%	6th, 7th element	acceleration and and strain	
13	10%	2nd element	acceleration with 5% noise	Impulsive excitation at node 4
14	5%, 10%	6th, 7th element		
15(a)	5%, 10%	6th, 7th element		Impulsive excitation at node 4 with −3% model error in all elements
15(b)	5%, 10%	6th, 7th element		

this will improve the final identified results from the above discussions. Also, a scheme for optimizing the best packets is needed.

Damage scenarios and their detection

Twelve damage scenarios were studied with different damage extents and from different types of response, as listed in Table 8.2. In fact, three patterns of damage (single damage, two adjacent damages and three damages) were considered and each damage pattern was identified from acceleration, strain and both types of response. The first nine scenarios were studied using the vertical response from node 7 when the beam is subject to the same impulsive excitation at one-third span as for last study. The results shown in Figure 8.13 show that either acceleration along or both acceleration and strain responses together can identify the damage accurately with a very small error. The bad results in Scenario 8 were also noticed. This gives numerical evidence that the strain response is less sensitive to local damage as observed in its formulation in Law et al., 2005.

Damage scenarios 10 to 12 were studied using the vertical measured responses from nodes 7 and 8 when the beam is subject to a sinusoidal excitation at one-third span at the frequency of 30 Hz close to the fundamental frequency of the beam. The magnitude of the force is 20N. Results in Figure 8.13 show that this arrangement could identify

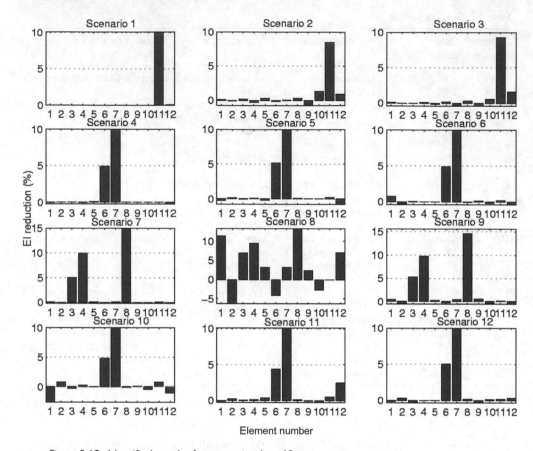

Figure 8.13 Identified results for scenarios I to I2

the damage very accurately with virtually no false alarms in other elements in the form of a large error. And it was noticed that the acceleration and strain responses, when used together, can give more accurate results than using either of the two separately as shown in Scenario 12 in Figure 8.13.

Effect of measurement noise and model error

Damage Scenarios 13 and 14, as listed in Table 8.2, were studied with 5% noise in the measured responses. Acceleration measurements from nodes 5 and 6 were used with impulsive excitation applied at node 4 as described previously, and the identified results are shown in Figure 8.14. The presence of noise seems not to adversely affect the identified result on the damaged elements. With the damage information distributed to a large number of time-scale wavelet packets, the solution of the respective system of equations implies an averaging of more samples, which is beneficial with respect to noise-contaminated data.

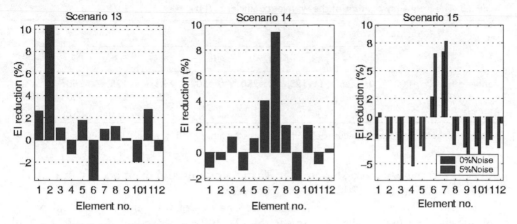

Figure 8.14 Identified results for scenarios 13 to 15

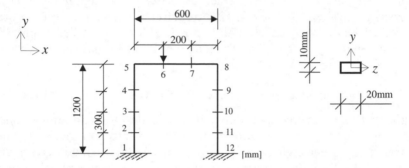

Figure 8.15 The one-storey frame structure

Scenarios 15(a) and (b) include a model error of an under-estimation in the flexural rigidity of the whole beam structure in the finite element model plus two adjacent damages in the beam with and without 5% noise in the measured acceleration. The excitation and measuring points are the same as for the last two scenarios. The identified results for the no noise case show a very accurate increase of approximately 3% in elements 3 to 5 and 9 to 11 and a reduction of 2% and 7.3% for the damaged elements 6 and 7. The identified 2% and 7.3% reductions are relative to the original modelled flexural rigidity, which is 97% of the intact value. However the identified results degenerate badly when 5% noise is included, but the damaged elements can still be localized with no false alarm in other elements.

8.3.5 Damage information from different wavelet coefficients

The one-storey plane frame structure as shown in Figure 8.15 and discussed in Section 7.2.3 serves for the numerical study. The structure is subject to a sinusoidal excitation $F(t) = 10\sin(12\pi t)$ N applied vertically at node 6. The columns are 1.2 m high and the cross-beam is 0.6 m long, and each member is 10 mm deep and 20 mm wide, with a

Table 8.3 The frequency content of the wavelet coefficients (Hz)

	Response	Wavelet							
		A4	A3	A2	A1	D4	D3	D2	D1
Bandwidth	0~1000	0~62.5	0~125	0~250	0~500	62.5~125	125~250	250~500	500~1000
Modes	f_0~f_{12}	f_0~f_2	f_0~f_3	f_0~f_6	f_0~f_8	f_3	f_4~f_6	f_7,f_8	f_9~f_{12}

Note: f_0 denotes the excitation frequency, f_1~f_{12} denotes the first 12 natural frequencies of the frame structure.

uniform rectangular cross-section. The elastic modulus and mass density of material are 69×10^9 N/m^2 and 2700 kg/m^3, respectively.

The finite element model of the structure consists of four and three equal beam-column elements in each vertical and horizontal member, respectively. The translational and rotational restraints at the supports are represented by large stiffnesses of 1.5×10^{10} kN/m and 1.5×10^9 kN-m/rad, respectively. Rayleigh damping is adopted for the system with $\xi = 0.01$. The first 12 natural frequencies of the structure are 13.09, 57.31, 76.7, 152.4, 196.5, 227.3, 374.7, 382.5, 580.2, 699.3, 765.3 and 983.3 Hz. The sampling frequency is 2000 Hz. The tolerance limits for both convergence criteria have been set equal to 1.0×10^{-6}.

8.3.6 *Frequency and energy content of wavelet coefficients*

The horizontal acceleration response computed at node 9 for a duration of one second after the application of the excitation is decomposed into four levels of Daubechies Db4 wavelets. The wavelets are divided into groups A and D with different bandwidth as shown in Table 8.3. Those in group A are the low-frequency wavelet coefficients, and those in group D are the high-frequency wavelet coefficients. Wavelets A1 and D1 are the largest scale wavelets and A4 and D4 are the smallest scale wavelets. The large-scale wavelets, D1 and A1, have better time resolution than the small-scale wavelets, D4 and A4, because of their wider bandwidths. The low-frequency wavelet coefficients were checked and were shown to be larger than those for the high-frequency wavelet coefficients, indicating a larger vibration energy in the low-frequency responses. This is because the low-frequency wavelets include the first few vibration modes of the structure, but the high-frequency wavelets include only some of the higher vibration modes of the structure.

Comparison with response sensitivity

The relative sensitivity of the wavelet coefficient to the parameter, S_{wc}, and the relative response sensitivity, S_r, are defined as

$$S_{wc} = \frac{\|\partial d_{j,k}/\partial \alpha_b\|}{\|d_{j,k}\|}, \quad S_r = \frac{\|\partial f(t)/\partial \alpha_b\|}{\|f(t)\|}$$

A comparison of the sensitivity of each set of wavelet coefficients and the response is compared and the results are shown in Table 8.4 for a perturbation in the flexural

Table 8.4 Comparison of response sensitivity to wavelet sensitivity

		Perturbation in the following finite element										
		1	2	3	4	5	6	7	8	9	10	11
S_r	Response	18.28	16.62	17.28	15.64	9.39	13.27	9.39	15.65	17.28	16.62	18.27
S_{wc}	A1	17.92	16.31	17.01	15.35	9.18	13.24	9.18	15.35	17.01	16.31	17.92
	A2	13.16	12.94	13.51	11.25	7.24	12.5	7.24	11.25	13.5	12.94	13.15
	A3	8.38	5.23	7.33	4.67	3.56	5.93	3.56	4.67	7.33	5.23	8.36
	A4	7.08	3.29	5.92	2.76	2.70	3.53	2.70	2.76	5.92	3.29	7.06
	D1	52.02	46.9	45.1	44.7	28.5	18.6	28.5	44.7	45.1	46.9	52.0
	D2	49.14	40.7	42.48	42.18	23.24	21.27	23.24	42.18	42.48	40.70	49.14
	D3	26.49	29.74	29.06	25.69	16.0	27.76	16.0	25.68	29.06	29.74	26.49
	D4	14.28	11.75	13.22	11.04	6.91	13.84	6.91	11.04	13.22	11.75	14.26

rigidity of each of the finite element. The response sensitivity is comparatively low compared with those from the wavelets. Large-scale wavelets are always more sensitive than small-scale wavelets, and high-frequency wavelet coefficients are more sensitive than low-frequency coefficients. The wavelet coefficient D1 has the highest sensitivity because of the contribution from the damage information carried by vibration modes 9 to 12, as shown in Table 8.3. While wavelet coefficient D2 has the second highest sensitivity, which is contributed to only by modes 7 and 8. This shows that modes 7 and 8 are the two more significant modes that carry much of the damage information.

The wavelet coefficient sensitivity is in general much higher than that of the response, apart from coefficients A3 and A4, which are contributed to by the first few modes. This shows that the lower vibration modes do not carry significant damage information.

Damage identification

The same plane frame structure as for last study is used. The excitation force $F(t) = 10 \sin(12\pi t)$N is applied at node 6 in the vertical direction. The horizontal acceleration response is computed at node 9 for a duration of a quarter of a second after the application of the excitation is used for the wavelet decomposition. The sampling rate is 2000 Hz and the following damage scenarios are with different percentage reductions in the flexural rigidity in an element.

- Scenario 1 – 5% reduction in element 2.
- Scenario 2 – 15% reduction in element 4.
- Scenario 3 – 5% and 10% reduction in elements 3 and 4, respectively.
- Scenario 4 – 10% reduction in element 1.
- Scenario 5 – 15%, 5% and 10% reduction in elements 3, 6 and 8, respectively.

The identified results obtained from the response sensitivity, each of the eight wavelet coefficients and a combination of wavelet coefficients A4, D1, D2, D3 and D4, which cover the whole frequency range of the response, are very close to the true value with a maximum error of identification in each scenario highlighted in Table 8.5. The number

Table 8.5 Error of identification in percentage

	Scenario 1	Scenario 2	Scenario 3	Scenario 4	Scenario 5
Response	−2.03/7	0.00/20	0.00, 0.00/20	−0.54/11	0.00, 0.00, 0.00/24
D1	0.00/17	−0.35/12	−0.44, −0.68/12	−0.31/9	0.04, −0.09, −0.52/12
D2	−0.41/9	−1.00/14	0.01, −2.88/14	−1.79/11	−0.09, 0.08, 0.10/13
D3	−0.57/9	0.02/16	−0.61, −0.24/16	−0.36/16	−0.25, −0.03, 0.16/11
D4	−0.01/20	0.00/28	−0.03, −0.01/28	−21.51/8	Fail
A1	−1.82/8	−0.31/13	−1.20, −0.09/13	−0.60/12	−0.85, −0.05, 0.12/16
A2	−0.57/7	0.00/10	−1.29, 0.08/10	−3.53/10	−0.49, −0.11, 0.18/14
A3	−1.07/8	−0.11/11	−0.53, 0.44/11	−3.33/12	1.08, 0.19, 0.11/66
A4	−0.47/9	−0.10/14	−2.60, 1.24/14	−1.84/18	Fail
A4 + D1 + D2 + D3 + D4	−2.12/7	−0.22/20	0.00, 0.00/20	−0.66/11	−0.65, 0.00, 0.03/16

Note: •/• denotes the error of identification (%)/required iteration number.

of iterations required for convergence in the different damage scenarios are also given in the table.

The performance of the combined group of wavelet coefficients is similar to the response sensitivity. Wavelet coefficients D4 and A4 perform badly in the cases of adjacent damages and with the damage adjacent to the support. More detailed inspection shows that component A4 consists of only the first two vibration modes of the structure and D4 consists of the 3rd vibration mode only. This again confirms the previous observation that the first few vibration modes do not carry significant information on the changes in the stiffness properties of the structure. In general, small-scale wavelet coefficients give less accurate results than the large-scale wavelet coefficients, and this is consistent with the observations from the last study.

Figure 8.16 shows the identified stiffness change in all the elements for Scenarios 3 to 5 using wavelet coefficients A1, A2, D1 and D2 in the identification. Both the damage location and severity are identified very accurately without any false alarm in other elements.

Effect of model error

The finite element model would not fully represent the real life structures with assumptions on the linearity, damping models, dynamic behaviour, joint flexibilities and constitutive laws of materials, etc. These assumptions are necessary to focus on the problem under study and to reduce the number of unknowns in the solution, or otherwise, including different models on the damping, damages, semi-rigid joints and different forms of finite element for the structure. The latter approach complicates the problem leading to computational errors in the identified results. However, the violation of the initial assumptions of the model would lead to bias errors which cannot be differentiated with those from computation and measurement noise. In many existing damage assessment techniques, engineering judgements were made on the structural behaviour and the standard finite element model of the structure is used. Parameters of the model are considered prone to model inaccuracy, and this effect is investigated in

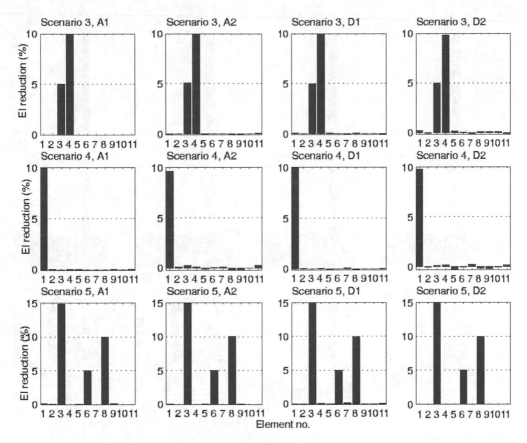

Figure 8.16 Identified results for scenarios 3 to 5

detail with the following list of model errors introduced into the finite element model of the plane frame. The same excitation force and sampling rate of signal as for last study is used. The horizontal response from node 9 is used for the decomposition. All the following scenarios are with 10% reduction in the flexural rigidity of element 3.

- Scenario 6 – both the support rotational and translational stiffnesses have been over-estimated ten times.
- Scenario 7 – 5% over-estimation in the flexural stiffness of all elements.
- Scenario 8 – 2% under-estimation in the density of material.
- Scenario 9 – the Rayleigh damping is over-estimated from 0.01 to 0.02.
- Scenario 10 – 10% under-estimation in the amplitude of excitation force.
- Scenario 11 – 5% over-estimation in the excitation frequency.
- Scenario 12 – includes all the model errors listed from scenarios 6 to 11.

The identified results for the stiffness changes in all the elements are shown in Figure 8.17 for Scenarios 6 to 8, where sensitivity of the response, wavelet coefficients

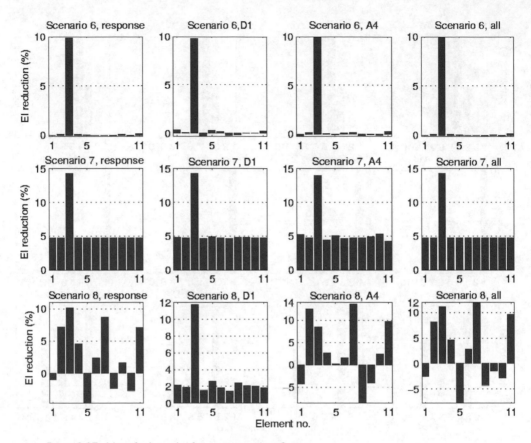

Figure 8.17 Identified results for scenarios 6 to 8

D1, A4 and a combination of A4, D1, D2, D3 and D4 are used in the identification. The results from the response and the combination of wavelet coefficients are similar; while wavelet coefficient A4 gives many false alarms, indicating that the model error of a mass density change cannot be detected using low frequency and low-scale wavelet coefficients. Wavelet coefficient D1 gives consistently very good results in these three damage scenarios. Results not shown indicate that wavelet D1 can also identify the damage element with similar accuracy with 3% under-estimation in the mass density, but there are also many false alarms in other elements with up to 7% stiffness reduction. All other wavelet coefficients fail in the identification of this damage scenario. In the case of Scenario 7, all the wavelets except D3 and D4 could identify the local damage in element 3 with similar results.

Figure 8.18 gives the results for Scenarios 9 to 12, using the response sensitivity and wavelet coefficient D1 sensitivity. The latter gives consistently very good results even in Scenario 12 when all the different types of model errors exist. It is noted that all the wavelets except wavelet D1 fail to give meaningful identified results for all the above scenarios.

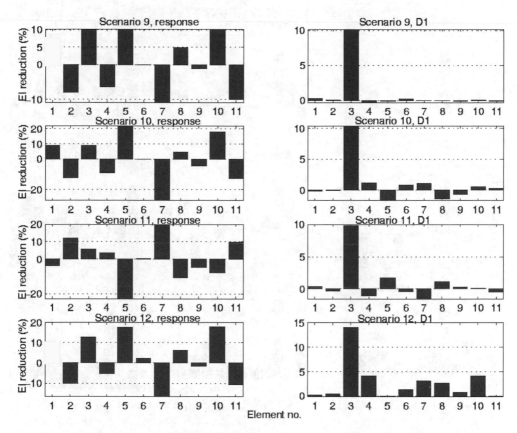

Figure 8.18 Identified results for scenarios 9 to 12

8.3.7 Noise effect

White noise is added to the calculated accelerations to simulate the polluted measurement time history. The same excitation force and sampling rate of the signal as for last study is used. The horizontal response from node 9 is used for the decomposition. Three damage scenarios as listed below are studied:

- Scenario 13 – 10% reduction in the flexural rigidity of element 3 with 5% noise.
- Scenario 14 – ditto, but with 10% noise.
- Scenario 15 – 10% reduction in element 3 with 1%, 2% and 3% noise, respectively, and with all the model errors as studied in Scenario 12.

The identified results from the response sensitivity, wavelet coefficient A1 and coefficient D1, are shown in Figure 8.19 for the first two scenarios, and only results from wavelet D1 are shown for Scenario 15. The 5% noise causes a smaller identified value in the damaged element and with false alarm in other elements. This effect becomes larger with noise, and high-frequency wavelet coefficient, D1, is found less resistant

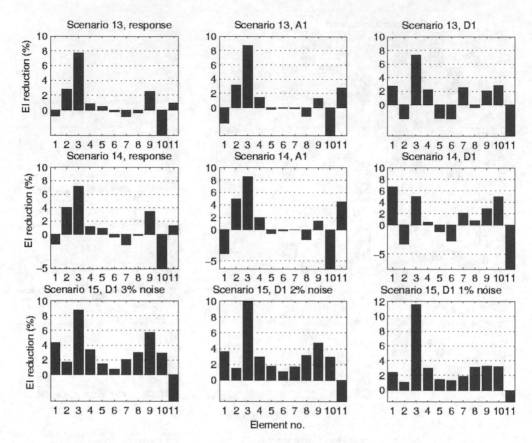

Figure 8.19 Identified results for scenarios 13 to 15

to the effect of random measurement noise than the response and the low-frequency wavelet coefficient, A1.

The random noise is noted to give random errors in the elements while the model errors lead to bias error in the finite elements as shown in Scenario 15. Acceptable results can be obtained only with 1% noise level. A small percentage of noise level can reduce the quality of identification greatly when model errors are included.

8.4 Approaches that are independent of input excitation

8.4.1 The unit impulse response function sensitivity

To reduce the dependence of the model, impulse response functions (IRFs) are considered instead of the response in the damage detection process because vibration responses are related to the excitation input. Impulse response functions are intrinsic functions of the system given the excitation location, and they can be extracted from the measured response. Existing impulse-response extraction techniques include Laplace transform-based extraction, conventional time-domain extraction and

FFT-based extraction (Robertson et al., 1998). FFT-based extraction is most commonly used to obtain the impulse response function. However, it has long been recognized that FFT-based vibration signal analysis exhibits several weakness.

In contrast to the FFT-based extraction procedure, which must process the data in both the time and frequency domains, the discrete wavelet transform (DWT)-based extraction procedure (Robertson et al., 1998) handles the experimental data only in the time domain. This involves the forward and inverse DWT plus an inversion operation, and is indeed preferable.

8.4.1.1 Wavelet-based unit impulse response

The equation of motion of a N DOFs damped structural system under the unit impulse excitation is

$$M\ddot{x} + C\dot{x} + Kx = D\delta(t) \tag{8.61}$$

where M, C and K are the $N \times N$ mass, damping and stiffness matrices, respectively; N is the number of DOFs of the system; D is the mapping matrix relating the force excitation location to the corresponding DOFs of the system; x, $\dot{x}$ and $\ddot{x}$ are the $N \times 1$ displacement, velocity and acceleration vectors, respectively; and $\delta(t)$ is the Dirac delta function. Assuming the system is in static equilibrium before the unit impulse excitation, then the forced vibration system with the unit impulse excitation can be converted to a free vibration system with the following initial conditions,

$$x(0) = 0, \quad \dot{x}(0) = M^{-1}D \tag{8.62}$$

Rewriting Equations (8.61) and (8.62), the unit impulse response function can be computed from the following:

$$\begin{cases} M\ddot{h} + C\dot{h} + Kh = 0 \\ h(0) = 0, \quad \dot{h}(0) = M^{-1}D \end{cases} \tag{8.63}$$

where h, $\dot{h}$ and $\ddot{h}$ are the unit impulse displacement, velocity and acceleration vectors, respectively. Using the Newmark method, the unit impulse response can easily be computed.

Expressing the local damage in the structural system in the form of $\Delta K = \sum_{i=1}^{ne} \alpha_i K_i$, where α_i is the fractional change in the stiffness of an element, and ne is the number of elements in the structure. Differentiating Equation (8.63) with respect to α_i, gives

$$\begin{cases} M\dfrac{\partial \ddot{h}}{\partial \alpha_i} + C\dfrac{\partial \dot{h}}{\partial \alpha_i} + K\dfrac{\partial h}{\partial \alpha_i} = -\dfrac{\partial K}{\partial \alpha_i}h \\ \dfrac{\partial h(0)}{\partial \alpha_i} = 0, \quad \dfrac{\partial \dot{h}(0)}{\partial \alpha_i} = \dfrac{\partial M^{-1}D}{\partial \alpha_i} \end{cases} \tag{8.64}$$

The sensitivities $\partial h/\partial\alpha_i$, $\partial\dot{h}/\partial\alpha_i$ and $\partial\ddot{h}/\partial\alpha_i$ can then be obtained from Equation (8.64) with Newmark Method.

A signal can be expressed by the Discrete Wavelet Transform (DWT) in terms of the local basis functions (Daubechies, 1992). Daubechies wavelets are used in the following studies as they satisfy the two crucial requirements: the orthogonality of local basis functions and second-order accuracy or higher, depending on the dilation expression adopted.

A function $f(t)$ can therefore be approximated in terms of its DWT as

$$f(t) = f_0^{DWT}\varphi(t) + f_1^{DWT}\psi(t) + f_2^{DWT}\psi(2t) + \cdots + f_{2^j+k}^{DWT}\psi(2^j t - k) \tag{8.65}$$

where $\varphi(t)$ and $\psi(t)$ are the scaling function and the mother wavelet function, respectively, and they satisfy the following relations,

$$\varphi(t) = \sqrt{2}\sum_{k=-\infty}^{+\infty}h(k)\varphi(2t-k); \quad \psi(t) = \sqrt{2}\sum_{k=-\infty}^{+\infty}g(k)\varphi(2t-k);$$

$$\psi_{j,k}(t) = 2^{-\frac{j}{2}}\psi(2^{-j}t - k) \tag{8.66}$$

where $h(k)$ and $g(k)$ are the low-pass and high-pass analysis filters, respectively, which are all constants; and $f_{2^j+k}^{DWT}$ is the wavelet transform coefficients. Because of the orthogonality on both the translation and scale of the Daubechies wavelets, the following is obtained,

$$\int \psi_{j,k}(t)\psi_{m,n}(t)dt = \delta_{j,m}\delta_{k,n} \tag{8.67}$$

Thus, for real wavelets from Equations (8.65) and (8.67),

$$f_0^{DWT} = \int f(t)\varphi(t)dt; \quad f_{2^j+k}^{DWT} = \int f(t)\psi_{j,k}(t)dt \tag{8.68}$$

The wavelet coefficient of the impulse-response function sensitivity has been shown (Law et al., 2006) equal to the first derivative of the wavelet coefficients of the impulse-response function with respect to the damage parameter, α_i, i.e.

$$DWT\left(\frac{\partial\ddot{h}}{\partial\alpha_i}\right) = \frac{\partial\ddot{h}^{DWT}}{\partial\alpha_i} \tag{8.69}$$

where $DWT(\partial\ddot{h}/\partial\alpha_i)$ is the discrete wavelet coefficient of $\partial\ddot{h}/\partial\alpha_i$. Therefore differentiating the second part of Equation (8.68) with respect to the stiffness parameter of an element gives

$$\left(\frac{\partial\ddot{h}}{\partial\alpha_i}\right)_{2^j+k}^{DWT} = \int \frac{\partial\ddot{h}}{\partial\alpha_i}\psi_{j,k}(t)dt \tag{8.70}$$

The sensitivity matrix can then be formed from Equation (8.70) as

$$S = \begin{bmatrix} \dfrac{\partial \ddot{h}_l^{DWT}}{\partial \alpha_1} & \dfrac{\partial \ddot{h}_l^{DWT}}{\partial \alpha_2} & \cdots & \dfrac{\partial \ddot{h}_l^{DWT}}{\partial \alpha_m} \end{bmatrix} \tag{8.71}$$

where $\ddot{h}_l^{DWT}$ is the DWT coefficient of $\ddot{h}_l$; $\ddot{h}_l$ denotes the acceleration impulse response function at location l; and m is the number of structural parameters.

8.4.1.2 Impulse response function via discrete wavelet transform

The equation of motion of an N DOFs damped structural system under general excitation is written as

$$M\ddot{x} + C\dot{x} + Kx = DF(t) \tag{8.72}$$

where $F(t)$ is the vector of excitation force. If the system has zero initial conditions, the solution of Equation (8.72) can be expressed as,

$$x(t) = \int_0^t h(t - \tau)F(\tau)d\tau \tag{8.73}$$

The acceleration response, $\ddot{x}_l(t_n)$, from location l at time t_n is,

$$\ddot{x}_l(t_n) = \int_0^{t_n} \ddot{h}_l(\tau) \cdot F(t_n - \tau)d\tau \tag{8.74}$$

Applying DWT to $\ddot{h}_l(\tau)$ and $F(t_n - \tau)$ respectively, gives

$$\ddot{h}_l(\tau) = \ddot{h}_{l,0}^{DWT}\,\varphi(\tau) + \ddot{h}_{l,1}^{DWT}\,\psi(\tau) + \cdots + \ddot{h}_{l,2^j+k}^{DWT}\psi(2^j\tau - k) \tag{8.75}$$

$$F(t_n - \tau) = F_0^{DWT}(t_n)\varphi(\tau) + F_1^{DWT}(t_n)\psi(\tau) + \cdots + F_{2^j+k}^{DWT}(t_n)\psi(2^j\tau - k) \tag{8.76}$$

Substituting Equations (8.75) and (8.76) into Equation (8.74),

$$\ddot{x}_l(t_n) = \int_0^{t_n} (\ddot{h}_{l,0}^{DWT}\,\varphi(\tau) + \ddot{h}_{l,1}^{DWT}\,\psi(\tau) + \cdots + \ddot{h}_{l,2^j+k}^{DWT}\psi(2^j\tau - k))$$
$$\times (F_0^{DWT}(t_n)\varphi(\tau) + F_1^{DWT}(t_n)\psi(\tau) + \cdots + F_{2^j+k}^{DWT}(t_n)\psi(2^j\tau - k))d\tau \tag{8.77}$$

The orthogonal condition shown in Equation (8.67) leads to

$$
\begin{cases}
\displaystyle\int_0^{t_n} \varphi(t)\varphi(t)dt = 1 \\[2mm]
\displaystyle\int_0^{t_n} \varphi(t)\psi(2^j t - k)dt = 0 \\[2mm]
\displaystyle\int_0^{t_n} \psi(2^j t - k)\psi(2^m t - n)dt = \delta_{j,m}\delta_{k,n}/2^j
\end{cases}
\tag{8.78}
$$

Substituting Equation (8.78) into Equation (8.77), gives

$$
\ddot{x}_l(t_n) = \overset{\bullet\bullet DWT}{h_{l,0}} F_0^{DWT}(t_n) + \overset{\bullet\bullet DWT}{h_{l,1}} F_1^{DWT}(t_n) + \cdots + \overset{\bullet\bullet DWT}{h_{l,2^j+k}} F_{2^j+k}^{DWT}(t_n)/2^j
\tag{8.79}
$$

Rewriting Equation (8.79) in matrix form,

$$
\ddot{x}_l(t_n) = F^{DWT}(t_n)\overset{\bullet\bullet DWT}{h_l}
\tag{8.80}
$$

where

$$
F^{DWT}(t_n) = [\, F_0^{DWT}(t_n) \quad F_1^{DWT}(t_n) \quad \cdots \quad F_{2^j+k}^{DWT}(t_n)/2^j \,],
$$

$$
\overset{\bullet\bullet DWT}{h_l} = \begin{bmatrix} \overset{\bullet\bullet DWT}{h_{l,0}} & \overset{\bullet\bullet DWT}{h_{l,1}} & \cdots & \overset{\bullet\bullet DWT}{h_{l,2^j+k}} \end{bmatrix}^T
$$

Equation (8.80) can be rewritten as follows for a time series

$$
\ddot{x}_l = F^{DWT}\overset{\bullet\bullet DWT}{h_l}
\tag{8.81}
$$

where $\ddot{x}_l = \begin{bmatrix} \ddot{x}_l(t_1) & \ddot{x}_l(t_2) & \cdots & \ddot{x}_l(t_n) \end{bmatrix}^T$, $F^{DWT} = \begin{bmatrix} F^{DWT}(t_1) \\ F^{DWT}(t_2) \\ \cdots \\ F^{DWT}(t_n) \end{bmatrix}$

Finally, $\overset{\bullet\bullet DWT}{h_l}$ can be computed in the form of a pseudo-inverse as

$$
\overset{\bullet\bullet DWT}{h_l} = (F^{DWT^T} F^{DWT})^{-1} F^{DWT^T} \ddot{x}_l
\tag{8.82}
$$

8.4.1.3   Solution algorithm

$\overset{\bullet\bullet DWT}{h_{l0}}$ and $\overset{\bullet\bullet DWT}{h_{ld}}$ are vectors of the DWT coefficient of the impulse response function from the two states of the structure, i.e. the intact and damaged states, respectively. $\partial \overset{\bullet\bullet DWT}{h_{l0}} /\partial\alpha$ is the sensitivity matrix of the DWT coefficient with respect to the local damage with reference to the intact state. $\Delta\alpha$ is the vector of parameter changes of the system. Thus the identification equation is

$$\ddot{h}_{ld}^{DWT} - \ddot{h}_{l0}^{DWT} = \frac{\partial \ddot{h}_{l0}^{DWT}}{\partial \alpha} \Delta \alpha$$

$$= S \Delta \alpha \tag{8.83}$$

The solution algorithm is similar to that for the wavelet-packet energy sensitivity with the following two criteria for convergence.

$$\left\| \frac{\{\ddot{h}_l^{DWT}\}_{i+1} - \{\ddot{h}_l^{DWT}\}_i}{\{\ddot{h}_l^{DWT}\}_{i+1}} \right\| \leq toler\ 1, \quad \left\| \frac{\{\Delta\alpha_{i+1}\} - \{\Delta\alpha_i\}}{\{\Delta\alpha_{i+1}\}} \right\| \leq toler\ 2$$

where i refers to the ith iteration. The tolerance limits for both convergence criteria have been set equal to 1.0×10^{-6}.

8.4.1.4 Simulation study

The 31-bar truss, shown in Figure 8.20, is modelled using 31 finite elements without internal nodes in the bars giving 28 DOFs. The cross-sectional area of the bar is $0.0025\,m^2$. Damage in the structure is introduced as a reduction in the stiffness of individual bars, but the inertial properties are unchanged. The translational restraints at the supports are represented by large stiffnesses of $1.0 \times 10^{10}\,kN/m$. Rayleigh damping is adopted for the system with $\xi_1 = 0.01$ and $\xi_2 = 0.01$. The first 12 natural frequencies of the structure are 36.415, 75.839, 133.608, 222.904, 249.323, 358.011, 372.509, 441.722, 477.834, 507.943, 538.1246 and 547.393 Hz. The sampling frequency is 2000 Hz. The tolerance limits for both convergence criteria have been set equal to 1.0×10^{-6}. The excitation is applied in the downward direction at node 5 while the vertical acceleration measurement at node 4 is recorded as shown in Figure 8.20.

Damage identification with model error and noise effect

Two different excitations are used in the different states of the structure. The excitation for the intact state is a triangular impulsive force with 320.4N peak value and it lasts for 0.005 seconds. The excitation for the damaged state is a sinusoidal excitation of $F = 20 \sin(40\pi t)$N. The sampling rate is 2000 Hz and the first 0.25 seconds of the

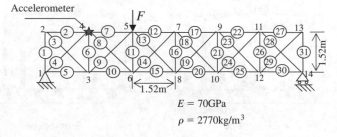

$E = 70GPa$

$\rho = 2770kg/m^3$

Figure 8.20 Thirty-one-bar truss structure

response is used for the damage identification. The impulse-response function wavelet coefficient sensitivity matrix is computed from the analytical model using Equations (8.63) and (8.71).

Different types of model errors are introduced into the finite element model of the truss structure as listed under Scenarios 1 to 4 in Table 8.6. Since the IRF is an intrinsic function of the structure that is independent of the input excitation, the acceleration response and excitation time histories are repeatedly 'measured' to obtain a set of IRFs from Equation (8.80). One sample and twenty samples of IRFs are used, respectively for scenarios without and with noise effect. The results in Figure 8.21 show that both the damage location and extent can be identified but with false positives in several other elements in the last scenario with the combined noise and model error effects. The presence of random noise in the 'measured' data is noted to amplify the erroneous effect due to the model errors.

8.4.1.5 Discussions

The method of damage detection using IRFs has been demonstrated with high accuracy in scenarios with multiple damages when there is no noise. This performance surpasses

Table 8.6 Damage scenarios

Damage scenario	Damage extent	Damage locations	Noise	Model error
1	5%, 10%	22nd, 26th element	no	2% reduction in the stiffness in all elements
2	5%, 10%	22nd, 26th element	no	50% increase in the support stiffness at two supports
3	5%, 10%	22nd, 26th element	no	2% decrease in the Rayleigh damping coefficients
4	5%, 10%	22nd, 26th element	5%	Include all the above model errors

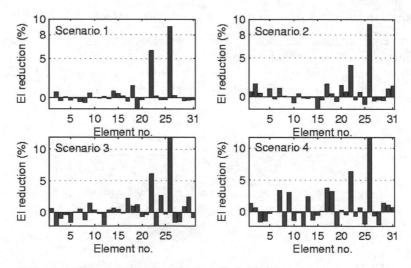

Figure 8.21   Identified results for scenarios 1 to 4

those from other existing damage detection methods. But the identified results suffer from the accompanying noise effect and initial model errors when dealing with real structures. The noise effect can be further reduced by averaging and taking a longer duration of measurement from more and better-located sensors. However, the initial model errors associated with a fixture in a joint, material nonlinearity, cracks and fissures in a member, etc. cannot be addressed with the present approach of modelling the damage with an averaged damage parameter. Separate damage models should be included in the initial model for proper identification. However, the wavelet is known to have a broad bandwidth of responses, and further benefit can be sought by employing sensitivity for wavelets from different bandwidths in the inverse analysis, as the noise effect and model error would have different contributions to the different bandwidth of the responses (Law et al., 2006).

8.4.2 The covariance sensitivity

This section presents one approach which is different to that shown in Section 8.4.1 for a general type of excitation. The excitation under consideration is ambient white noise excitation, which usually exists in the environment with large-scale infrastructures. The covariance of responses is expressed explicitly as a function of the impulse response function of the system under single or multiple ambient excitations. The sensitivity of the covariance of responses with respect to the physical parameters of the structure is derived analytically.

8.4.2.1 Covariance of measured responses

The equation of motion of the NDOFs damped structural system under general excitation is

$$M\ddot{x} + C\dot{x} + Kx = D_k F_k(t) \tag{8.84}$$

where $F_k(t)$ is the excitation force at the kth DOF. If the system has zero initial conditions, the solution of Equation (8.84) can be expressed as,

$$x_k(t) = \int_{-\infty}^{t} h_k(t - \tau)F_k(\tau)d\tau \tag{8.85}$$

where $h_k(t)$ is the vector of impulse response of the system under a unit impulse excitation at the kth DOF. The measured acceleration responses $\ddot{x}_{pk}(t)$ from location p at time t and $\ddot{x}_{qk}(t+\tau)$ from location q at time $t + \tau$ are, respectively,

$$\ddot{x}_{pk}(t) = \int_{-\infty}^{t} \ddot{h}_{pk}(t - \sigma)F_k(\sigma)d\sigma, \quad \ddot{x}_{qk}(t+\tau) = \int_{-\infty}^{t+\tau} \ddot{h}_{qk}(t + \tau - \sigma)F_k(\sigma)d\sigma \tag{8.86}$$

The cross-correlation function, $R_{pqk}(\tau)$, relating the two measured responses is given by Bendat and Piersol (1980) as

$$R_{pqk}(\tau) = E\{\ddot{x}_{pk}(t)\ddot{x}_{qk}(t + \tau)\} \tag{8.87}$$

where $E\{\ \}$ indicates the expectation operator. Equation (8.87) can further be written as

$$R_{pqk}(\tau) = \lim_{T \to \infty} \frac{1}{T} \int_0^T \ddot{x}_{pk}(t)\ddot{x}_{qk}(t + \tau)dt$$

$$= \lim_{NN \to \infty} \frac{1}{NN} \sum_{n=0}^{NN} \ddot{x}_{pk}(n\Delta t)\ddot{x}_{qk}(n\Delta t + \tau) \qquad (8.88)$$

where NN is the number of data points within the duration T under studied. The cross-correlation function $R_{pqk}(\tau)$ is then obtained from two measured responses of a structure.

8.4.2.2 When under single random excitation

Alternatively, $R_{pqk}(\tau)$ can be formulated in terms of the physical structural parameters of the system. The accelerations $\ddot{x}_{pk}(t)$ and $\ddot{x}_{qk}(t + \tau)$ are calculated for the N DOFs damped structural system defined by Equation (8.84) using the Newmark method and direct differentiation. Substituting the calculated accelerations into Equation (8.87), gives

$$R_{pqk}(\tau) = E\left\{ \int_{-\infty}^{t} \ddot{h}_{pk}(t - \sigma_1)F_k(\sigma_1)d\sigma_1 \int_{-\infty}^{t+\tau} \ddot{h}_{qk}(t + \tau - \sigma_2)F_k(\sigma_2)d\sigma_2 \right\} \qquad (8.89)$$

With random excitation, $F_k(\bullet)$ is random in Equation (8.89), and the equation can be rewritten as

$$R_{pqk}(\tau) = \int_{-\infty}^{t} \int_{-\infty}^{t+\tau} \ddot{h}_{pk}(t - \sigma_1)\ddot{h}_{qk}(t + \tau - \sigma_2)E(F_k(\sigma_1)F_k(\sigma_2))d\sigma_1 d\sigma_2 \qquad (8.90)$$

F_k is assumed to be of white noise distribution, and the autocorrelation function of F_k is (Bendat and Piersol, 1980)

$$E(F_k(\sigma_1)F_k(\sigma_2)) = S_k \delta(\sigma_1 - \sigma_2) \qquad (8.91)$$

where S_k is a constant and $\delta(t)$ is the Dirac delta function. Substituting Equation (8.91) into (8.90) with $\int_{-\infty}^{+\infty} f(t)\delta(t)dt = f(0)$, gives

$$R_{pqk}(\tau) = S_k \int_{-\infty}^{t} \ddot{h}_{pk}(t - \sigma_1)d\sigma_1 \int_{-\infty}^{t+\tau} \ddot{h}_{qk}(t + \tau - \sigma_2)\delta(\sigma_1 - \sigma_2)d\sigma_2$$

$$= S_k \int_{-\infty}^{t} \ddot{h}_{pk}(t - \sigma_1)\ddot{h}_{qk}(t + \tau - \sigma_1)d\sigma_1 \qquad (8.92a)$$

or,

$$R_{pqk}(\tau) = S_k \int_0^{+\infty} \ddot{h}_{pk}(t)\ddot{h}_{qk}(\tau + t)dt \qquad (8.92b)$$

It should be noted that $\overset{\bullet\bullet}{h}_{pk}(t)$ is an intrinsic function of the structure and is dependent only on the excitation location. Equation (8.92b) also shows that $R_{pqk}(\tau)$ has the same property as $\overset{\bullet\bullet}{h}_{pk}(t)$. The auto-correlation function can also be derived in the same way by putting $q = p$ in Equation (8.92). Equation (8.92b) can then be computed directly by integration. This formulation relates the cross-correlation function with the physical structural system via the impulse response function.

8.4.2.3 When under multiple random excitations

When the N DOFs damped structural system is under multiple excitations, Equation (8.84) can be written as,

$$M\overset{\bullet\bullet}{x} + C\overset{\bullet}{x} + Kx = D_1 F_1(t) + D_2 F_2(t) + \cdots D_N F_N(t) \tag{8.93}$$

where $D_i = [0, 0, \ldots, 1, \ldots, 0]^T$, where the ith element in D_i equals one and others equal zero. If there is no excitation at the ith DOF of the structure, $F_i(t) = 0$. Based on linear superposition theory and from zero initial conditions, Equation (8.86) gives

$$
\begin{aligned}
\overset{\bullet\bullet}{x}_p(t) &= \int_{-\infty}^{t} \overset{\bullet\bullet}{h}_{p1}(t-\sigma)F_1(\sigma)d\sigma + \int_{-\infty}^{t} \overset{\bullet\bullet}{h}_{p2}(t-\sigma)F_2(\sigma)d\sigma + \cdots \\
&\quad + \int_{-\infty}^{t} \overset{\bullet\bullet}{h}_{pN}(t-\sigma)F_N(\sigma)d\sigma \\
\overset{\bullet\bullet}{x}_q(t) &= \int_{-\infty}^{t} \overset{\bullet\bullet}{h}_{q1}(t-\sigma)F_1(\sigma)d\sigma + \int_{-\infty}^{t} \overset{\bullet\bullet}{h}_{q2}(t-\sigma)F_2(\sigma)d\sigma + \cdots \\
&\quad + \int_{-\infty}^{t} \overset{\bullet\bullet}{h}_{qN}(t-\sigma)F_N(\sigma)d\sigma
\end{aligned}
\tag{8.94}
$$

where $\overset{\bullet\bullet}{x}_p(t)$ and $\overset{\bullet\bullet}{x}_q(t)$ are the acceleration responses from locations p and q, respectively; and $\overset{\bullet\bullet}{h}_{pi}(t)$ and $\overset{\bullet\bullet}{h}_{qi}(t)$ are unit impulse acceleration responses at time t from locations p and q, respectively. Then the cross-correlation functions of $\overset{\bullet\bullet}{x}_p(t)$ and $\overset{\bullet\bullet}{x}_q(t+\tau)$ can be obtained similar to Equation (8.89) as

$$R_{pqk}(\tau) = E\left\{ \sum_{i=1}^{N} \int_{-\infty}^{t} \overset{\bullet\bullet}{h}_{pi}(t-\sigma_1)F_i(\sigma_1)d\sigma_1 \sum_{j=1}^{N} \int_{-\infty}^{t+\tau} \overset{\bullet\bullet}{h}_{qj}(t+\tau-\sigma_2)F_j(\sigma_2)d\sigma_2 \right\} \tag{8.95a}$$

Equation (8.95a) can be rewritten similar to Equation (8.90) as

$$R_{pqk}(\tau) = \sum_{i=1}^{N}\sum_{j=1}^{N} \int_{-\infty}^{t+\tau} \int_{-\infty}^{t} \overset{\bullet\bullet}{h}_{pi}(t+\tau-\sigma_1)\overset{\bullet\bullet}{h}_{qj}(t-\sigma_2)E(F_i(\sigma_1)F_j(\sigma_2))d\sigma_1 d\sigma_2 \tag{8.95b}$$

With random excitation, $F_i(\sigma_1)$ and $F_j(\sigma_2)$ are white noise functions, and

$$E(F_i(\sigma_1)F_j(\sigma_2)) = S_i\delta(i-j)\delta(\sigma_1-\sigma_2) \tag{8.96}$$

where S_i is a constant determined from the amplitude level of the random excitation. Substituting Equation (8.96) into Equation (8.95b), gives

$$R_{pq}(\tau) = \sum_{i=1}^{N}\sum_{j=1}^{N}\int_{-\infty}^{t+\tau}\int_{-\infty}^{t}\ddot{h}_{pi}(t+\tau-\sigma_1)\ddot{h}_{qj}(t-\sigma_2)S_i\delta(i-j)\delta(\sigma_1-\sigma_2)d\sigma_1 d\sigma_2$$

$$\tag{8.95c}$$

Considering the property of the Dirac delta function, Equation (8.95c) can be written as

$$R_{pq}(\tau) = \sum_{k=1}^{N}S_k\int_{0}^{+\infty}\ddot{h}_{pk}(t)\ddot{h}_{qk}(t+\tau)dt \tag{8.95d}$$

Comparing Equation (8.95d) with Equation (8.92b), the former reduces into Equation (8.95d) for $(S_k = 0, (i \neq k))$ for single excitation. Equation (8.95d) can further be simplified as

$$R_{pq} = \sum_{k=1}^{N}S_k\ddot{h}_{qk}^{DWT}\ddot{h}_{pk}^{DWT} \tag{8.97}$$

where

$$R_{pq} = \begin{bmatrix} R_{pq}(\tau_0) \\ R_{pq}(\tau_1) \\ \cdots \\ R_{pq}(\tau_n) \end{bmatrix}, \ddot{h}_{pk}^{DWT} = \begin{bmatrix} \ddot{h}_{pk,0}^{DWT} \\ \ddot{h}_{pk,1}^{DWT} \\ \cdots \\ \ddot{h}_{pk,2^j+l}^{DWT} \end{bmatrix}, \ddot{h}_{qk}^{DWT} = \begin{bmatrix} \ddot{h}_{qk}^{DWT}(\tau_0) \\ \ddot{h}_{qk}^{DWT}(\tau_1) \\ \cdots \\ \ddot{h}_{qk}^{DWT}(\tau_n) \end{bmatrix},$$

$$\ddot{h}_{qk}^{DWT}(\tau_i) = \begin{bmatrix} \ddot{h}_{qk,0}^{DWT}(\tau_i) \\ \ddot{h}_{qk,1}^{DWT}(\tau_i) \\ \cdots \\ \ddot{h}_{qk,2^j+l}^{DWT}(\tau_i)/2^j \end{bmatrix}^{T}$$

8.4.2.4 Sensitivity of the cross-correlation function

The sensitivity of the cross-correlation function $R_{pq}(\tau)$ can be obtained from Equation (8.95d) as

$$\frac{\partial R_{pq}(\tau)}{\partial \alpha_i} = \sum_{k=1}^{N} S_k \left(\int_0^{+\infty} \frac{\partial \ddot{h}_{pk}(t)}{\partial \alpha_i} \ddot{h}_{qk}(t+\tau)dt + \int_0^{+\infty} \ddot{h}_{pk}(t)\frac{\partial \ddot{h}_{qk}(t+\tau)}{\partial \alpha_i}dt \right) \qquad (8.98)$$

where $\partial \ddot{h}_{pk}(t)/\partial \alpha_i$ and $\partial \ddot{h}_{qk}(t+\tau)/\partial \alpha_i$ are obtained numerically from Equation (8.64) in Section 8.4.1.

8.5 Condition assessment including the load environment

8.5.1 Sources of external excitation

All structures are subject to ambient excitations like wind, rain, ground micro-tremor and temperature effects as well as the operating loads. The environmental forces are usually small, while the operating load is significant. Most existing methods of condition assessment of a structure require an input which may be the ambient environmental forces or artificial forced excitation. The operation loads are usually treated as random forces or just ignored. This practice may be appropriate for a large structure such as a suspension bridge but not for the usual types of infrastructure such as a box-section bridge deck. The provision of sufficiently large energy input for the identification of such a structure is formidable. Methods that include the moving operating load in the system identification are scarce but the work by the authors on the moving load identification should be referred to. The following sections present two methods making use of ground excitation and ambient excitation as external excitation for the condition assessment of a structure.

8.5.2 Under earthquake loading or ground-borne excitation

This section presents an approach where an earthquake loading or micro-tremor is used as an excitation force for the condition assessment. The wavelet-packet transform coefficient sensitivity given in Section 8.3.3 is studied particularly with a short duration of data sampled at a low frequency. Although the simulation studies shown below are discussed with reference to ground micro-tremor excitation, the approach can be used for the case with ground-borne blast excitation or traffic-induced excitations, which are easily available in practice.

8.5.2.1 Simulation studies

The five-bay three-dimensional frame structure studied in Section 7.4 and shown in Figure 8.22 serves for the numerical study. The finite element model consists of 37 three-dimensional Euler beam elements and 17 nodes. The length of all the horizontal, vertical and diagonal tube members between the centres of two adjacent nodes is exactly 0.5 m. The material and geometrical properties of the frame member are referred to Table 7.1. The structure orients horizontally and is fixed into a rigid support

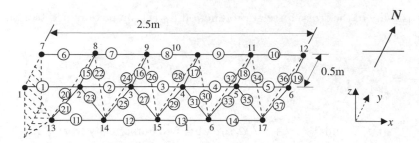

Figure 8.22 A five-bay three-dimensional frame structure

at three nodes at one end. Each node has six DOFs, and there are 102 DOFs for the whole structure. The elastic modulus of material of all the elements is taken as unknown in the identification.

The translational and rotational restraints at the supports are represented by large stiffnesses of 1.5×10^{11} kN/m and 1.5×10^{10} kN-m/rad, respectively, in six directions. Rayleigh damping is adopted for the system with $\xi_1 = 0.01$ and $\xi_2 = 0.005$. The first 12 natural frequencies of the structure are 9.21, 28.26, 33.71, 49.01, 49.72, 71.02, 89.80, 153.93, 194.33, 209.80, 256.51 and 274.82 Hz from the eigenvalue analysis of the structure. The sampling frequency is 200 Hz.

The structure is subject to the El-Centro ground motion in three directions with the magnitude reduced 100 times as plotted in Figure 8.23 to simulate a micro-tremor. The response of the structure is computed at all the DOFs and is subsequently re-sampled in the ratio 1 in 4, corresponding to a sampling rate of 50 Hz to form a subset of the response. The vertical acceleration response at node 5 of the structure is recorded for a duration of ten seconds after the excitation begins. The response is decomposed into four levels of Daubechies Db4 wavelet packets, which are 16 signal components represented by their WPT coefficients, and each has the same frequency bandwidth of 1.5625 Hz.

8.5.2.2 *The sensitivities*

The sensitivities of the calculated response with respect to the elastic modulus of material are also computed from Equation (7.2) and re-sampled again in the ratio 1 in 4, and the corresponding WPT coefficients with respect to the elastic modulus of material of the 4th and 20th elements of the structure are shown in Figure 8.24. The WPT coefficient sensitivities are arranged in a vector in their order of frequency band. The number of WPT coefficients is seen to be more than the number of original data because in the extraction of the wavelets, zero padding is performed at the end of the data series in case the number of original data cannot be equally divided. It can be seen from Figure 8.24 that some of the WPT coefficients are more sensitive than the response (Lu and Law, 2006a). Hence, the WPT coefficients from the 6th to 16th sets of wavelets are used to detect damage in the following simulation study for a better performance (Law et al., 2006). These coefficients correspond to a higher frequency range and are more sensitive to a stiffness change.

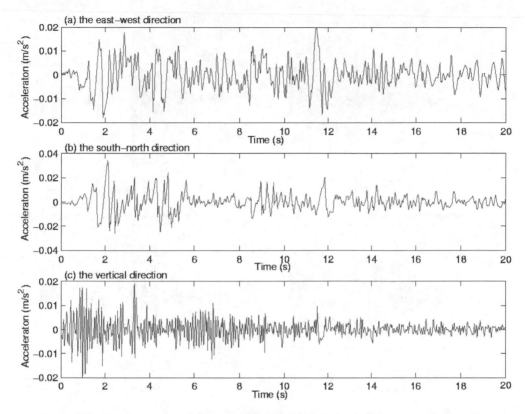

Figure 8.23 Excitation at the support of the structure.

8.5.2.3   *Damage identification from WPT sensitivity and response sensitivity*

Damage Scenarios 1 to 5, as listed in Table 8.7, are studied encompassing the cases with single and multiple damages, with and without noise and initial model error. Ten seconds of the response are used, apart from Scenario 5 where the first 50 seconds of response is used with the low sampling frequency of 10 Hz. The tolerance limits for both convergence criteria have been set equal to 1.0×10^{-8}, apart from Scenario 1 where 1.0×10^{-6} is adopted. The 'measured' acceleration responses are obtained through computation using the Newmark method with a time step of 0.005 seconds between two consecutive time instances. The same excitation force, sampling rate of signal, convergence criteria, measured location and WPT coefficients are used for all the studies, unless otherwise stated.

The identified results from WPT coefficients for Scenario 1 without noise and model errors are shown in Figure 8.25, and they are very close to the true values. The results obtained from the response sensitivity are also close to the true values at the damage locations, but there are a number of false alarms in other elements, such as element 28. The response sensitivities are obtained from Equation (7.2) using the Newmark method. It is noted that the convergence criteria is 1.0×10^{-6} for this scenario. If the

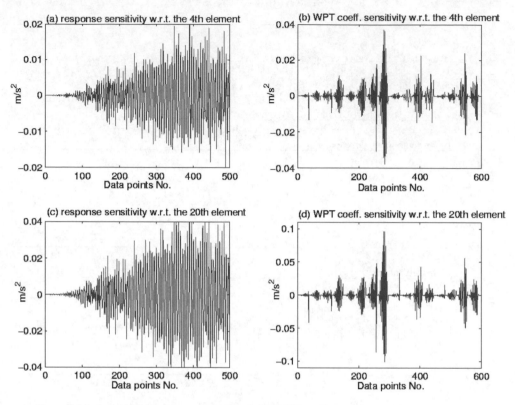

Figure 8.24   Sensitivities of the response and the WPT coefficients

Table 8.7   Damage scenarios

Scenario	Damage	Noise	Model error	Sampling rate (Hz)
1	5% in element 16 10% in element 26 10% in element 27	–	–	50
2	5% in element 16 10% in element 26 10% in element 27	–	1% increase in the elastic modulus of all elements	50
3	10% in element 7	5%	–	50
4	5% in element 26 5% in element 27	5%	1% increase in the elastic modulus of all elements	50
5	5% in element 16 10% in element 26 10% in element 27	–	–	10
6	5% in element 16 10% in element 26 10% in element 27	–	–	100, 200

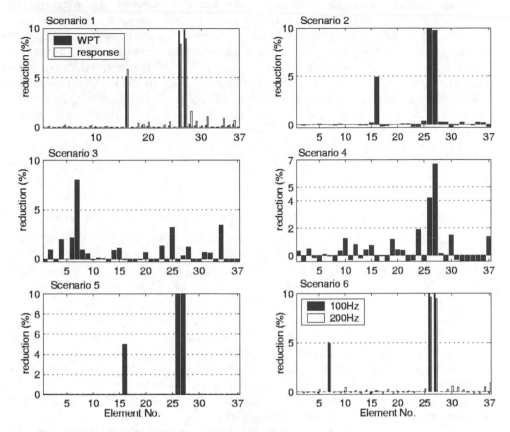

Figure 8.25 Identified results for scenarios 1 to 6

iteration for the identification is allowed to continue with a convergence criteria of 1.0×10^{-8}, the false alarms disappear and the results obtained are very similar to those from the WPT coefficients. This indicates that the response sensitivity can also be used to identify damage accurately but converge more slowly than the WPT coefficients sensitivity. This can be explained by the fact that the selected WPT coefficients obtained from the higher frequency response components are more sensitive to damage than the response sensitivity (Law et al., 2006). This also demonstrates that by a proper selection of the WPT coefficients, the non-sensitive components can be removed, resulting in a higher accuracy and faster convergence of the results.

Effect of model error and noise

In this study, only an error in the elastic modulus of material of an element is considered. Scenario 2 has 1% over-estimation in the elastic modulus of material of each member of the structure. The identified results in Figure 8.25 show that the method can tolerate some model error with good accuracy.

Damage Scenario 3 has 5% noise in the 'measured' acceleration. The location and extent of the damage can be identified with some accuracy as shown in Figure 8.25, but there are also alarms in several adjacent elements. The random noise is noted to give error in the identified damage of the damaged element, although a good indication of the damage location is still achieved. This shows that the identified results are easily corrupted by measurement noise, and noise reduction prior to identification is needed.

Scenario 4 is studied with damage in two adjacent elements to test the resolution capability of the approach with close proximity of two damages. The results in Figure 8.25 indicate that WPT coefficients can identify the damage elements quite accurately but with some error in the damage magnitude. It is believed that more spatial information of the structure can be captured from using responses from more locations of the structure to mitigate the noise effect.

Performance from a subset of the measured response

In practice, most of the on-line health monitoring systems operate with a high sampling frequency, sufficient to cover a wide range of the structural response. Scenario 5 is on the use of a subset of the original response re-sampled in the ratio 1 in 20. All sixteen sets of WPT coefficients are used in the identification. The identified results in Figure 8.25 show that this subset of the 'measured' response yields very accurate results.

It should be noted that for all the Scenarios in Table 8.7, the response obtained from the equation of motion of the structure contains components from all the structural vibration modes with frequencies below 100 Hz, and the re-sampling of the response only results in a subset of the response measurement. The time-domain approach does not depend mainly on the frequency information of the measured response as the frequency-domain method does, and it also makes use of the vibration amplitude in the solution process as in the following discussions.

The amplitude of the structural response, when expressed in the form of a Duhamel integral of $\sum_{i=1}^{n} \int \frac{e^{-\xi_i \omega_i \tau}}{\omega_{id}} \sin(\omega_{id}\tau) \cdot F(t-\tau)d\tau$, includes the structural information in the two terms of $\sin(\omega_{id}\tau)$ and $e^{-\xi_i \omega_i \tau}/\omega_{id}$. The Nyquist frequency has to be satisfied if the structural vibration information (both the amplitude and frequency) is sought from the term $\sin(\omega_{id}\tau)$. However, the term $e^{-\xi_n \omega_n \tau}$ is an exponential function and is independent of the sampling frequency. The time-domain method is therefore less dependent on the Nyquist frequency and the damage detection can be performed successfully with a low sampling rate.

Another comparison is made in Scenario 6 with two sets of responses. The first set is computed from Equation (7.1) with a sampling rate of 200 Hz, and the other set is obtained from re-sampling the first set of response in the ratio 1 in 2. The first eight sets of WPT coefficients from the first set and all sixteen sets of WPT coefficients from the second set of responses are used for the identification covering the same bandwidth of $0 \sim 50.0$ Hz. Accurate identified results, shown in Figure 8.25, provide further evidences to the above discussions.

8.5.3   Under normal randvom support excitation

The formulation of the condition assessment of a structure under ambient excitation was given in Section 8.4.2. This section gives an experimental example of the use of such a technique in the assessment of a laboratory structure. The sensitivity of wavelet-packet component energy described in Section 8.3.1 is adopted in the identification equation, while a damage localization stage is added before the damage quantification stage to reduce the candidate set of unknowns for better convergence of the computation.

8.5.3.1 Damage localization based on mode shape changes

The structural mode shapes can be identified from the ambient vibration responses using the Natural Excitation Technique (James et al., 1995) in conjunction with the Eigensystem Realization Algorithm (ERA) (Juang and Pappa, 1985), or using the conventional peak-picking technique. Since mode shapes of higher order are usually obtained with difficulties in real measurement, only the first mode shape is employed for the damage localization.

The modal strain energy of the ith element corresponding to the first mode shape from the healthy structure is defined as

$$s_{1i} = \Phi_1^T K_i \Phi_1 \tag{8.99}$$

where K_i is the ith elemental stiffness matrix and Φ_1 is the first mode shape of the system. Then the total modal strain energy of the structure corresponding to the first mode is obtained as

$$s_1 = \Phi_1^T K \Phi_1 \tag{8.100}$$

The fractional contribution to the modal strain energy from the ith member is denoted as

$$F_{1i} = \frac{s_{1i}}{s_1} \tag{8.101}$$

Similarly, for a damaged structure, the corresponding F_{1i}^d is defined as

$$F_{1i}^d = \frac{s_{1i}^d}{s_1^d} \tag{8.102}$$

$$s_{1i}^d = (\Phi_1^d)^T K_i^d \Phi_1^d \tag{8.103}$$

$$s_1^d = (\Phi_1^d)^T K^d \Phi_1^d \tag{8.104}$$

where K_i^d and K^d are the ith elemental stiffness matrix and the global stiffness matrix of the damaged structure, respectively; and Φ_1^d is the first mode shape of the damaged structure. Equation (8.103) can be expressed as

$$s_{1i}^d = \alpha_i (\Phi_1^d)^T K_i \Phi_1^d \tag{8.105}$$

For the case with relatively small damage in the structure, it can be assumed that

$$F_{1i} \approx F_{1i}^d \tag{8.106}$$

Based on the assumption of small damage in the structure, $K^d \approx K$, thus

$$s_1^d = (\Phi_1^d)^T K^d \Phi_1^d \approx (\Phi_1^d)^T K \Phi_1^d \qquad (8.107)$$

Substituting Equations (8.99) to (8.104) into Equation (8.106) yields a damage index as

$$\Delta \alpha_i = 1 - \frac{(\Phi_1^T K_i \Phi_1)((\Phi_1^d)^T K \Phi_1^d)}{((\Phi_1^d)^T K_i \Phi_1^d)(\Phi_1^T K \Phi_1)} \qquad (8.108)$$

It is noted that the above formulation of the damage index needs full measurement to obtain the mode shape. In the case of incomplete measurement, the measured information can be expanded to the full mode shape, or, only the measured information is taken into account in Equation (8.108). Both of these practices would lead to errors of different extents, which is a feature with the mode shape-based approach. However, it is noted that this localization stage serves to reduce the candidate set of possible damage elements for the next stage of damage quantification. Other frequency-based or time-based methods can also be used for this purpose. The solution algorithm and convergence criteria described in Section 8.3.1 are adopted in the identification.

8.5.3.2 Laboratory experiment

A nine-bay three-dimensional frame structure shown in Figure 8.26 is fabricated in the laboratory using the Meroform M12 construction system. It consists of 69, 22 mm diameter alloy steel tubes jointed together with 29 standard Meroform ball nodes. Each tube is fitted with a screwed end connector which, when tightened into the node, also clamps the tube by means of an internal compression fitting. All the connection bolts are tightened with the same torsional moment to avoid asymmetry or nonlinear effects caused by man-made assembly errors. The experimental setup is also shown in Figure 8.26(a) and the support is shown in Figure 8.26(b). The finite element model of the structure is shown in Figure 8.27. The structure has the material and geometric properties as shown in Table 7.1.

Modelling of the structure

This paper adopts the hybrid beam model including the semi-rigid end connections (Law et al., 2001a) for the model improvement of the structure. The initial model assumes a large fixity factor, p, for the rotational stiffness of the joints which is taken equal to 0.999 with 1.0 equal to that for a rigid joint.

The total weight of the ball and half of the weight of the bolt connecting the ball with the frame element are placed at each node as a lump mass. The other half of the weight of the bolt is considered as part of the finite element. In additional, another lump mass of 72g is added to each node to represent the weight of the moving accelerometers. The natural frequencies calculated from the finite element model are found very close to the measured values.

(b) The rigid concrete support and the actuator

(a) The cantilevered 3D steel tube frame structure

(c) The damaged elements

Figure 8.26 Experimental set-up and the damaged elements of the frame structure

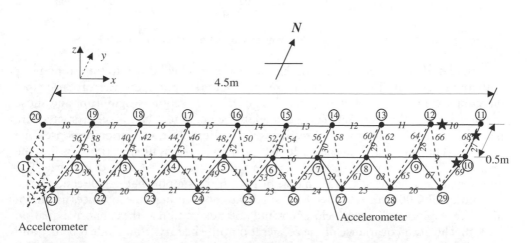

Figure 8.27 A nine-bay three-dimensional frame structure (★- a damaged element)

Ambient vibration test for damage detection

The structure is excited with a random white noise signal through a LING PO 300 exciter approximately at the centroid of the support in the *y*-direction. The support is

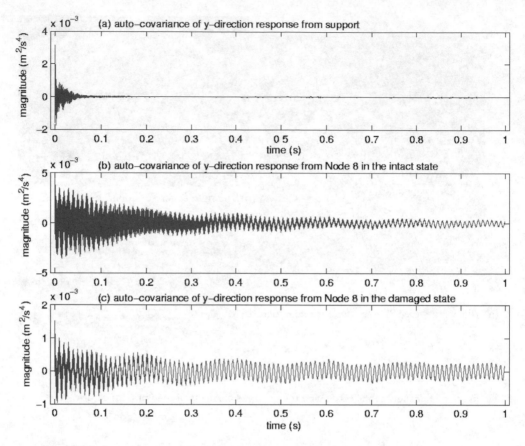

Figure 8.28 Auto-covariance of accelerations at support and structure

very rigid and heavy compared with the frame and it is held down to the strong floor with four steel bolts. The acceleration responses of the support along the three principal directions are measured. Only the response in the y-direction is significant and those in the other two directions are small enough to be neglected. Nine accelerometers are placed at nodes 2 to 10 in the y-direction for recording the acceleration response-time histories. The sampling frequency is 2000 Hz with a low-pass filter at 1000 Hz, and the responses for a duration of 800 seconds are used to calculate the auto-covariance. The auto-covariance of the acceleration time history at the support in the y-direction has a magnitude of 0.0031 m^2/s^4 at $t = 0$ and small values for other time instances as shown in Figure 8.28(a), indicating a close to white noise excitation at the support (Equation (8.96)). The auto-covariance of the acceleration time history from Node 8 before and after the damage occurrence is shown in Figures 8.28(b) and 8.28(c), respectively. The covariances are noted to be relatively smooth with a low noise level.

Damage scenarios

Local faults are introduced by replacing three intact members with damaged ones. The artificial damage is of two types. Type I is a perforated slot cut in the central length of the member. The slot is 13.7 cm long, and the remaining depth of the tube in the cut cross-section is 14.375 mm. Type II is the removal of a layer of material from the surface of the member. The external diameter of the tube is reduced from 22.02 mm to 21.47 mm, and the length of the weakened section is 202 mm, located in the middle of the beam leaving 99 mm and 75 mm length of original tube cross-section on both sides. Figure 8.26(c) gives a close up view of the damaged frame members. Type I damage is located in element 10 and Type II damage is located in elements 27 and 68. The equivalent damages computed by the Guyon method are 5% and 9.5% reduction in the elastic modulus of material of element 10 and elements 27 and 68, respectively.

Model improvement for damage detection

The two-stage approach is adopted for the damage detection. The first stage updates the rotational stiffness at the joints to obtain an improved analytical model for the intact structure. Nine y-direction acceleration responses obtained from nodes 2 to 10 are measured and the corresponding auto-covariance WPT energy components are computed for use in Equation (8.43) in Section 8.3.1. There are $69 \times 6 = 414$ unknown rotational stiffnesses in the identification. The first two thousand data points of auto-covariance of each acceleration response are decomposed into six levels of Daubechies Db4 wavelet packets to have 64 covariance WPT energy components, and each packet has the same frequency bandwidth of 15.625 Hz. All 64 WPT energy components are used in the experimental identification procedure and there are $9 \times 64 = 576$ equations in Equation (8.43). The measured modal damping ratios from the intact structure have been used in the equation of motion for the computation of the analytical auto-covariance. The updated rotational stiffness do not differ too much from the original value with the largest change in member 20 at node 22 with the updated $p = 0.86$ for the x-axis rotational stiffness.

The second stage updates the local faults in all the members of the structure in terms of their elastic modulus. The damage localization is performed first. The measured first mode shape with 52 translational DOFs from the intact and the damaged states of the structure are expanded to the full DOFs by the dynamic condensation method with Gauss–Jordan elimination (Mario, 1997). The obtained damage localization vector is shown in Figure 8.29(a), indicating that the real damaged members can be identified but with false alarms in other undamaged members because of the error due to measurement noise and the mode shape expansion.

The auto-covariance calculated from the y-direction response from nodes 2 and 10 are used in Equation (8.43) for the identification. There are 26 unknowns, which are the suspected damaged elements shown in Figure 8.29(a) in the identification. The first two thousand data points of the auto-covariance of each response are used and all $64 \times 9 = 576$ WPT energy components are used in Equation (8.43). The identified damage extent for all the elements are shown in Figure 8.29(b). The identified reduction in the elastic modulus in the damaged members are 4.76%, 7.09% and 6.73% for elements 10, 27 and 68, respectively, which are fairly close to the values of 5%, 9.5% and 9.5%. There are false alarms in elements 9, 50 and 67 with a reduction of more than 2% even though they are in fact undamaged.

It should be noted that the selection of damage elements at the end of the cantilever is due to the consideration of experimental ease and assembly repeatability. It also carries

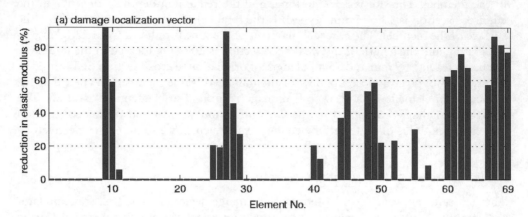

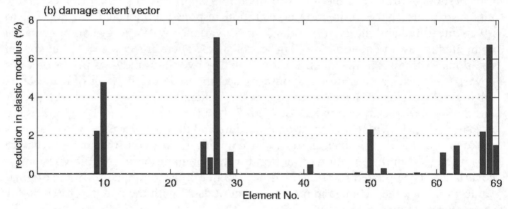

Figure 8.29 Identified results for the experimental structure

an intention to identify a damage scenario with damage elements experiencing a small change in the stress and strain, which is usually difficult to be identified with existing techniques.

8.6 Conclusions

The wavelet-based approach provides further opportunity to use information from different bandwidths for the condition assessment of a structure. The time-scale representation of the dynamic response gives a more detailed description of the system characteristic properties than the information from either a time series or its Fourier transform. For an impulsive excitation, a low sampling rate has been shown to give similar identified results compared to that obtain from data at a high sampling rate. Different types of excitations have also been shown to be useful in identifying local damages via the algorithms present in this chapter.

Chapter 9

Uncertainty analysis

9.1 Introduction

Structural condition assessment involves many uncertainties, which come from the modelling procedure, measurements and environmental variations (Ang and Tang, 1984; 2007). Ignoring the effects of these uncertainties could lead to an unsuccessful assessment with damage undetected or intact elements identified as damaged. It is important to locate the sources of these uncertainties and how they evolve throughout the damage identification process.

Uncertainty in models and computer simulations generally arises from the measurement, environmental variability, parameter variability and modelling with a lack of knowledge. Measurement variability mainly comes from experimental procedures and equipment. It includes measurement errors and data processing errors. Environmental variability includes the uncertainty associated with environmental variables, such as temperature, humidity, wind and traffic loads. Parameter variability is the uncertainty associated with the input parameters of a particular model. Modelling uncertainty, or lack of knowledge, refers to the uncertainty associated with the functional form of the models implemented or simplified assumptions. In summary, the uncertainties in structural condition assessment may be attributed to (Hao and Xia, 2005):

- Inaccuracy in finite element model discretization;
- Uncertainties in geometry and boundary conditions;
- Variations in materials properties;
- Errors associated with the measured signals;
- Environmental variability (such as temperature, humidity, wind and traffic).

These uncertainties can be further classified into three groups according to their sources, namely as, the modelling, measurement and environmental uncertainties. They propagate through the damage identification process, and an integrated procedure is required to estimate the probability that the failure criterion is met or not with this effect.

Ang and Leon (2005) have tried to model and analyze these uncertainties for risk-informed decisions, while Ellingwood (2005) reported on the state of practice of risk-informed condition assessment of civil infrastructure for their maintenance, management (Frangopol and Liu, 2007) and safety evaluation (Liu et al., 2009a; 2009b). The effect on the maintenance of civil structures (Frangopol and Messervey, 2009a) and the life-cycle cost estimation of deteriorating structures has also been reported (Frangopol et al., 1997; Frangopol and Messervey, 2009b). However, the integration of the reliability analysis with structural health monitoring has not been attempted. The reliability analysis results of a structure are known to be a function of the health monitoring or condition assessment results, and yet the inclusion of such a relationship in the reliability analysis has not been performed due to the lack of knowledge of the uncertainty propagation in the assessment process.

The three types of uncertainties and their influence to the structural condition assessment are discussed in this chapter. Some of the existing approaches to formulate their effect on the dynamic response and on the condition assessment results are also given. Finally, a box-section bridge deck structure is taken to illustrate how these uncertainties are considered in the integration of structural condition assessment with the subsequent reliability analysis to yield an updated set of indices.

9.2 System uncertainties

The different stages of the computational modelling and simulation process relevant to the damage identification are depicted in Figure 9.1 (Oberkampf et al., 2002). Uncertainty is presented in all these stages, including the conceptual modelling of the physical system, its symbolic representation as a mathematical model and the subsequent numerical computation and solution. In this section, some of the general considerations associated with uncertainty quantification and special issues related to model uncertainty are discussed.

9.2.1 Modelling uncertainty

Modelling uncertainty or uncertainty from the model can be introduced during either the conceptual or the mathematical modelling processes. During the conceptual modelling stage, no mathematical equations are written, but the fundamental assumptions regarding possible events and physical processes are made. All possible factors that

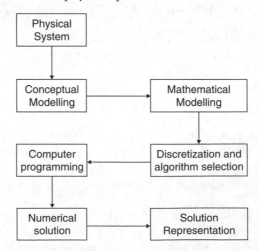

Figure 9.1 Phases of computational modelling and simulation

could affect the set of requirements for the modelling and simulation are considered. Also, the possible physical couplings of these factors are identified. Since it is not possible to model complex and realistic systems at the level of elementary particles, most real-world mathematical models are comprised of a set of field equations and associated boundary and initial conditions. These equations are approximations of the complicated physics of the real world. Thus, modelling uncertainty can arise from approximations or errors in the form of the model equations, the applied loads and the boundary and initial conditions.

The discretization phase converts the continuum mathematics model into a discrete mathematics problem suitable for numerical solution. In general, a finite element model is a discrete numerical model of a continuous structural system. When a finite element model is constructed to predict the dynamic properties of a structure, it usually involves some simplifications, where complicated parts of the structure are represented approximately by standard elements in the model. Discretization errors are introduced when there is a difference between the actual dynamic properties of the system and those predicted by the finite element model.

Methods are available for the systematic treatment of uncertainty in physical models. If the uncertainty is due to physics that are known, but not modelled, either the model can be revised to include these physics or an attempt can be made to characterize the uncertainty included. However, if the physics are not known, little can be done to characterize the accompanying uncertainty.

9.2.2 *Parameter uncertainty*

The parameter uncertainty refers simply to uncertainty in the input parameters of a model. Once the form of a model has been specified by a complete set of mathematical equations, all that remains for a complete mathematical statement of the problem is numerical quantification of the various constants. These constants may be geometric or material properties, or they may be those that characterize the applied loads, the boundary conditions or the initial conditions. Parameter uncertainties are caused by the differences between the actual parameters and the analytical ones estimated from the drawing. It is common for it to be impossible to construct a real structure exactly the same way as it is designed. Parameter uncertainty is perhaps the most widely studied type of uncertainty. The characterization of the effect of this type of uncertainty on the model prediction is analyzed by a wide variety of techniques, such as sensitivity analysis, Monte Carlo methods, reliability-based methods, fuzzy set, interval propagation methods and stochastic finite elements.

The parameter uncertainty can be reduced by model updating if the vibration properties of the real system are available. It has become an important topic in the development of model updating algorithms and many procedures for model verification or validation.

9.2.3 *Measurement and environmental uncertainty*

Measurement uncertainty refers to the uncertainties arising from the measurement process itself. The inherent variability may come from noise in the measurement process due to factors such as thermal instability, internal or external electromagnetic fields and the quantization error. Error and potential uncertainty may arise from electrical bias and the gain errors of the signal processing tools. Measurement uncertainty forms a significant part of the overall uncertainty. In many situations, measured data are taken as the indisputable truth. This is obviously not correct because the accuracy of measurements depends on the characteristics of the hardware, and on the physical or empirical models underlying the design of the data collection system. An honest quantification of measurement uncertainty is crucial to a meaningful characterization of uncertainty as a whole.

The treatment of measurement uncertainties has a history almost as long as measurement itself. The most common method of treating bias, gain and temporal or frequency errors is by calibrating each piece of measurement equipment with respect to known standards. Less common is the calibration of the entire measurement system.

Environmental variability includes the uncertainty associated with environmental variables such as temperature and humidity. For the purpose of this discussion, it also includes measurement errors, data processing errors and uncertainty with the description of loads and boundary conditions associated with environmental variables. Most environmental variability can be dealt with as a parametric uncertainty. Uncertainty that cannot be modelled easily is generally accounted for by perturbing the available information with a random process as shown in the following sections.

9.3 System identification with parameter uncertainty

The sources of uncertainty were studied in Section 9.2., These mainly come from the modelling procedure and measurement and environmental variations. Another task in the uncertainty analysis is to study the effect of uncertainty on the condition assessment and to estimate the statistical distribution or bounds of the identified results, which is termed the uncertainty propagation in system identification.

While there are many studies on the effect of uncertainty in the modal analysis, reliability analysis, model updating, etc., other researchers try to quantify the uncertainty in identified modal parameters using the singular value decomposition technique (SVD) (Ruotolo and Surace, 1999) or using the principal component analysis of a matrix of estimated features (Kullaa, 2002). This is successful as the damage features are outliers. However, it only works well when the SVD of the data matrix from intact observations can be computed. In another application of the SVD, the parameter variations are decomposed into principal components that are weighted based on the sensitivity of the performance matrix to the parameter variations (Vanlanduit et al., 2005). Parameter bounds in the form of an interval model are generated and each interval corresponds to one identified bounded uncertainty parameter with its associated principal direction.

Two commonly used numerical approaches to study the uncertainty propagation are the Monte Carlo simulation and the perturbation method. In the Monte Carlo simulation, a large number of samples with a given probability distribution are generated (Kottegoda and Rosso, 1997). A specified process is then applied to these data samples to estimate the statistics of the output. The Bootstrap method (Efron and Tibshirani, 1994) is somewhat similar to Monte Carlo simulation. It has the advantage of requiring no assumptions about the distribution functions or covariance matrices. However, this technique requires that every data sample be reused in the generation of the next sample. An alterative approach is the perturbation method. This approach expands a nonlinear function with a truncated Taylor series expansion at a known point and then proceeds to the approximation of the moments of solutions from the expansion. Certainly, the Monte Carlo simulation method can be combined with the perturbation method for the uncertainty propagation study. The minimum variance method making use of the above combined techniques was first studied by Collins et al. (1974) to obtain the statistical properties of updating parameters. Beck and his co-workers (Beck and Katafygiotis, 1998; Katafygiotis and Beck, 1998) have also reformulated this method in a more general framework where the Bayesian method is used to minimize the variance of the updated structural parameters.

9.3.1 Monte Carlo simulation

The Monte Carlo simulation has been widely used in statistical analysis. Different sets of input data are obtained based on the statistics and an assumed probability distribution of the input data. A specified process is then applied to the simulated data to obtain the output for each set of inputs, from which, the distribution and the interval bounds of the output can be estimated. It is particularly useful when a closed-form solution is impossible or very difficult to obtain.

To perform a Monte Carlo simulation, one needs to specify the probability distribution of the variables involved, which must be known or assumed. The generation of outcomes from a prescribed probability distribution is a fundamental task in the simulation process. In designing the Monte Carlo simulation, one must determine how many simulations are required to assess the system behaviour with the error of assessment decreasing as $n^{-1/2}$, where n is the number of simulations. To estimate the probability, p, of an event within $100\,\varepsilon\%$ (where ε is the acceptable tolerance, $0 \leq \varepsilon \leq 1$) of its true value with $100(1-\alpha)\%$ confidence, n must satisfy (Kottegoda and Rosso, 1997)

$$n \geq z_{\alpha/2}^2 (1-p)/\varepsilon^2 p \tag{9.1}$$

where $z_{\alpha/2}$ denotes a standard normal variate that is exceeded with the probability of $\alpha/2$. Since n is a function of p, which is unknown before the experiment is performed, the value of p must be estimated before the experiment.

While increasing the sample size is one way to reduce the standard error of a Monte Carlo analysis, the fact of doing so can be computationally expensive. A better solution is to employ some technique of variance reduction based on the properties of correlated samples. Standard techniques of variance reduction including antithetic variates, control variates, importance sampling and stratified sampling (Kottegoda and Rosso, 1997) may be used for this purpose.

9.3.2 Integrated perturbed and Bayesian method

Before the integrated perturbed and Bayesian methods are introduced, the perturbation method is first presented. The identified distribution of updated parameters is then combined with the prior distributions of updated parameters via Bayesian updating to achieve the posterior distributions of the updated parameters.

The perturbation method generally takes a small disturbance into the undisturbed system. Here the uncertainty is taken as a perturbation from the noise-free data. It is closely related to the Taylor series expansion and is useful when the uncertainty level is not high, such that the first few terms can approximate the important features of the solution and the remaining ones can be truncated. In model updating, the relationship between the measured vibration characteristics and the structural parameters can be expressed as:

$$\{e\} = [S]\{\Delta\alpha\} \tag{9.2}$$

where $\{e\}$ is the modal data change vector containing the differences between the eigenvalues and mode shapes at the corresponding measured degrees-of-freedom (DOF) of the structure before and after updating; $\{\Delta\alpha\} = \{\tilde{\alpha}\} - \{\alpha\}$ is the vector of parameter change, where $\{\alpha\}$ and $\{\tilde{\alpha}\}$ are structural elemental parameters of the initial and updated finite element model; and $[S]$ is the corresponding sensitivity matrix.

Different accuracies of approximation to the solution statistics can be achieved depending on the order of truncation of the Taylor series expansion used in the perturbation method. In practical applications, either the first-order, second moment approach or the second-order, second moment approach can be used to approximate the solution moments. The first-order, second moment approach approximates the nonlinear function with a linear expansion at a point of the space of the random variables, and the obtained mean and covariance are of first-order accuracy. The limitations of the first-order perturbation method are that uncertainties must not be too large and that the nonlinearity is not significant. Quadratic accuracy can be achieved by approaching the nonlinear function with a second-order Taylor series expansion as illustrated with the following example.

Applying the perturbation technique, Equation (9.2) can be expanded as a second-order Taylor series in terms of the uncertainty, X_i, (Liu, 1995), which is taken as the selected uncertainty here.

$$[S] = [S]^0 + \sum_{i=1}^{N} \frac{\partial[S]}{\partial X_i} X_i + \frac{1}{2} \sum_{i=1}^{N} \sum_{j=1}^{N} \frac{\partial^2[S]}{\partial X_i \partial X_j} X_i X_j \tag{9.3}$$

$$\{\Delta\alpha\} = \{\Delta\alpha\}^0 + \sum_{i=1}^{N} \frac{\partial\{\Delta\alpha\}}{\partial X_i} X_i + \frac{1}{2} \sum_{i=1}^{N} \sum_{j=1}^{N} \frac{\partial^2\{\Delta\alpha\}}{\partial X_i \partial X_j} X_i X_j \tag{9.4}$$

$$\{e\} = \{e\}^0 + \sum_{i=1}^{N} \frac{\partial\{e\}}{\partial X_i} X_i + \frac{1}{2} \sum_{i=1}^{N} \sum_{j=1}^{N} \frac{\partial^2\{e\}}{\partial X_i \partial X_j} X_i X_j \tag{9.5}$$

It is noted that superscript '0' represents the respective noise-free value. Substituting the above equations into Equation (9.2) and comparing the terms of $1, X_i, X_i$ and X_j, then $\{\Delta\alpha\}^0, \partial\{\Delta\alpha\}/\partial X_i, \partial^2\{\Delta\alpha\}/\partial X_i \partial X_j$ $(i, j = 1, 2, \ldots, N)$ can be solved as follows

$$\{\Delta\alpha\}^0 = ([S]^0)^+ \{e\}^0 \tag{9.6}$$

$$\frac{\partial\{\Delta\alpha\}}{\partial X_i} = ([S]^0)^+ \left(\frac{\partial\{e\}}{\partial X_i} - \frac{\partial[S]}{\partial X_i} \{\Delta\alpha\}^0 \right) \tag{9.7}$$

$$\frac{\partial^2 \{\Delta\alpha\}}{\partial X_i \, \partial X_j} = ([S]^0)^+ \left(\frac{\partial^2 \{e\}}{\partial X_i \, \partial X_j} - \frac{\partial^2 [S]}{\partial X_i \, \partial X_j} \{\Delta\alpha\}^0 - 2\frac{\partial [S]}{\partial X_i} \frac{\partial \{\Delta\alpha\}}{\partial X_j} \right) \tag{9.8}$$

where $^+$ denotes the general inverse of a matrix. Noting that $E(X_i) = 0$ and the mean values and covariance matrix of $\{\Delta\alpha\}$ approximating the nonlinear function can be obtained from Equation (9.4) as follows

$$E(\{\Delta\alpha\}) = E(\{\Delta\alpha\}^0) + \frac{1}{2} \sum_{i=1}^{N} \frac{\partial^2 \{\Delta\alpha\}}{\partial^2 X_i} Cov(X_i, X_i) \tag{9.9}$$

$$[Cov(\Delta\alpha, \Delta\alpha)] = \left[\frac{\partial \{\Delta\alpha\}}{\partial \{X\}} \right] [Cov(X, X)] \left[\frac{\partial \{\Delta\alpha\}}{\partial \{X\}} \right]^T \tag{9.10}$$

In practice, structural parameters are estimated from sources such as the design drawing, which are subjected to uncertainties. Such an estimate of structural parameters is termed as prior information. The Bayesian approach is very useful when one is faced with two sets of uncertain information and one needs to know which set to believe. Both the prior information and the newly obtained information are used to account for the relative uncertainty associated with each other. The newly obtained information can be derived from the perturbation approach. The prior information is given by the structural analysis prior to the testing of a structure. The Bayesian approach is then applied to obtain a new (posterior) distribution with the above information.

Assuming there is a random structural parameter, α, with a probability density function, $f_1(\alpha)$, the identified probability density function is obtained as $f_2(\alpha)$ from the above perturbation method. The posterior probability density function of the structural parameter can be expressed as the following with the Bayesian probabilistic framework (Sohn and Law, 1997; Vanik et al., 2000),

$$f(\alpha) = kL(\alpha)f_2(\alpha) \tag{9.11}$$

where $L(\alpha)$ represents the likelihood function and k is the normalized constant.

For the case where the probability density functions of both $f_1(\alpha)$ and $f_2(\alpha)$ are normally distributed, the posterior distribution function of the structural parameter, $f(\alpha)$, also complies with a normal distribution, where the mean and standard deviation are obtained as (Hua et al., 2007)

$$\mu = \frac{\mu_1 \sigma_2^2 + \mu_2 \sigma_1^2}{\sigma_1^2 + \sigma_2^2} \tag{9.12}$$

$$\sigma = \sqrt{\frac{\sigma_1^2 \sigma_2^2}{\sigma_1^2 + \sigma_2^2}} \tag{9.13}$$

where μ_1 and μ_2 are the means of prior and the identified distribution function, respectively; and σ_1 and σ_2 are the corresponding standard deviations, respectively.

It is clear that both the mean and standard deviation of the posterior distribution functions are weighted averages of the prior and the identified distribution functions of the structural parameters. By using the Bayesian theory, knowledge of the analyst and experience of the experimentalist can be rationally incorporated.

Examples of using the first-order perturbation with respect to parameter uncertainties can be found using the measured modal data (Xia et al., 2002) and using directly the measured acceleration of the structure (Li and Law, 2008). The propagation of uncertainty within the structure in the iterative updating process has also been studied (Li and Law, 2008).

9.4 Modelling the uncertainty

Since the uncertainties and their effects cannot be easily removed or quantified in the structural condition assessment, some sort of mathematical tool can be used to model them and include their effects in the assessment process. If the uncertainty $u(x, \theta)$ is a normally distributed random process, it is a function of the position vector defined over the domain D, with θ belonging to the space of random events Ω. It can be very conveniently modelled by the Karhunen–Loéve expansion (Ghanem and Spanos, 1991) where proofs have been given on its equivalence to the Proper Orthogonal Decomposition Methods (Wu et al., 2003) with,

$$u(x, \theta) = \overline{u}(x) + \sum_{n=1}^{m} \xi_n(\theta)\sqrt{\lambda_n}\varphi_n(x) \qquad (9.14)$$

where $\overline{u}(x)$ denotes the expected value of $u(x, \theta)$ over all possible realizations of the process; and λ_n and $\varphi_n(x)$ are the eigenvalue and the eigenvector of the covariance kernel, respectively. This representation introduces just m independent standard normal random variables $\xi_n(\theta)$ with the property of zero mean and unit standard deviation as

$$E(\xi_n(\theta)) = 0, \quad E(\xi_k(\theta)\xi_l(\theta)) = \delta_{kl} \qquad (9.15)$$

where δ_{nm} is the Kronecker delta. An explicit expression for $\xi_n(\theta)$ can be written as

$$\xi_n(\theta) = \frac{1}{\sqrt{\lambda_n}} \int_D \widetilde{u}(x, \theta)\varphi_n(x)\, dx \qquad (9.16)$$

The Karhunen–Loéve expansion is related to the required specification of the covariance, $C(x_1, x_2)$ by the following relation:

$$C(x_1, x_2) = \sum_{n=0}^{\infty} \lambda_n \varphi_n(x_1)\varphi_n(x_2) \qquad (9.17)$$

which is bounded, symmetric and positive definite.

Lanata and Del Grosso (2006) give an example of modelling the uncertainties using the Karhunen–Loéve expansion for the structural condition assessment; Wu and Law, (2009) give an example of the modelling of the Gaussian uncertainties within the bridge-vehicle interaction problem. The following gives an example of the study of propagation of uncertainties in a structural system during the condition assessment process in which the uncertainties of the system parameters, excitation forces and measured responses from the perturbed state of the structure are discussed (Li and Law, 2008). A further application of the assessment results in updating the reliability of a box-section bridge deck is also presented (Li and Law, 2008).

9.5   Propagation of uncertainties in the condition assessment process

9.5.1   Theoretical formulation

An approximate model of the local damage is adopted in which the damaged structural system is expressed as $K_d = \sum_{i=1}^{ne} \alpha_i K_i$, where α_i is the fractional stiffness of an element typically with $1.0 \geq \alpha_i \geq 0.0$; and ne is the number of finite elements. The damage identification equation is based on the acceleration response sensitivity method described in Chapter 7 as

$$S \cdot \Delta\alpha = \Delta\ddot{x} = \ddot{x}_d - \ddot{x}_u \qquad (9.18)$$

where $\ddot{x}_d$ and $\ddot{x}_u$ are vectors of the acceleration response at the lth DOF of the damaged and intact structures, respectively; $S = \begin{bmatrix} \frac{\partial \ddot{x}_l}{\partial \alpha_1} & \frac{\partial \ddot{x}_l}{\partial \alpha_2} & \cdots & \frac{\partial \ddot{x}_l}{\partial \alpha_m} \end{bmatrix}$ is the acceleration sensitivity matrix; and $\Delta\ddot{x}$ is the vector of changes in the acceleration response. An iterative computation is adopted to get $\Delta\alpha$ with the least-squares method from

$$\Delta\alpha = (S^T S)^{-1} S^T (\ddot{x}_d - \ddot{x}_u) \qquad (9.19)$$

with the specific tolerance of acceptance equal to 1.0×10^{-6}.

9.5.1.1   Uncertainties of the system

Random errors exist in the structural parameters of an initial analytical model. The force excitation needs to be measured from the damaged structure with noise included in the measurement. The measured acceleration response also contains measurement noise. Since a model updating technique is used in the damage identification procedure, all the above random errors will propagate in the computation system with iterations. They are analyzed with an example below to check on how these errors erode the identification results.

The different random variables above are all assumed to have zero mean and normally distributed. They are also assumed independent with no coupling effect. Assuming X_p denotes the random variables associated with the structural parameters of the initial analytical model – these may be the mass density, geometric parameters and the material elastic modulus of the initial analytical model. However, only the parameters associated with the mass (the material density) and stiffness (the elastic modulus of material) are studied for illustration of the analysis and they are denoted by X_{ρ_i} and X_{E_i}, respectively, for the ith element as

$$\tilde{\rho}_i = \rho_i(1 + X_{\rho_i}), \quad \tilde{E}_i = E_i(1 + X_{E_i}) \qquad (9.20)$$

where $\widetilde{\bullet}$ denotes the measured value including the uncertainty; ρ_i and E_i denote the true values; and i denotes the ith element. Each finite element is assigned these random variables representing the uncertainties in its mass and stiffness properties of the element. The uncertainties with other structural parameters can be similarly defined.

The second type of uncertainty arising from the measured exciting force is denoted by the random variable X_f, and the measured force excitation vector including the uncertainty is related to the random variables as

$$\widetilde{f}_i = f_i + X_{f_i} \tag{9.21}$$

where f_i denotes the true excitation value at the ith time instance.

Similarly, the third type of uncertainty arising from the measured acceleration response is denoted by the random variable, $X_{\ddot{x}}$, and it is related to the measured acceleration response vector including the uncertainty as

$$\widetilde{\ddot{x}}_{di} = \ddot{x}_{di} + X_{\ddot{x}_{di}} \tag{9.22}$$

where $\ddot{x}_{di}$ denotes the true value of the measured response at the ith time instance from the damaged structure.

9.5.1.2 Derivatives of local damage with respect to the uncertainties

The first derivation of Equation (9.18) with respect to the general random variables, X, is given as

$$\frac{\partial S}{\partial X} \cdot \Delta\alpha + S \cdot \frac{\partial \Delta\alpha}{\partial X} = \frac{\partial \widetilde{\ddot{x}}_d}{\partial X} - \frac{\partial \ddot{x}_u}{\partial X} \tag{9.23}$$

Substituting Equation (9.18) into Equation (9.23), $\frac{\partial \Delta\alpha}{\partial X}$ can be obtained as

$$\frac{\partial \Delta\alpha}{\partial X} = (S^T S)^{-1} S^T \left(\frac{\partial \widetilde{\ddot{x}}_d}{\partial X} - \frac{\partial \ddot{x}_u}{\partial X} - \frac{\partial S}{\partial X} \cdot (S^T S)^{-1} S^T (\widetilde{\ddot{x}}_d - \ddot{x}_u) \right) \tag{9.24}$$

Equation (9.24) gives the general relationship between an uncertainty and the identified local damage vector. Matrix S and response vector $\ddot{x}_u$ can be obtained from the analytical model. Response vector $\widetilde{\ddot{x}}_d$ is obtained from measurement. The unknown terms $\frac{\partial \widetilde{\ddot{x}}_d}{\partial X}$, $\frac{\partial \ddot{x}_u}{\partial X}$ and $\frac{\partial S}{\partial X}$ are obtained below for each type of uncertainty for the solution of Equation (9.24).

9.5.1.3 Uncertainty in the system parameter

Considering the uncertainty in the system structural parameter, $X = X_p$, gives $\frac{\partial \widetilde{\ddot{x}}_d}{\partial X_p} = 0$, since the measured response is independent of the system parameter. Equation (9.24) can be rewritten as,

$$\frac{\partial \Delta\alpha}{\partial X_p} = -(S^T S)^{-1} S^T \left(\frac{\partial \ddot{x}_u}{\partial X_p} + \frac{\partial S}{\partial X_p} (S^T S)^{-1} S^T (\widetilde{\ddot{x}}_d - \ddot{x}_u) \right) \tag{9.25a}$$

The sensitivity of the analytical response with respect to X_p, $\frac{\partial \ddot{x}_u}{\partial X_p}$, can be obtained by taking the first derivation of the equation of motion with respect to X_p as

$$M\frac{\partial \ddot{x}}{\partial X_p} + C\frac{\partial \dot{x}}{\partial X_p} + K\frac{\partial x}{\partial X_p} = -\frac{\partial M}{\partial X_p}\ddot{x} - \frac{\partial C}{\partial X_p}\dot{x} - \frac{\partial K}{\partial X_p}x \qquad (9.26)$$

The sensitivity, $\frac{\partial \ddot{x}_u}{\partial X_p}$, can be computed using the Newmark method (Lu and Law, 2007a).

The derivation of the equation of motion with respect to the stiffness fractional change, α, is

$$M\frac{\partial \ddot{x}}{\partial \alpha} + C\frac{\partial \dot{x}}{\partial \alpha} + K\frac{\partial x}{\partial \alpha} = -\frac{\partial K}{\partial \alpha}x \qquad (9.27)$$

from which the sensitivity matrix, S, can be obtained. Further differentiation of Equation (9.27) with respect to the random variable, X_p, gives

$$M\frac{\partial^2 \ddot{x}}{\partial \alpha\, \partial X_p} + C\frac{\partial^2 \dot{x}}{\partial \alpha\, \partial X_p} + K\frac{\partial^2 x}{\partial \alpha\, \partial X_p} = -\frac{\partial M}{\partial X_p}\frac{\partial \ddot{x}}{\partial \alpha} - \frac{\partial C}{\partial X_p}\frac{\partial \dot{x}}{\partial \alpha} - \frac{\partial K}{\partial X_p}\frac{\partial x}{\partial \alpha}$$
$$- \frac{\partial K}{\partial \alpha}\frac{\partial x}{\partial X_p} - \frac{\partial^2 K}{\partial \alpha\, \partial X_p}x \qquad (9.28)$$

Since $\frac{\partial x}{\partial X_p}$, $\frac{\partial \dot{x}}{\partial X_p}$ and $\frac{\partial \ddot{x}}{\partial X_p}$ have been obtained from Equation (9.26), and $\frac{\partial x}{\partial \alpha}$, $\frac{\partial \dot{x}}{\partial \alpha}$ and $\frac{\partial \ddot{x}}{\partial \alpha}$ have been obtained from Equation (9.27), $\frac{\partial^2 \ddot{x}}{\partial \alpha\, \partial X_p}$ can finally be computed from Equation (9.28) to form the sensitivity matrix, $\frac{\partial S}{\partial X_p}$,

$$\frac{\partial S}{\partial X_p} = \left[\begin{array}{cccc} \frac{\partial^2 \ddot{x}_l}{\partial \alpha_1\, \partial X_p} & \frac{\partial^2 \ddot{x}_l}{\partial \alpha_2\, \partial X_p} & \cdots & \frac{\partial^2 \ddot{x}_l}{\partial \alpha_m \partial X_p} \end{array}\right] \qquad (9.29)$$

and $\frac{\partial \Delta \alpha}{\partial X_p}$ can finally be computed from Equation (9.25a).

9.5.1.4 Uncertainty in the exciting force

Considering the uncertainty in the exciting force, X_f, gives $\frac{\partial \tilde{\ddot{x}}_d}{\partial X_f} = 0$, and Equation (9.24) can be rewritten as

$$\frac{\partial \Delta \alpha}{\partial X_f} = -(S^T S)^{-1}S^T\left(\frac{\partial \ddot{x}_u}{\partial X_f} + \frac{\partial S}{\partial X_f}\cdot(S^T S)^{-1}S^T(\tilde{\ddot{x}}_d - \ddot{x}_u)\right) \qquad (9.25b)$$

The derivation of the equation of motion with respect to X_f is given as

$$M\frac{\partial \ddot{x}}{\partial X_f} + C\frac{\partial \dot{x}}{\partial X_f} + K\frac{\partial x}{\partial X_f} = D\frac{\partial \tilde{f}}{\partial X_f} - \frac{\partial K}{\partial X_f}x \qquad (9.30)$$

Thus, from Equation (9.21)

$$\frac{\partial \tilde{f}(t)}{\partial X_{f_i}} = [0 \quad \cdots \quad 0 \quad 1 \quad 0 \quad \cdots \quad 0]^T \tag{9.31}$$

where only the ith component of the vector is one with zeros in all the other components. For the initial analytical model, $\frac{\partial K}{\partial X_f} = 0$, but in the updated analytical model, the effect of force uncertainty has propagated in the updated model with $\alpha = \alpha^0 + \Delta\alpha^1 + \Delta\alpha^2 + \ldots$, where α is the updated vector of the fractional parameter; α^0 is the initial vector of the fractional parameter; and $\Delta\alpha^i$ is the vector of the fractional parameter change identified after the ith iteration. Since the initial vector of the fractional parameter is independent of the force uncertainty, $\frac{\partial \alpha^0}{\partial X_f} = 0$, and

$$\frac{\partial K}{\partial X_f} = \frac{\partial K}{\partial \alpha}\frac{\partial \alpha}{\partial X_f} = \frac{\partial K}{\partial \alpha}\frac{\partial(\alpha^0 + \Delta\alpha^1 + \Delta\alpha^2 + \cdots)}{\partial X_f}$$

$$= \frac{\partial K}{\partial \alpha}\left(\frac{\partial \alpha^0}{\partial X_f} + \frac{\partial(\Delta\alpha^1)}{\partial X_f} + \frac{\partial(\Delta\alpha^2)}{\partial X_f} + \cdots\right)$$

$$= \frac{\partial K}{\partial \alpha}\left(\frac{\partial(\Delta\alpha^1)}{\partial X_f} + \frac{\partial(\Delta\alpha^2)}{\partial X_f} + \cdots\right) \tag{9.32}$$

Since $\frac{\partial \Delta\alpha^i}{\partial X_f} \neq 0$, then $\frac{\partial K}{\partial X_f} \neq 0$. The sensitivity, $\frac{\partial \ddot{x}_u}{\partial X_f}$, can be computed from Equations (9.30) to (9.32) by the Newmark method.

Differentiating Equation (9.27) with respect to the random variable, X_f, gives

$$M\frac{\partial^2 \ddot{x}}{\partial \alpha \, \partial X_f} + C\frac{\partial^2 \dot{x}}{\partial \alpha \, \partial X_f} + K\frac{\partial^2 x}{\partial \alpha \, \partial X_f} = -\frac{\partial K}{\partial \alpha}\frac{\partial x}{\partial X_f} - \frac{\partial K}{\partial X_f}\frac{\partial x}{\partial \alpha} - \frac{\partial^2 K}{\partial X_f \, \partial \alpha}x \tag{9.33}$$

which is similar to Equation (9.24). $\frac{\partial S}{\partial X_f}$ can then be computed from Equations (9.30) to (9.33) using the Newmark method and the sensitivity, $\frac{\partial \Delta\alpha}{\partial X_f}$, can then be obtained from Equation (9.25b). The effect of the uncertainty is seen to be propagating with iterations in the structural system through Equations (9.32), (9.33) and (9.25b).

9.5.1.5　Uncertainty in the structural response

The sensitivity of $\Delta\alpha$ with respect to the random variable $X_{\ddot{x}}$ can also be obtained similar to that for the structural parameters. Since $\ddot{x}_u$ and S are not related to $X_{\ddot{x}}$ in the initial analytical model, $\frac{\partial \ddot{x}_u}{\partial X_{\ddot{x}}} = 0$ and $\frac{\partial S}{\partial X_{\ddot{x}}} = 0$. Equation (9.24) then gives

$$\frac{\partial \Delta\alpha}{\partial X_{\ddot{x}}} = (S^T S)^{-1}S^T \frac{\partial \tilde{\ddot{x}}_d}{\partial X_{\ddot{x}}} \tag{9.34a}$$

However, in the updated analytical model, both $\ddot{x}_u$ and S are related to $X_{\ddot{x}}$ with the uncertainties propagated in the system. The derivation of the equation of motion with respect to $X_{\ddot{x}}$ gives

$$M\frac{\partial \ddot{x}}{\partial X_{\ddot{x}}} + C\frac{\partial \dot{x}}{\partial X_{\ddot{x}}} + K\frac{\partial x}{\partial X_{\ddot{x}}} = -\frac{\partial K}{\partial X_{\ddot{x}}}x \tag{9.35}$$

The sensitivities, $\frac{\partial x}{\partial X_{\ddot{x}}}$ and $\frac{\partial \ddot{x}}{\partial X_{\ddot{x}}}$, can be obtained from Equation (9.35) using the Newmark method. $\frac{\partial K}{\partial X_{\ddot{x}}}$ can be obtained in a manner similar to Equation (9.32). The derivation of Equation (9.26) with respect to $X_{\ddot{x}}$ gives

$$M\frac{\partial^2 \ddot{x}}{\partial \alpha\, \partial X_{\ddot{x}}} + C\frac{\partial^2 \dot{x}}{\partial \alpha\, \partial X_{\ddot{x}}} + K\frac{\partial^2 x}{\partial \alpha\, \partial X_{\ddot{x}}} = -\frac{\partial K}{\partial \alpha}\frac{\partial x}{\partial X_{\ddot{x}}} - \frac{\partial^2 K}{\partial \alpha\, \partial X_{\ddot{x}}}x - \frac{\partial K}{\partial X_{\ddot{x}}}\frac{\partial x}{\partial \alpha} \tag{9.36}$$

and $\frac{\partial \Delta \alpha}{\partial X_{\ddot{x}}}$ can be written as follows, incorporating Equations (9.35) and (9.36) as

$$\frac{\partial \Delta \alpha}{\partial X_{\ddot{x}}} = (S^T S)^{-1} S^T \left(\frac{\partial \widetilde{\ddot{x}}_d}{\partial X_{\ddot{x}}} - \frac{\partial \ddot{x}_u}{\partial X_{\ddot{x}}} - \frac{\partial S}{\partial X_{\ddot{x}}}(S^T S)^{-1} S^T (\widetilde{\ddot{x}}_d - \ddot{x}_u) \right) \tag{9.34b}$$

Equation (9.32) is obtained from Equation (9.22) as

$$\frac{\partial \widetilde{\ddot{x}}_d(t)}{\partial X_{\ddot{x}}} = [0 \quad \cdots \quad 0 \quad 1 \quad 0 \quad \cdots \quad 0]^T \tag{9.37}$$

and the sensitivity, $\frac{\partial \Delta \alpha}{\partial X_{\ddot{x}}}$, is obtained from Equations (9.34) and (9.37). The effect of the uncertainty is seen to be propagating with iterations in the structural system through a modified version of Equation (9.32) and Equations (9.35) and (9.34b).

In summary, the vector of sensitivity, $\frac{\partial \Delta \alpha}{\partial X}$, can be obtained from Equations (9.25a), (9.25b) and (9.24) as

$$\frac{\partial \Delta \alpha}{\partial X} = \left[\frac{\partial \Delta \alpha}{\partial X_p} \quad \frac{\partial \Delta \alpha}{\partial X_f} \quad \frac{\partial \Delta \alpha}{\partial X_{\ddot{x}}} \right] \tag{9.38}$$

9.5.1.6 Statistical characteristics of the damage vector

The mean value of the damage vector, $\Delta \alpha$, can be obtained directly from Equation (9.19) as

$$E(\Delta \alpha) = (S^T S)^{-1} S^T (\ddot{x}_d - \ddot{x}_u) \tag{9.39}$$

since the uncertainties considered in this paper are with zero means. The damage vector, $\Delta \alpha$, can be regarded as a function of the random variables, and it can be expressed as a truncated second-order Taylor series as

$$\Delta \alpha(X) = \Delta \alpha(0) + \sum_{i=1}^{mt} \frac{\partial \Delta \alpha(0)}{\partial X_i} X_i + \frac{1}{2} \sum_{i=1}^{mt} \sum_{j=1}^{mt} \frac{\partial^2 \Delta \alpha(0)}{\partial X_i\, \partial X_j} X_i X_j \tag{9.40}$$

where X_i and X_j denote the ith and jth variables, respectively; and mt is the number of random variables in the statistical analysis. The covariance matrix of $\Delta\alpha$ may be obtained as (Papadopoulos and Garcia, 1998)

$$[cov(\Delta\alpha, \Delta\alpha)]_{m\times m} \approx \left[\frac{\partial\Delta\alpha}{\partial X}\right]_{m\times mt} [cov(X, X)]_{mt\times mt} \left[\frac{\partial\Delta\alpha}{\partial X}\right]_{m\times mt}^T \tag{9.41}$$

Since the random variables, X_i and X_j $(i \neq j)$, are assumed independent,

$$[cov(X, X)]_{mt\times mt} = \begin{bmatrix} cov(X_1, X_1) & 0 & \cdots & 0 \\ 0 & cov(X_2, X_2) & \cdots & 0 \\ \cdots & \cdots & \cdots & \cdots \\ 0 & 0 & \cdots & cov(X_{mt}, X_{mt}) \end{bmatrix} \tag{9.42}$$

It is noted that the covariance matrix of the random variables, $cov(X, X)$, can be computed separately.

Random variables, X_p, are usually assumed to take the following form as

$$X_p = Ep \cdot N \tag{9.43}$$

where Ep is a constant defining the level of variation and N is a normally distributed vector with zero mean and unit standard deviation. This gives

$$cov(X_p, X_p) = \begin{bmatrix} (Ep)^2 & 0 & \cdots & 0 \\ 0 & (Ep)^2 & \cdots & 0 \\ \cdots & \cdots & \cdots & \cdots \\ 0 & 0 & \cdots & (Ep)^2 \end{bmatrix} \tag{9.44}$$

For the random variable, X_f, with the excitation force, is modelled as

$$X_f = Ep \cdot N \cdot \sigma_{\tilde{f}} \tag{9.45}$$

where $\tilde{f}$ is the vector of polluted force excitation; and $\sigma_{\tilde{f}}$ is the standard deviation of the measured force time history. Thus,

$$cov(X_f, X_f) = \begin{bmatrix} (Ep)^2 \cdot var(\tilde{f}) & 0 & \cdots & 0 \\ 0 & (Ep)^2 \cdot var(\tilde{f}) & \cdots & 0 \\ \cdots & \cdots & \cdots & \cdots \\ 0 & 0 & \cdots & (Ep)^2 \cdot var(\tilde{f}) \end{bmatrix} \tag{9.46}$$

where $var(\bullet)$ is the variance of the time history. For the random variables, $X_{\ddot{x}}$, which are the measurement noise in the acceleration response

$$X_{\ddot{x}} = Ep \cdot N \cdot \sigma_{\tilde{\ddot{x}}_d} \tag{9.47}$$

Then the covariance of $X_{\ddot{x}}$ is

$$
cov(X_{\ddot{x}}, X_{\ddot{x}}) = \begin{bmatrix} (Ep)^2 \cdot var(\tilde{\ddot{x}}_d) & 0 & \cdots & 0 \\ 0 & (Ep)^2 \cdot var(\tilde{\ddot{x}}_d) & \cdots & 0 \\ \cdots & \cdots & \cdots & \cdots \\ 0 & 0 & \cdots & (Ep)^2 \cdot var(\tilde{\ddot{x}}_d) \end{bmatrix} \tag{9.48}
$$

Hence, Equations (9.44), (9.46) and (9.48) give the covariance matrix of all the random variables in the present study.

9.5.1.7 *Statistical analysis in damage identification*

Statistical analysis of the identified results can be performed using Equations (9.39) and (9.41). Assuming that α^0 corresponds to the set of initial parameter changes in the analytical model, the updated set of parameter changes, α^1, will be given as

$$
\alpha^1 = \alpha^0 + \Delta\alpha^1 \tag{9.49}
$$

with an expectation value of

$$
E(\alpha^1) = E(\alpha^0) + E(\Delta\alpha^1) \tag{9.50}
$$

and a covariance of

$$
\begin{aligned}
[cov(\alpha^1, \alpha^1)]_{m \times m} &= cov(\alpha^0 + \Delta\alpha^1, \alpha^0 + \Delta\alpha^1) \\
&= cov(\alpha^0, \alpha^0) + cov(\alpha^0, \Delta\alpha^1) + cov(\Delta\alpha^1, \alpha^0) + cov(\Delta\alpha^1, \Delta\alpha^1)
\end{aligned} \tag{9.51}
$$

It is noted that Equations (9.50) and (9.51) will remain valid during the whole process of convergence of the identified results.

9.5.2 *Numerical example*

9.5.2.1 *The structure*

The three-dimensional five-bay cantilever steel frame structure studied in Sections 7.4 and 8.5 is shown in Figure 9.2. The finite element model consists of 37 three-dimensional Euler beam elements and 17 nodes. A summary of the main material and geometrical properties of the members of the frame structure is given in Table 7.1. The support conditions and dynamic characteristics of the structure are referred to Section 8.5.2.1.

A sinusoidal excitation is applied onto the structure at the 8th node in the z-direction with an amplitude of 3 N and at a frequency of 30 Hz. The acceleration response computed at the 5th node in the z-direction is taken as the 'measured' response and the first 500 data points are used to identify the damage. The sampling frequency is 2000 Hz. A damage case with 5% and 10% reduction of the flexural stiffness in the 7th member and the 26th member, respectively, is studied. The effect of each type of uncertainty on the damage identification is discussed below.

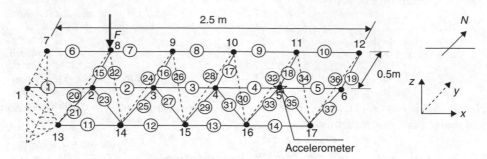

Figure 9.2 A five-bay three-dimensional frame structure

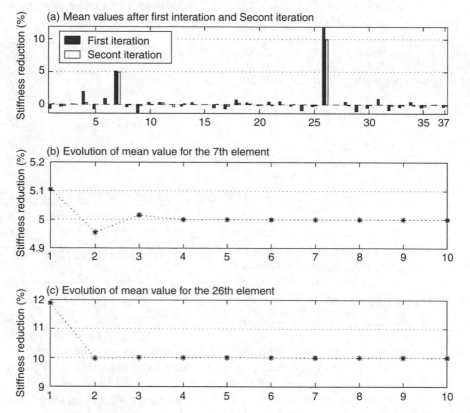

Figure 9.3 Mean value of identified results and its evolution with iterations due to uncertainty in mass density

9.5.2.2 Uncertainty with the mass density

The uncertainty of mass density of material is assumed to have a 1% amplitude with $Ep = 0.01$ in Equation (9.43), and damage identification is performed on the structure using the proposed approach. The mean stiffness change, $\mu_{\Delta\alpha}$, obtained from Equation (9.19) for all the elements in the structure are shown in Figure 9.3(a) after the first and

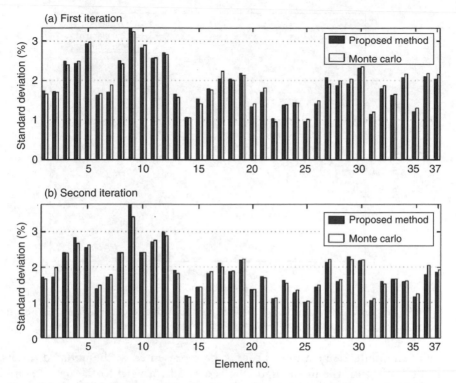

Figure 9.4 Standard deviation of identified results due to uncertainty in mass density

second iteration of the computation. The standard deviation of the stiffness change, $\sigma_{\Delta\alpha}$, for all the elements obtained from the proposed method after the first and second iterations are shown in Figure 9.4. They are compared with those computed from the Monte Carlo method (Robert and Casella, 1999) from 1000 samples of data. The two sets of standard deviations are very close to each other, indicating that the statistical method is correct. The standard deviation from the first iteration ranges from 0.96% to 3.34% of the identified flexural stiffness of the member when there is only 1% amplitude in the variation of the mass density of the analytical model. This indicates that the uncertainty amplifies the error in the identified results. The standard deviation in Figure 9.4(b) after the second iteration ranges from 1.01% to 3.77% and they are of similar magnitude to those after the first iteration. This indicates that the amplifying effect of the variation in the mass density is not significant in the second iteration, since the bulk of the damage vector has been updated in the first iteration, as shown in Figure 9.3(a).

To study the effect of the variation throughout the iterations, 10 iterations were performed and the mean values of the damaged members are shown in Figures 9.3(b) and 9.3(c). The standard deviations for the 7th, 18th and 26th members are shown in Figure 9.5. The damage parameters converge quickly to the true values with only four iterations. The standard deviation in Figure 9.5 also converges quickly to a constant with increasing iterations in all the elements. Since the uncertainty with the mass

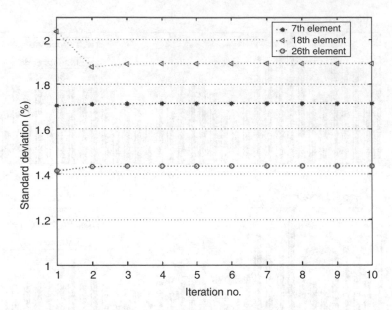

Figure 9.5 Evolution of standard deviation with iterations due to uncertainty in mass density

density of the finite element model cannot be represented in the updated results, its detrimental effect on the model updating is carried forward to the next iteration in the form of variation in the identified result. It is only after four iterations that the correct reductions in the 7th and 26th members are updated, and the variation of the identified results becomes stabilized with a constant standard deviation.

Equation (9.41) shows that the standard deviation of the damage vector is linearly related to the amplitude of the uncertainty, Ep. Only the computation for the case with a unit variation is necessary, as it can easily be extended to other cases with different variation amplitudes. It should be noted that all the statistical analysis in this study is restricted to a small variation such that the linear approximation assumption in Equation (9.18) is valid.

9.5.2.3   Uncertainty with the elastic modulus of material

When a 1% variation is included in the elastic modulus of material of the initial analytical model, the mean values of the identified flexural stiffness are the same as those shown in Figure 9.3, as they are also computed from Equation (9.19). The standard deviation of the identified results is shown in Figure 9.6 alongside those obtained from the Monte Carlo technique. The standard deviation for all members has a maximum value of 1.68% after the first iteration, and it drops to 0.26% after the second iteration. This shows that (a) the effect of random variation in the elastic modulus is comparable to that of the mass density after the first iteration; and (b) the significant reduction after the second iteration shows that the flexural stiffness, which is closely related to the elastic modulus, has been updated with the mitigation of the associated variation effect in the subsequent iterations.

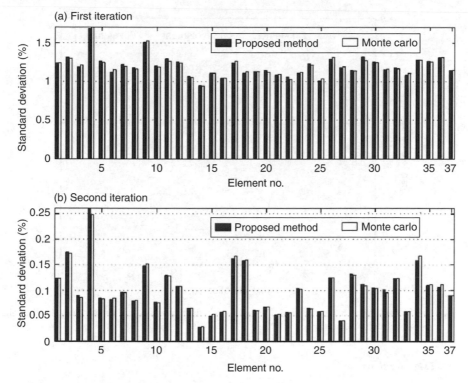

Figure 9.6 Standard deviation of identified results due to uncertainty in the elastic modulus of elasticity

To study the effect of variation with iterations, 10 iterations are performed and the mean values of the identified results are shown in Figures 9.3(b) and 9.3(c), while the standard deviations for the 7th, 18th and 26th members are shown in Figure 9.7. The flexural stiffness converges quickly to the true values with only four iterations and the standard deviations in Figure 9.7 also reduce quickly to zero with about four iterations. The standard deviation for other undamaged members also exhibits this behaviour. This indicates that the effect of variation in the stiffness parameter can be fully represented in the updated model after a few iterations with no residual uncertainties in the updated finite element model.

9.5.2.4 *Uncertainty with the excitation force and the measured response*

For the case of including the 1% variation as defined in Equation (9.45) for the excitation force, the observations and the statistics of the identified results are similar to those found for the case of the mass density variation and the details are not discussed any further. The case with 1% random noise, as defined in Equation (9.47), added into the 'measured' acceleration responses from the damaged structures to simulate the measurement noise is not discussed for the same reason.

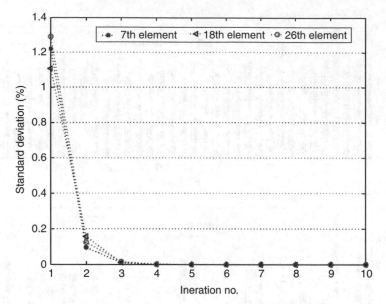

Figure 9.7 Evolution of standard deviation with iterations due to uncertainty in the elastic modulus of material

9.5.3 *Discussions*

As a summary of the above discussions, the effect of variation in the flexural stiffness of a member is least significant and can be mitigated with increasing iterations of model updating. The effect of variation in the mass density of the finite element is more significant, and those for the excitation force and the structural response are highly significant.

Results show that the uncertainties in the system parameters remain in the updating process in the form of a coefficient of variation in the identified results. The standard deviation does not diverge when the bulk of the damage vector is obtained after a few iterations. It should be noted that the assumptions of zero mean and normal distribution for the uncertainties are the basis of the present study. If the zero mean assumption is not valid, Equation (9.39) is also not valid. If the normal random distribution is not satisfied, the distribution of the identified parameter will not be normal distribution, and this will add difficulties to the subsequent reliability analysis with the identified results.

9.6 Integration of system uncertainties with the reliability analysis of a box-section bridge deck structure

The conventional capacity rating of a bridge structure has been largely based on the Working Stress Rating criteria, which do not systematically take into account any information on the variability of strength and loading, the extent of damage or deterioration and the characteristics of actual response or redundancy specific to the structure. Thus, the nominal rating load or reserve capacity evaluated by the conventional code-related formula, in general, fails to predict the realistic carrying capacity or reserve capacity of deteriorated or damaged bridges. Reliability-based capacity rating has since been suggested to predict a realistic relative reserve safety by incorporating actual bridge conditions and uncertainties.

The structural reliability of several types of pre-stressed concrete beams has been compared when designed to the code requirements of different countries (Nowak et al., 2001; Du and Au, 2005). Different live load models and resistance models are used for different countries. The ultimate limit state of flexural and shear capacities and serviceability limit states (e.g. tension stress in concrete) are considered in similar studies. The uncertainties in the analysis are expressed in terms of a bias factor and coefficient of variation. Haukaas and Kiureghian (2005) have attempted to rank the relative significance of different types of uncertainties using the reliability sensitivity measures derived by Bjerager and Krenk (1989). However, this reliability sensitivity is not directly related to the real conditions of the structure when under structural health monitoring. Such a link is made possible by Xia et al. (2002), who related the uncertainty with the structural modal parameter in terms of a sensitivity matrix. However, this work remained one step short of the integration of the reliability analysis with structural health monitoring until Hosser et al. (2008) proposed a framework for such study. Frangopol et al. (2008a; 2008b) and Catbas et al. (2008) later carried out the reliability analysis of bridge structures using the long-term structural health monitoring data.

Section 9.5 discussed the system uncertainties that are related explicitly to the identified local anomalies, such that the latter has a variance distribution. This statistical feature of the identified result thus carries the information of the system uncertainty of the initial system. This characteristic of the statistics enables the integration of structural health monitoring with the subsequent reliability analysis of the structure. The reliability index can then be updated from this improved set of statistics.

9.6.1 Numerical example

The bridge-vehicle system studied in Section 7.5.3 is again investigated for an updating of the reliability of a bridge structure with additional information from the structural condition assessment results, including the system uncertainties. Excitation generated by the moving vehicle shown in Figure 7.21 serves as an input to the structure while the acceleration responses recorded at the top of the bridge deck are used for the assessment using the response sensitivity-based method described in Chapter 7. The uncertainties in the elastic modulus of concrete and the measurement noise are included in the condition assessment procedure to obtain the statistics for the identified results with propagation in the condition assessment process, as discussed in Section 9.5. These statistics are then included in the reliability analysis to give an updated set of safety predictions on the bridge structure.

A damage zone involving a group of elements in the bridge deck is simulated with 20% reduction in the elastic modulus of material in the web elements of the central span section – i.e. in the 18th and 19th elements in the left web, the 54th and 55th elements in the middle web and the 90th and 91st elements in the right web, as indicated in Figure 9.8. In the structural condition assessment process, the uncertainties included are the elastic modulus of concrete with a coefficient of variation of 0.08 and 1% random noise in the measured responses.

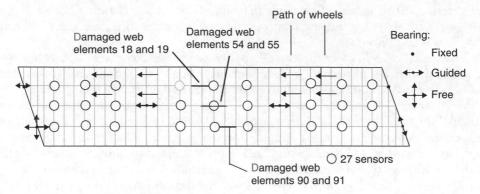

Figure 9.8 Plan of bridge and sensor configuration

Damage detection is performed with the two-axle three-dimensional vehicle crossing the bridge deck along the travel path shown in Figure 9.8. It is noted that the vehicle travels on top of the left web. The velocity of the vehicle is 20 m/s and the sampling rate is 100 Hz. Road surface roughness Class C (ISO, 1995) is included in the analysis. It is noted that Class C road roughness corresponds to the average road pavement conditions, which is modelled as a noise distribution in the measured responses in addition to the 1% measurement noise. The parameters of the vehicle are taken from Marchesiello et al. (1999) with a mass of 17000 kg.

9.6.2 *Condition assessment*

Only the 108 web elements are included as candidates in the assessment. Each inter-action force is modelled with 600 terms in its orthogonal expansions. Since the interaction forces of the four wheels and local damages in the structural elements are all unknowns, acceleration responses recorded at 27 locations on top of the deck as shown in Figure 9.8 are used in the identification to ensure that the identifica-tion equation is over-determined. The vehicle passes over the deck in 3.75 seconds, and there are 375 time instances with the vehicle moving on top of the bridge deck. There are $27 \times 375 = 10125$ equations for solving the $4 \times 600 + 108 = 2508$ number of unknowns in the inverse problem.

The interaction forces calculated are kept as reference for comparison, while the calculated responses of the deck are used for the inverse identification. Figures 9.9 and 9.10 give the identified interaction force-time histories and the identified local damages in the bridge deck. The identified wheel-load-time histories from 'unpol-luted' measured responses overlap with the true values, and the identified results from measurement with 1% noise are very accurate with a small relative error less than 5%. Table 9.1 gives the associated information for convergence. There are some false positives in elements adjacent to the damages in Figure 9.10.

Figure 9.11 shows the coefficient of variation (COV) of the elastic constant in all the web elements obtained after three iterations using the true or identified interaction forces with a COV of 0.08 in the elastic modulus and 1% noise in the measured

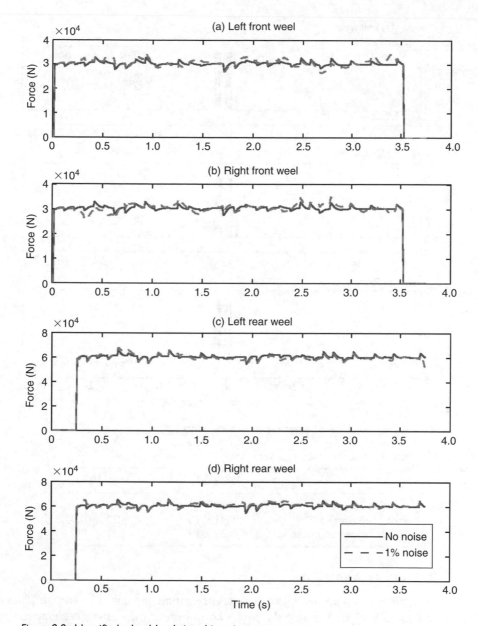

Figure 9.9 Identified wheel-load-time histories

responses. The two sets of COV results are close to each other. Since the computation of the standard deviations including uncertainties is very time consuming, only three iterations are performed here. It is found that the results obtained after the second and third iterations are converging and therefore the standard deviations after the third iteration are adopted as the final result.

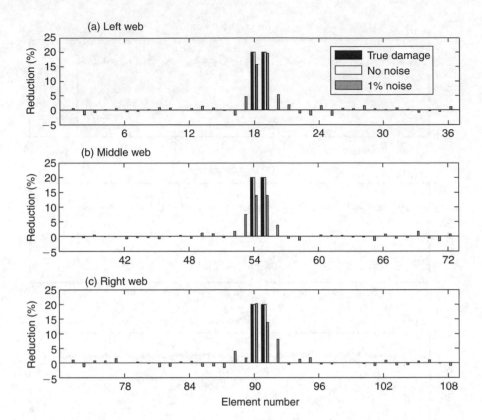

Figure 9.10 Identified local damages in the web elements

Table 9.1 Information on convergence

	No noise	With 1% noise
Required iterations	5	3
Error of convergence	1.41×10^{-5}	4.32×10^{-11}
Regularization parameter λ	4.22×10^{-17}	2.31×10^{-4}

9.6.3 Reliability analysis

The component and system reliability analyses are conducted based on the statistics in Figures 9.10 and 9.11 to evaluate the safety condition of the damaged elements and the bridge system. The former is assessed based on the most severe bending tensile stress effect that occurs in the elements when the live load is on the central span. According to the design code, the bridge carriageway was divided into three notional lanes, and 25 units of HB loading are placed on one notional lane of the carriageway with HA loading on the other two lanes, as shown in Figure 9.12. The working tensile stress under the damaged state is calculated using the 'equivalent strain' assumption (Voyiadjis and Kattan, 2005). For the system reliability analysis, the HB loading of the above load combination is incrementally increased, such that the deflection of

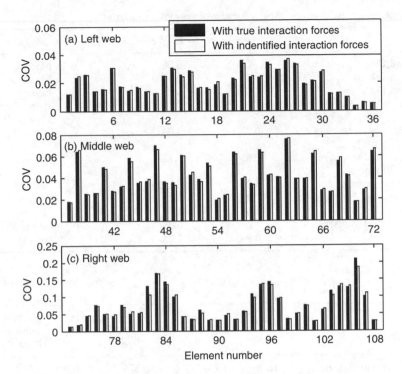

Figure 9.11 Coefficient of variation of identified results from different forces

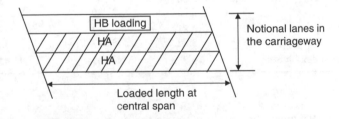

Figure 9.12 Distribution of HA and HB loading in reliability analysis

the bridge deck approaches the limiting value. The maximum deflection criterion is defined in this study as 8/100 of the span length under a general live load according to the AASHTO LRFD code. This criterion has been widely adopted for bridge structures.

The statistical parameters of the basic variables of the loading and structure are shown in Table 9.2. Three types of reliability indices are calculated to study the effect of uncertainties with different combinations of damage extent and COVs. They are:

Type I: intact structure with initial COV for all the elements;
Type II: structure with true damage in the damaged elements and initial COV for all the elements;
Type III: structure with identified damage and identified COV, as shown in Figures 9.10 and 9.11.

Table 9.2 Statistical parameters of basic variables in reliability analysis

Variable	Bias factor	Coefficient of variation	Distribution type
Dead load, D1	1.03	0.08	Normal
Dead load, D2	1.05	0.10	Normal
Dead load, D3	1.0	0.25	Normal
Live load	1.15	0.18	Type I Extreme value Distribution
Pre-stress – new structure (age < 5 years)	0.97	0.04	Normal
Pre-stress – old structure (age > 30 years)	0.85	0.12	Normal
Area of pre-stress strands	1.0	0.0125	Normal
Tensile strength of concrete (f_t)	1.0	0.23	Normal
Modulus of elasticity	1.0	0.08	Normal
Dimensions (thickness)	1.0	0.05	Normal
Model coefficient	0.945	0.03	Normal

Table 9.3 Reliability index

		Intact with initial COV (Type I)	True damage with initial COV (Type II)	Identified damage with identified COV (Type III)
Component reliability index	18th	1.268	0.876	1.050
	19th	2.481	2.124	2.128
	54th	2.069	1.747	1.881
	55th	2.736	2.420	2.526
	90th	2.814	2.367	2.397
	91st	3.723	3.517	3.687
System reliability index		4.953	4.909	4.901

The Type I index gives the baseline before the damage occurrence and is the set of indices corresponding to the design stage. The Type II index is associated with the damage occurrence, but without the uncertainty effect from the condition assessment. This is the conventional reliability index computed from an estimated damage pattern. The Type III index is associated with the assessment results incorporating the propagation of uncertainty in the structure. This is the set of updated reliability indices incorporating the effect from the structural condition assessment.

The component and system reliability indices are computed and they are shown in Table 9.3. The HB loading is placed approximately on top of the left web (Figure 9.8) creating high tensile stresses in elements 18 and 19 compared to other elements and the original component reliability indices (Type I) are smaller than the others. The effect of local damage reduces the corresponding Type II and Type III indices. When the statistics of the identified damages are further included in the analysis, the component reliability indices (Type III) have a notable difference compared with those of Type II. (It cannot, however, be concluded that the Type III component reliability indices will be larger than those of Type II, as it depends on the values of both the identified mean and COVs of the variables). It is also noted that the system reliability does not decrease dramatically since the damage in a few elements has a comparatively small effect on the whole system. The Type III reliability analysis results listed in Table 9.3 give the updated safety predictions on the bridge deck including the effect of structural condition assessment of the structure.

9.7 Conclusions

Uncertainty plays a significant role in structural damage identification and condition assessment. Much effort has been made to quantify their effect in the health monitoring of a structure. However, their effect is always mixed with those from the local damages under study. The Karhunen–Loéve expansion is a convenient approach to model these uncertainties with an attempt to include their effect in the assessment, and this approach is becoming more popular with the health monitoring community. The modelling of the different random variables in the structural condition monitoring can be further integrated with the reliability analysis of a structure, whereby the statistics of the system parameters are altered resulting in an improved set of reliability indices.

References

Abraham, M.A., Park, S.Y. and Stubbs, N. (1995) Loss of prestress prediction on nondestructive damage location algorithms. *SPIE, Smart Structures and Materials*, 2456: 60–67.

Abramovich, H. (1992) Natural frequencies of Timoshenko beams under compressive axial loads. *Journal of Sound and Vibration*, 157(1): 183–189.

Ahmadian, H., Mottershead, J.E. and Friswell, M.I. (1998) Regularization methods for finite element model updating. *Mechanical Systems and Signal Processing*, 12(1): 47–64.

Andelfinger, U. and Ramm, E. (1992) *An Assessment of Hybrid-mixed Four Node Shell Elements. Nonlinear Analysis of Shells by Finite Elements*. Edited by F.G. Rammerstorfer. CISM Courses and Lectures No. 328. International Centre for Mechanical Sciences. Springer-Verlag.

Ang, A.H-S. and De Leon, D. (2005) Modeling and analysis of uncertainties for risk-informed decisions in infrastructure engineering. *Structure and Infrastructure Engineering*, 1(1), 19–32. Taylor & Francis.

Ang, A.H-S. and Tang, W.H. (1984) *Probability Concepts in Engineering Planning and Design Volume II*. New York: John Wiley & Sons.

Ang, A.H-S. and Tang, W.H. (2007) *Probability Concepts in Engineering Planning and Design Volume II*, 2nd ed. New York: Wiley.

Bagarello, F. (1995) Multiplication of distribution in one dimension: possible approaches and applications to δ-function and its derivatives. *Journal of Mathematic Analysis and Application*, 196(3): 885–901.

Bakhtiari-Nejad, F., Rahai, A. and Esfandiari, A. (2005) A structural damage detection method using static noisy data. *Engineering Structures*, 27(12): 1784–1793.

Banan, M.R. and Hjelmstad, K.D. (1994) Parameter estimation of structures from static response. I. Computational aspects. *Journal of Structural Engineering ASCE*, 120(11): 3243–3258.

Barbero, E.J. and Reddy, J.N. (1991) Modelling of delamination in composite laminates using a layer-wise plate theory. *International Journal of Solids and Structure*, 28(3): 373–388.

Baschmid, A., Dizua, G. and Pizzigoni, B. (1984) The influence of unbalance on cracked rotors. *Proceedings of the International Conference on Vibration of Rotating Machinery*, 193–198.

Beck, J.L. and Katafygiotis, L.S. (1998) Updating models and their uncertainties. I: Bayesian statistical frameworks. *Journal of Engineering Mechanics ASCE*, 124(4): 455–461.

Bendat, J.S. and Piersol, A.G. (1980) *Engineering Applications of Correlation and Spectral Analysis*. New York: John Wiley.

Berman, A. and Flannely, W.G. (1971) Theory of incomplete models of dynamic structures. *AIAA Journal*, 9(8): 1481–1487.

Bjerager, P. and Krenk, S. (1989) Parametric sensitivity in first order reliability theory. *Journal of Engineering Mechanics ASCE*, 115(7): 1577–1582.

Blair, M.A., Camino, T.S. and Dickens, J.M. (1991) An iterative approach to a reduced mass matrix. *Proceedings of the 8th International Modal Analysis Conference*, Florence, Italy, 510–515.

Briggs, J.C. and Tse, M.K. (1992) Impact force identification using extracted modal parameters and pattern matching. *International Journal of Impact Engineering*, 12(3): 361–372.

British Standard Institution. (1990) *British Standard-Structural Use of Steelwork in Building*. BS5950.

Buda, G. and Caddemi, S. (2007) Identification of concentrated damages in Euler-Bernoulli beams under static loads. *Journal of Engineering Mechanics ASCE*, 133(8): 942–956.

Bui, H.D. (1994) Inverse problems in engineering mechanics. *Proceedings of the Second International Symposium on Inverse Problems—ISIP '94*, Paris, France, 2–4 November 1994. Rotterdam; Brookfield, Vt.: A.A. Balkema.

Caddemi, S. and Greco, A. (2006) The influence of instrumental errors on the static identification of damage parameters for elastic beams. *Computers and Structures*, 84(26–27): 1696–1708.

Catbas, F.N., Susoy, M. and Frangopol, D.M. (2008) Structural health monitoring and reliability estimation: long span truss bridge application with environmental monitoring data. *Engineering Structures*, 30(9): 2347–2359.

Cawley, P. and Adams, R.D. (1979) The location of defects in structures from measurement of natural frequencies. *Journal of Strain Analysis*, 14: 49–57.

Chan, H.C., Cheung, Y.K. and Huang, Y.P. (1992) Crack analysis of reinforced concrete tension members. *Journal of Structural Engineering ASCE*, 118(8): 2118–2132.

Chan, H.C., Cheung, Y.K. and Huang, Y.P. (1993) Analytical crack model for reinforced-concrete structures. *Journal of Structural Engineering ASCE*, 119(5): 1339–1357.

Chan, S.L. (1994) Vibration and modal analysis of steel frames with semi-rigid connections. *Engineering Structures*, 16(1): 25–31.

Chan, S.L. and Chui, P.P.T. (2000) *Non-linear Static and Cyclic Analysis of Steel Frames with Semi-rigid Connections*. Elsevier.

Chan, S.L. and Zhou, Z.H. (1998) On the development of a robust element for second-order non-linear integrated design and analysis. *Journal of Constructional Steel Research*, 47(1–2): 169–190.

Chang, C.C. and Chen, L.W. (2003) Vibration damage detection of a Timoshenko beam by spatial wavelet based approach. *Applied Acoustics*, 64(12): 1217–1240.

Chaudhry, Z. and Ganino, A.J. (1994) Damage detection using neutral network: an initial experimental study on debonded beams. *Journal of Intelligent Material Systems and Structures*, 5(4): 585–589.

Chen, J.C. and Garba, J.A. (1980) Analytical model improvement using modal testing results. *AIAA Journal*, 18(6): 684–690.

Chen, J. and Li, J. (2004) Simultaneous identification of structural parameters and input time history from output-only measurements. *Computational Mechanics*, 33(5): 365–374.

Chen, W.F., Takanashi, K., Nakashima, M. and White, D. (1998) US-Japan seminar on innovations in stability concepts and methods for seismic design in structural steel. *Engineering Structures*, 20(4–6): 239–569.

Choi, C.K., Kim, K.H. and Hong, H.S. (2002) Spline finite strip analysis of prestressed concrete box-girder bridges. *Engineering Structures*, 24(12): 1575–1586.

Chui, P.P.T. and Chan, S.L. (1997) Vibration and deflection characteristics of semi-rigid jointed frames. *Engineering Structures*, 19(12): 1001–1010.

Coifman, R.R. and Wickerhauser, M.V. (1992) Entropy-based algorithms for best basis selection. *IEEE Transactions on Information Theory*, 38(2): 713–718.

Collins, J.D. and Hart, G.C., Hasselman, T.K. and Kennedt, B. (1974) Statistical identification of structures. *AIAA Journal*, 12: 185–190.

Cornwell, P., Doebling, S.W. and Farrar, C.R. (1999) Application of the strain energy damage detection method to plate-like structures. *Journal of Sound and Vibration*, 224(2): 359–374.

Dallard, P., Fitzpatrick, A.J., Flint, A., Bourva, S.L.e, Low, A., Ridsdill, R.M. and Willford, M. (2001) The London Millennium Footbridge. *The Structural Engineer*, 79(22): 17–33.

Daubechies, I. (1988) Orthonormal bases of compactly supported wavelets. *Communications on Pure and Applied Mathematics*, 41(7): 909–996.

Daubechies, I. (1992) *Ten Lectures on Wavelets*. Philadelphia, PA: SIAM.

Devriendt, C. and Fontul, M. (2005) Transmissibility in coupled structures: identification of the sufficient set of coupling forces. *First International Operational Modal Analysis Conference*, Copenhagen, Denmark, April, 2005. Edited by Rune Brincker and Nis Møller. Published by Aalborg University: 445–453.

Diaz Valdes, S.H. and Soutis, C. (1999a) Application of the rapid frequency sweep technique for delamination detection in composite laminates. *Advanced Composite Letters*, 8(1): 19–23.

Diaz Valdes, S.H. and Soutis, C. (1999b) Delamination detection in composite laminates from variations of their modal characteristics. *Journal of Sound and Vibration*, 228(1): 1–9.

Dimarogonas, A.D. (1996) Vibration of cracked structures: a state of the art review. *Engineering Fracture Mechanics*, 55(5): 831–857.

Di Paola, M. and Bilello, C. (2004) An integral for damage identification of Euler-Bernoulli beams under static loads. *Journal of Engineering Mechanics ASCE*, 130(2): 225–234.

Doebling, S.W. (1995) *Measurement of Structural Flexibility Matrices for Experiments with Incomplete Reciprocity*. PhD Dissertation, Aerospace Engineering Sciences Dept., University of Colorado, Boulder, CO.

Doebling, S.W. (1996) Damage detection and model refinement using elemental stiffness perturbations with constrained connectivity. *Proceedings of the AIAA/ASME/AHS Adaptive Structures Forum*: 360–370.

Doebling, S.W., Peterson, L.D. and Alvin, K.F. (1998a) Experimental determination of local structural stiffness by disassembly of measured flexibility matrices. *Journal of Vibration and Acoustics ASME*, 120(4): 949–957.

Doebling, S.W., Farrar, C.R. and Prime, M.B. (1998b) A summary review of vibration-based damage identification methods. *Shock and Vibration Digest*, 30(2): 91–105.

Douka, E., Loutridis, S. and Trochidis, A. (2003) Crack identification in beams using wavelet analysis. *International Journal of Solids and Structures*, 40(13): 3557–3569.

Doyle, J.F. (1994) A genetic algorithm for determining the location of structural impacts. *Experimental Mechanics*, 34(1): 37–44.

Du, J.S. and Au, F.T.K. (2005) Deterministic and reliability analysis of prestressed concrete bridge girders: comparison of the Chinese, Hong Kong and AASHTO LRFD Codes. *Structural Safety*, 27(3): 230–245.

Efron, B. and Tibshirani, R.J. (1994) *An Introduction to the Bootstrap*. New York: Chapman & Hall.

Ellingwood, B.R. (2005) Risk-informed condition assessment of civil infrastructure: state of practice and research needs, *Structure and Infrastructure Engineering*, 1(1), 7–18. Taylor & Francis.

Farrar, C.R., Duffey, T.A., Doebling, S.W. and Nix, D.A. (1999) A statistical pattern recognition paradigm for vibration-based structural health monitoring. *Proceedings of the 2nd International Workshop on Structural Health Monitoring*, 764–773.

Figueiras, J.A. and Póvoas, H.H.C.F. (1994) Modelling of prestress in non-linear analysis of concrete structures. *Computers & Structures*, 53(1): 173–187.

Frangopol, D.M. and Liu, M. (2007) Maintenance and management of civil infrastructure based on condition, safety, optimization, and life-cycle cost. *Structure and Infrastructure Engineering*, 3(1), 29–41.

Frangopol, D.M., Lin, K-Y. and Estes, A.C. (1997) Life-cycle cost design of deteriorating structures. *Journal of Structural Engineering ASCE*, 123(10), 1390–1401.

Frangopol, D.M., and Messervey, T.B. (2009a). "Maintenance principles for civil structures," Chapter 89 in *Encyclopedia of Structural Health Monitoring*, C. Boller, F-K. Chang, and Y. Fujino, eds., John Wiley & Sons Ltd, Chicester, UK, Vol. 4, 1533–1562.

Frangopol, D.M., and Messervey, T.B. (2009b). "Life-cycle cost and performance prediction: Role of structural health monitoring," in *Frontier Technologies for Infrastructures Engineering*, Chapter 16, S-S, Chen and A.H-S. Ang, eds., CRC Press, Taylor & Francis, London, 361–381.

Frangopol, D.M., Strauss, A. and Kim, S.Y. (2008a) Bridge reliability assessment based on monitoring. *Journal of Bridge Engineering ASCE*, 13(3): 258–270.

Frangopol, D. M., Strauss, A., and Kim, S. (2008b). "Use of monitoring extreme data for the performance prediction of structures: General approach" *Engineering Structures*, Elsevier, 30 (12), 3644–3653.

Fregolent, A., D'ambrogio, W., Salvini, P. and Sestieri, A. (1996) Regularisation techniques for dynamic model updating using input residual. *Inverse Problems in Engineering*, 2: 171–200.

Friswell, M.I. and Mottershead, J.E. (1995) *Finite Element Model Updating in Structural Dynamics*. Kluwer Academic Publisher: Dordrecht, The Netherlands.

Friswell, M.I. and Mottershead, J.E. (1996) *Identification in engineering systems: Proceedings of the conference held at Swansea*, March 1996. Swansea [England]: University of Wales Swansea.

Friswell, M.I., Garvey, S.D. and Penny, J.E.T. (1995) Model reduction using dynamic and iterated IRS techniques. *Journal of Sound and Vibration*, 186(2): 311–323.

Friswell, M.I., Mottershead, J.E. and Lees, A.W. (1999) *Identification in engineering systems: proceedings of the second international conference held in Swansea*, March 1999. Publisher [Swansea : University of Wales].

Fox, R.L. and Kappor, M.R. (1968) Rates of change of eigenvalues and eigenvectors. *AIAA Journal*, 6(12): 2426–2429.

Gao, L. and Haldar, A. (1995) Nonlinear seismic analysis of space structures with partially restrained connections. *Microcomputer Civil Engineering*, 10(1): 27–37.

Ge, L. and Soong, T.T. (1998) Damage identification through regularization method. I: Theory. *Journal of Engineering Mechanics ASCE*, 124(1): 103–108.

Gentile, A. and Messina, A. (2003) On the continuous wavelet transforms applied to discrete vibrational data for detecting open cracks in damaged beams. *International Journal of Solids and Structures*, 40(2): 295–315.

Ghanem, R. and Spanos, P.D. (1991) *Stochastic Finite Elements: A Spectral Approach*, New York: Springer-Verlag, Inc.

Gladwell, G.M.L. and Ahmadian, H. (1995) Generic element matrices suitable for finite element model updating. *Mechanical Systems and Signal Processing*, 9(6): 601–614.

Griffin, S.F. and Sun, T.C. (1991) Health monitoring of dumb and smart structures. *Proceedings of the 28th Annual Meeting of SES*, November 6–8, Gainsville, Florida.

Grigorian, C.E., Yang, T.S. and Popov, E.P. (1992) Slotted bolted connection energy dissipators. *Report of National Science Foundation*, University of California, Berkeley.

Goldfarb, D. and Idnani, A. (1983) A numerically stable dual method for solving strictly convex quadratic problems. *Mathematical Programming*, 27(1): 1–33.

Golub, G.H. (1996) *Matrix Computation*. Johns Hopkins University Press.

Golub, G.H., Heath, M. and Wahba, G. (1979) Generalized cross-validation as a method for choosing a good ridge parameter. *Technometrics*, 27(2): 215–223.

Gontier, C. and Bensaibi, M. (1995) Time domain identification of a substructure from in situ analysis of the whole structure. *Mechanical Systems and Signal Processing*, 9(4): 379–396.

Gordis, J.H. (1992) An analysis of the improved reduced system (IRS) model reduction procedure. *Proceedings of the 7th International Modal Analysis Conference*, San Diego, California, February: 471–479.

Guo, H.Y. (2006) Structural damage detection using information fusion technique. *Mechanical Systems and Signal Processing*, 20(5): 1173–1188.

Guo, L., Li, Z.X. and Chen, H.T. (2006) Multi-stage model updating method via substructure analysis. *Engineering Science*, 8(9): 42–48.

Gurley, K. and Kareem, A. (1999) Application of wavelet transform in earthquake, wind and ocean engineering. *Engineering Structures*, 21(2): 149–167.

Guyan, R.J. (1965) Reduction of stiffness and mass matrices. *AIAA Journal*, 3(2): 380.

Hansen, P.C. (1990) The discrete Picard condition for discrete ill-posed problems. *BIT*, 30: 658–672.

Hansen, P.C. (1992) Analysis of discrete ill-posed problems by means of the L-curve. *SIAM Review*, 34(4): 561–580.

Hansen, P.C. (1994) Regularization tools: A Matlab package for analysis and solution of discrete ill-posed problems. *Numerical Algorithms*, 6: 1–35.

Hao, H. and Xia, Y. (2005) A review of uncertainty effect in structural condition assessment. *Proceedings of the 9th International Conference on Inspection Appraisal Repairs and Maintenance of Structures*, Singapore, CI Premier Pte Ltd. 29–44.

Haukaas, T. and Der Kiureghian, A. (2005) Parameter sensitivity and importance measures in nonlinear finite element reliability analysis. *Journal of Engineering Mechanics ASCE*, 131(10): 1013–1026.

Hemez, F.M. and Farhat, C. (1994) An energy based optimum sensor placement criterion and its application to structural damage detection. *Proceedings of the 12th IMAC*, 1568–1575.

Herrera, I. (1965) Dynamic models for Masing type materials and structure. *Boletin Sociedad Mexicana De Ingenieria Sismica*, 3(1): 1–8.

Hjelmstad, K.D. and Banan, M.R. (1995) Time-domain parameter estimation algorithm for structures I: Computational aspects. *Journal of Engineering Mechanics ASCE*, 121(3): 424–434.

Hjelmstad, K.D. and Shin, S. (1997) Damage detection and assessment of structures from static response. *Journal of Engineering Mechanics ASCE*, 123(6): 568–576.

Hoff, C. and Natke, H.G. (1989) Correction of a finite element model by input-output measurements with application to a radar tower. *The International Journal of Analytical and Experimental Modal Analysis*, 4(1): 1–7.

Holland, J. (1975) *Adaptation in Natural and Artificial Systems*. Cambridge, Mass.: MIT Press.

Hong, J.C., Kim, Y.Y., Lee, H.C. and Lee, Y.W. (2002) Damage detection using the Lipschitz exponent estimated by the wavelet transform: applications to vibration modes of a beam. *International Journal of Solids and Structures*, 39(7): 1803–1816.

Hosser, D., Klinzmann, C. and Schnetgoke, R. (2008) A framework for reliability-based system assessment based on structural health monitoring. *Structure and Infrastructure Engineering*, 4(4): 271–285.

Hoyos, A. and Aktan, A.E. (1987) *Regional Identification of Civil Engineered Structures Based on Impact Induced Transient Responses*. Research Report 87-1, Louisiana State University, Baton Rouge.

Hua, X.G., Ni, Y.Q., Chen, Z.Q. and Ko, J.M. (2007) Structural reliability as a measure for assessing the quality of stochastically updated finite element model. *Proceedings of the 2nd International Conference of Structural Condition Assessment, Monitoring and Improvement*. Beijing, China: Science Press: 65–72.

Irons, B. (1965) Structural eigenvalue problems elimination of unwanted variables. *AIAA Journal*, 3(5): 961–962.

Islam, A.S. and Craig, K.C. (1994) Damage detection in composite structures using piezoelectric materials. *Smart Materials and Structures*, 3(3): 318–328.

ISO 8606:1995(E). (1995) Mechanical vibration—road surface profiles—reporting of measured data.

James, III G.H., Carne, T.G. and Lauffer, J.P. (1995) The natural excitation technique (Next) for modal parameters extraction from operating structures. Modal analysis. *The International Journal of Analytical and Experimental Modal Analysis*, 10(4): 260–277.

Jiang, S.F., Chan, G.K. and Zhang, C.M. (2005) Data fusion technique and its application in structural health monitoring. *Structural Health Monitoring and Intelligent Infrastructure*. Ou, Li & Duan (eds). London: Taylor & Francis Group. ISBN 0 415 39652 2: 1125–1130.

Jones, S.W., Kerby, P.A. and Nethercot, D.A. (1982) Columns with semi-rigid joints. *Journal of Structural Engineering ASCE*, 108(2): 361–372.

Juang, J.N. and Pappa, R.S. (1985) An eigensystem realization algorithm for modal parameters identification and model reduction. *Journal of Guidance, Control and Dynamics*, 8(5): 620–627.

Kammer, D.C. (1991) Sensor placement for on-orbit modal identification and correlation of large space structures. *Journal of Guidance, Control and Dynamics*, 14(2): 251–259.

Kammer, D.C. (1997) Estimation of structural response using remote sensor locations. *Journal of Guidance, Control and Dynamics*, 20(3): 501–508.

Katafygiotis, L.S. and Beck, J.L. (1998) Updating models and their uncertainties. II: Model identifiability. *Journal of Engineering Mechanics ASCE*, 124(4): 463–467.

Kijewski, T. and Kareem, A. (2003) Wavelet transform for system identification in civil engineering. *Computer-Aided Civil and Infrastructure Engineering*, 18(5): 339–355.

Kim, J.T., Yun, C.B., Ryu, Y.S. and Cho, H.M. (2004) Identification of prestress-loss in PSC beams using modal information. *Structural Engineering and Mechanics*, 17(3–4): 467–482.

Krawczuk, M. (1993) A rectangular plate finite element with an open crack. *Computers and Structures*, 46(3): 487–493.

Krawczak, M. and Ostachowicz, W. (1994) A finite plate element for dynamic analysis of a crack plate. *Computer Methods in Applied Mechanics and Engineering*, 115: 67–78.

Krawczuk, M. and Ostachowicz, W. (2002) Identification of delamination in composite beams by genetic algorithm. *Science and Engineering of Composite Materials*, 10(2): 147–155.

Koh, C.G., Hong, B. and Liaw, C.Y. (2003) Substructural and progressive structural identification methods. *Engineering Structures*, 25: 1551–1563.

Koh, C.G. and Shankar, K. (2003) Substructural identification method without interface measurement. *Journal of Engineering Mechanics ASCE*, 129(7): 769–776.

Kottegoda, N.T. and Rosso, R. (1997) *Statistics, Probability, and Reliability for Civil and Environmental Engineers*. McGraw-Hill Companies, Inc.

Kuhar, E.J. and Stahle, C.V. (1974) Dynamic transformation method for modal synthesis. *AIAA Journal*, 12(5): 672–678.

Kukreti, A.R. and Abolmaali, A.S. (1999) Moment-rotation hysteresis behavior of top and seat angle steel frame connections. *Journal of Structural Engineering ASCE*, 125(8): 810–820.

Kullaa, J. (2002) Elimination of environment influences from damage-sensitive features in a structural health monitoring system. *Proceedings of the 1st European Workshop on Structural Health Monitoring*: 742–749.

Kwon, Y.W. and Aygunes, H. (1996) Dynamic finite element analysis of laminated beams with delamination cracks using contact-impact conditions. *Computers and Structures*, 58(6): 1160–1169.

Lanata, F. and Del Grosso, A. (2006) Damage detection and localization for continuous static monitoring of structures using a proper orthogonal decomposition of signals. *Smart Materials and Structures*, 15(6): 1811–1829.

Lantau fixed crossing: wind and structural health monitoring master plan Task 7.12 — Report No. 2 – Assessment strategy of possible structural damage in the Tsing Ma Bridge, Kap Shui Mun Bridge, and Ting Kau Bridge. (1998) Department of Civil and Structural Engineering, Hong Kong Polytechnic University, Hung Hom, Kowloon, Hong Kong.

Law, S.S. and Lu, Z.R. (2005) Time domain responses of a prestressed beam and prestress identification. *Journal of Sound and Vibration*, 288(4–5): 1011–1125.

Law, S.S. and Zhu, X.Q. (2004) Dynamic behaviour of damaged concrete structures under moving vehicular loads. *Engineering Structures*, 26(9): 1279–1293.

Law, S.S., Shi, Z.Y. and Zhang, L.M. (1998) Structural damage detection from incomplete and noisy modal test data. *Journal of Engineering Mechanics ASCE*, 124(11): 1280–1288.

Law, S.S., Chan, T.H.T. and Wu, D. (2001a) Super-element with semi-rigid joints in model updating. *Journal of Sound and Vibration*, 239(1): 19–39.

Law, S.S., Chan, T.H.T., Zhu, X.Q. and Zeng, Q.H. (2001b) Regularization in moving force identification. *Journal of Engineering Mechanics ASCE*, 127(2): 136–148.

Law, S.S., Wu, D. and Shi, Z.Y. (2001c) Model updating of semi-rigid jointed structures using generic parameters theory. *Journal of Engineering Mechanics ASCE*, 127(11): 1174–1183.

Law, S.S., Wu, Z.M. and Chan, S.L. (2003) Hybrid finite element with frictional joints for dynamic analysis of steel frames. *Journal of Engineering Mechanics ASCE*, 129(5): 564–570.

Law, S.S., Bu, J.Q., Zhu, X.Q. and Chan, S.L. (2004a) Vehicle axle loads identification using finite element method. *Engineering Structures*, 26: 1143–1153.

Law, S.S., Wu, Z.M. and Chan, S.L. (2004b) Vibration control study of a suspension footbridge using hybrid slotted bolted connection elements. *Engineering Structures*, 26(1): 107–116.

Law, S.S., Li, X.Y., Zhu, X.Q. and Chan, S.L. (2005) Structural damage detection from wavelet packet sensitivity. *Engineering Structures*, 27(9): 1339–1348.

Law, S.S., Li, X.Y. and Lu, Z.R. (2006) Structural damage detection from wavelet coefficient sensitivity with model errors. *Journal of Engineering Mechanics ASCE*, 132(10): 1077–1087.

Law, S.S., Zhang, K. and Duan, Z.D. (2008) Structural damage detection from coupling forces between substructures. *Proc. of 15th International Congress on Sound and Vibration*, Daejoen, Korea. 6–10 July 2008, Paper #T0489.

Lee, H. and Park, Y.S. (1995) Error analysis of indirect force determination and a regularization method to reduce force determination error. *Mechanical Systems and Signal Processing*, 9(6): 615–633.

Lee, I.W. and Jung, G.H. (1997a) An efficient algebraic method for the computation of natural frequency and mode shape sensitivities—Part I. Distinct natural frequencies. *Computers and Structures*, 62(3): 429–435.

Lee, I.W. and Jung, G.H. (1997b) An efficient algebraic method for the computation of natural frequency and mode shape sensitivities—Part II. Multiple natural frequencies. *Computers and Structures*, 62(3): 437–443.

Lee, J.W., Kim, J.D., Yun, C.B., Yi, J.H. and Shim, J.M. (2002) Health-monitoring method for bridges under ordinary traffic loadings. *Journal of Sound and Vibration*, 257(2): 247–264.

Lee, U., Lesieutre, G.A. and Fang, L. (1997) Anisotropic damage mechanics based on strain energy equivalence and equivalent elliptical microcracks. *International Journal of Solids and Structures*, 34(33): 4377–4397.

Lee, U., Cho, K. and Shin, J. (2003) Identification of orthotropic damages within a thin uniform plate. *International Journal of Solids and Structures*, 40(9): 2195–2213.

Lemaitre, J. (1985) A continuous damage mechanics model for ductile fracture. *Journal of Engineering Materials and Technology ASME*, 107(1): 83–89.

Leung, Y.T. (1978) An accurate method of dynamic condensation in structural analysis. *International Journal for Numerical Methods in Engineering*, 12: 1705–1715.

Leung, Y.T. (1979) An accurate method of dynamic substructuring with simplified computation. *International Journal for Numerical Methods in Engineering*, 14: 1241–1256.

Lewitt, C.W., Chesson, E. and Munse, W.H. (1969) Restraint characteristics of flexible riveted and bolted beam-to-column connections. *Engineering Bulletin of University of Illinois 500.*

Li, H.J., Yang, H.Z. and Hu, S.L.J. (2006) Modal strain energy decomposition method for damage localization in 3D frame structures. *Journal of Engineering Mechanics ASCE*, 132(9): 941–951.

Li, J. and Chen, J. (1999) A statistical average algorithm for the dynamic compound inverse problem. *Computational Mechanics*, 30(2): 88–95.

Li, J. and Law, S.S. (2008) Reliability analysis for a bridge structure based on condition assessment results. *Eleventh East Asia-Pacific Conference on Structural Engineering & Construction (EASEC-11) "Building a Sustainable Environment"*, November 19–21, 2008, Taipei, TAIWAN. Paper 605.

Li, X.Y. and Law, S.S. (2008) Damage identification of structures including system uncertainties and measurement noise. *AIAA Journal*, 46(1): 263–276.

Li, Y.Y., Cheng, L., Yam, L.H. and Wong, W.O. (2002) Identification of damage locations for plate-like structures using damage sensitive indices: strain modal approach. *Computers and Structures*, 80(25): 1881–1894.

Liew, K.M. and Wang, Q. (1998) Application of wavelet theory for crack identification in structures. *Journal of Engineering Mechanics ASCE*, 142(2): 152–157.

Limkatanyu, S. and Spacone, E. (2002) Reinforced concrete frame element with bond interfaces. I: Displacement-based, force-based, and mixed formulations. *Journal of Structural Engineering ASCE*, 128(3): 346–355.

Lin, Y.H. and Trethewey, M.W. (1990) Finite element analysis of elastic beams subjected to moving dynamic loads. *Journal of Sound and Vibration*, 136(2): 323–342.

Ling, X. and Haldar, A. (2004) Element level system identification with unknown input with Rayleigh damping. *Journal of Engineering Mechanics ASCE*, 130(8): 877–885.

Liu, M., Frangopol, D.M., and Kim, S. (2009a). "Bridge system performance assessment from structural health monitoring: A case study," *Journal of Structural Engineering*, ASCE 135 (6), 733–742.

Liu, M., Frangopol, D.M., and Kim, S. (2009b). "Bridge safety evaluation based on monitored live load effects," *Journal of Bridge Engineering*, ASCE, 14(4), 257–269.

Liu, P.L. (1995) Identification and damage detection of trusses using modal data. *Journal of Structural Engineering ASCE*, 121(4): 599–608.

Liu, P.L. and Chian, C.C. (1997) Parametric identification of truss structures using static strains. *Journal of Structural Engineering ASCE*, 123(7): 927–933.

Lu, Q., Ren, G. and Zhao, Y. (2002) Multiple damage location with flexibility curvature and relative frequency change for beam structures. *Journal of Sound and Vibration*, 253(5): 1101–1114.

Lu, Z.R. (2005) *Dynamic response sensitivity for structural condition assessment*, PhD Thesis. Civil and Structural Engineering Department, Hong Kong Polytechnic University.

Lu, Z.R. and Law, S.S. (2005) System identification including the load environment. *Journal of Applied Mechanics ASME*, 71(5): 739–741.

Lu, Z.R. and Law, S.S. (2006a) Force identification based on sensitivity in time domain. *Journal of Engineering Mechanics ASCE*, 132(10): 1050–1056.

Lu, Z.R. and Law, S.S. (2006b) Identification of prestress force from measured structural responses. *Mechanical System and Signal Processing*, 20(8): 397–412.

Lu, Z.R. and Law S.S. (2007a) Features of dynamic response sensitivity and its application in damage detection. *Journal of Sound and Vibration*, 303(1–2): 305–29.

Lu, Z.R. and Law, S.S. (2007b) Identification of system parameters and input force from output only. *Mechanical Systems and Signal Processing*, 21(5): 2099–2111.

Luo, H. and Hanagud, S. (1995) Delamination detection using dynamic characteristics of composite plates. *Proceedings of AIAA/ASME/ASCE/AHS Structures, Structural Dynamics & Materials Conference*: 129–139.

Maeck, J., Abdel Wahab, M., Peeters, B., De Roeck, G., De Visscher, J., De Wilde, W.P., Ndambi, J.M. and Vantomme, J. (2000) Damage identification in reinforced concrete structures by dynamic stiffness determination. *Engineering Structures*, 22, 1339–1349.

Mahmoud, M.A. (2001) Effect of cracks on the dynamic response of a simple beam subject to a moving load. *Proceedings of the Institute of Mechanical Engineers Part F: Journal of Rail and Rapid Transit*, 215(3): 207–215.

Majumdar, P.M. and Suryanarayan, S. (1988) Flexural vibrations of beams with delaminations. *Journal of Sound and Vibration*, 125(3): 441–461.

Majumder, L. and Manohar, C.S. (2003) A time-domain approach for damage detection in beam structures using vibration data with a moving oscillator as an excitation source. *Journal of Sound and Vibration*, 268(4): 699–716.

Mallat, S. and Hwang, W.L. (1992) Singularity detection and processing with wavelets. *IEEE Transactions on Information Theory*, 38(2): 617–643.

Mammone, R.J. (1992) *Computational Methods of Signal Recovery and Recognition*. New York: Wiley.

Marchesiello, S., Fasana, A., Garibaldi, L. and Piombo, B.A.D. (1999) Dynamics of multi-span continuous straight bridges subject to multi-degrees of freedom moving vehicle excitation. *Journal of Sound and Vibration*, 224(3): 541–561.

Mario, P. (1997) *Structural Dynamics Theory and Computation*. International Thomson Publishing.

Mason, J.C. and Handscomb, D.C. (2003) *Chebyshev Polynomials*. Boca Raton, FL: Chapman & Hall/CRC: 145–163.

Matta, K.W. (1987) Selection of degrees of freedom for dynamic analysis. *Journal of Pressure Vessel Technology*, 109(1): 65–69.

Mazurek, D.F. and Dewolf, J.T. (1990) Experimental study of bridge monitoring techniques. *Journal of Structural Engineering ASCE*, 115(9): 2532–2549.

Mindlin, R.D. (1949) Compliance of elastic bodies in contact. *Journal of Applied Mechanics ASME*, 16: 259–268.

Miller, C.A. (1980) Dynamic reduction of structural models. *Journal of the Structural Division ASCE*, 106(10): 2097–2108.

Miyamoto, A., Tei, K., Nakamura, H. and Bull, J.W. (2000) Behavior of prestressed beam strengthened with external tendons. *Journal of Structural Engineering ASCE*, 126(9): 1033–1044.

Moller, P.W. and Friberg, O. (1998) An approach to mode pairing problem. *Mechanical System and Signal Processing*, 12(4): 515–523.

Mottershead, J.E. and Friswell, M.I. (1993) Model updating in structural dynamics: a survey. *Journal of Sound and Vibration*, 167(2): 347–375.

Morozov, V.A. (1984) *Methods for Solving Incorrectly Posed Problems*. Berlin: Springer-Verglag: 1–64.

Nelson, R.B. (1976) Simplified calculations of eigenvector derivatives. *AIAA Journal*, 14: 1201–1205.

Nethercot, D.A. (1985) *Steel Beam-to-Column Connections—A Review of Test Data*. Construction Industry Research and Information Association, London, England.

Newmark, N.W. (1959) A method of computation for structural dynamics. *Journal of Engineering Mechanics Division ASCE*, 85(3): 67–94.

Nowak, A.S., Park, C.H. and Casas, J.R. (2001) Reliability analysis of prestressed concrete bridge girders: comparison of Eurocode, Spanish Norma IAP and AASHTO LRFD. *Structural Safety*, 23(4): 331–344.

Oberkampf, W.L., Deland, S., Rutherford, B.M., Diegert, K.V. and Alvin, K.F. (2002) Error and uncertainty in modelling and simulation. *Reliability Engineering and System Safety*, 75(3): 333–357.

O'Callahan, J.C. (1989) A procedure for an improved reduced system (IRS) model. *Proceedings of the 7th International Modal Analysis Conference*, Las Vegas: 17–21.

Oh, B.H. and Jung, B.S. (1998) Structural damage assessment with combined data of static and modal test. *Journal of Structural Engineering ASCE*, 124(8): 956–965.

Okafor, A.C., Chandrashekhara, K. and Jiang, Y.P. (1996) Delamination prediction in composite beams with built-in piezoelectric devices using modal analysis and neutral network. *Smart Materials and Structures*, 5(3): 338–347.

Pai, P.F. and Young, L.G. (2001) Damage detection of beams using operational deflection shapes. *International Journal of Solids and Structures*, 38(18): 3161–3192.

Pandey, A.K., Biswas, M. and Samman, M.M. (1991) Damage detection from changes in curvature mode shapes. *Journal of Sound and Vibration*, 145(2): 321–332.

Pandey, A.K. and Biswas, M. (1994) Damage detection in structures using changes in flexibility. *Journal of Sound and Vibration*, 169(1): 3–17.

Pandey, A.K. and Biswas, M. (1995) Experimental verification of flexibility difference method for locating damage in structures. *Journal of Sound and Vibration*, 184(2): 311–328.

Papadopoulos, L. and Garcia, E. (1998) Structural damage identification: a probabilistic approach. *AIAA J.*, 36(11): 2137–2145.

Pearl, J. (1988) *Probabilistic Reasoning in Intelligent Systems: networks of plausible inference*. Morgan Kaufmann.

Perel, V.Y. and Palazotto, A.N. (2002) Finite element formulation for dynamics of delaminated composite beams with piezoelectric actuators. *International Journal of Solids and Structure*, 39(17): 4457–4483.

Peters, R.J. (1986) An unmanned and undetectable highway speed vehicle weighting system. *Proceedings of 13th ARRB and 5th REAAA Combined Conference*, Part 6: 70–83.

Piombo, B.A.D., Fasana, A., Marchesiello, S. and Ruzzene, M. (2000) Modelling and identification of the dynamic response of a supported bridge. *Mechanical Systems and Signal Processing*, 14(1): 75–89.

Popov, E.O. (1983) *Seismic Moment Connections for Moment-resisting Steel Frames*, Report No. UCB/EERC-83/02, Earthquake Engineering Research Center, University of California, Berkeley, CA.

Qian, G.L., Gu, S.N. and Jiang, J.S. (1991) A finite element model of cracked plates and application to vibration problems. *Computers and Structures*, 39(5): 483–487.

Qu, Z.Q. and Fu, Z.F. (2000) An iterative method for dynamic condensation of structural matrices. *Mechanical Systems and Signal Processing*, 14(4): 667–678.

Ratcliffe, C.P. and Bagaria, W.J. (1998) A vibration technique for locating delamination in a composite beam. *AIAA Journal*, 36(6): 1074–1077.

Ren, W.X., De Roeck, D. and Harik, I.E. (2000) Singular value decomposition based truncation algorithm in structural damage identification through modal data. *Proceedings of 14th Engineering Mechanics Conference*: 21–24.

Ren, Y. (1992) *The Analysis and Identification of Friction Joint Parameters in the Dynamic Response of Structures*. PhD Thesis, Imperial College of Science, Technology and Medicine, London University, UK.

Richard, R.M. and Abbott, B.J. (1975) Versatile elastic-plastic stress-strain formula. *Journal of Engineering Mechanics Division ASCE*, 101(4): 511–515.

Robert, C.P. and Casella, G. (1999) *Monte Carlo Statistical Methods*. New York: Springer.

Robertson, A.N., Peterson, L.D., James, G.H. and Doebling, S.W. (1996) Damage detection in aircraft structures using structures using dynamically measured static flexibility matrices.

Proceedings of the 12th International Modal Analysis Conference (Dearborn, MI), Society of Experimental Mechanics, Bethel, CT: 857–865.

Robertson, A.N., Park, K.C. and Alvin, K.F. (1998) Extraction of impulse response data via wavelet transform for structural system identification. *Journal of Vibration and Acoustics ASME*, 120(1): 252–260.

Rucka, M. and Wilde, K. (2006) Crack identification using wavelets on experimental static deflection profiles. *Engineering Structures*, 28(2): 279–288.

Rudisill, C.S. (1974) Derivatives of eigenvalues and eigenvectors for a general matrix. *AIAA Journal*, 1(5): 412.

Ruotolo, R. and Surace, C. (1999) Using SVD to detect damage in structures with different operating conditions. *Journal of Sound and Vibration*, 226(3): 425–439.

Saiidi, N., Douglas, B. and Feng, S. (1994) Prestress force effect on vibration frequency of concrete bridges. *Journal of Structural Engineering ASCE*, 120(7): 2233–2241.

Sanayei, M. and Saletnik, M.J. (1996) Parameter estimation of structures from static strain measurements. I: Formulation. *Journal of Structural Engineering ASCE*, 122(5): 555–562.

Sanayei, M., Imbaro, G.R., Mcclain, J.A.S. and Brown, L.C. (1997) Structural model updating using experimental static measurements. *Journal of Structural Engineering ASCE*, 123(6): 792–798.

Shan, V.N. and Raymund, M. (1982) Analytical selection of masters for the reduced eigenvalue problem. *International Journal for Numerical Method in Engineering*, 18(1): 89–98.

Shen, M.H.H. and Grady, J.E. (1992) Free vibration of delaminated beams. *AIAA Journal*, 30(5): 1361–1370.

Shen, M.H.H. and Pierre, C. (1990) Natural modes of Bernoulli-Euler beams with symmetric cracks. *Journal of Sound and Vibration*, 138(1): 115–134.

Shenton, III H.W. and Hu, X.F. (2006) Damage identification based on dead load redistribution: methodology. *Journal of Structural Engineering ASCE*, 132(8): 1254–1263.

Shi, G. and Atluri, S.N. (1989) Static and dynamic analysis of space frames with nonlinear flexible connections. *International Journal for Numerical Methods in Engineering*, 28(11): 2635–2650

Shi, G. and Atluri, S.N. (1992) Nonlinear dynamic response of frame-type structures with hysteretic damping at the joints. *AIAA Journal*, 30(1): 234–240.

Shi, T., Jones, N.P. and Ellis, J.H. (2000a) Simultaneous estimation of system and input parameters from output measurements. *Journal of Engineering Mechanics ASCE*, 126(7): 746–753.

Shi, Z.Y., Law, S.S. and Zhang, L.M. (1998) Structural damage localization from modal strain energy change. *Journal of Sound and Vibration*, 218(5): 825–844.

Shi, Z.Y., Law, S.S. and Zhang, L.M. (2000b) Structural damage detection from modal strain energy change. *Journal of Engineering Mechanics ASCE*, 126(12): 1216–1223.

Shi, Z.Y., Law, S.S. and Zhang, L.M. (2000c) Optimum sensor placement for structural damage detection. *Journal of Engineering Mechanics ASCE*, 126(11): 1173–1179.

Shi, Z.Y., Law, S.S. and Zhang, L.M. (2002) Improved damage quantification from elemental strain energy change. *Journal of Engineering Mechanics ASCE*, 128(5): 521–529.

Shoukry, S.N. (1985) A mathematical model for the stiffness of fixed joints between machine parts. *Proceedings of the NUMETA'85 Conference*, Swansea: 851–858.

Sih, G.C. and Liebowitz, H. (1968) *Mathematical Theories of Brittle Fracture*. Volume II, Chapter 2. Ed. Liebowitz, H. New York: Academic Press.

Simo, J.C. and Ju, J.W. (1987) Strain- and stress-based continuum damage models. *International Journal of Solids and Structures*, 23(7): 821–869.

Sjövall, P. and Abrahamsson, T. (2007) Substructure system identification from coupled system test data. *Mechanical Systems and Signal Processing*, 22: 15–33.

Smyth, A. and Wu, L.L. (2007) Multi-rate Kalman filtering for the data fusion of displacement and acceleration response measurements in dynamic system monitoring. *Mechanical Systems and Signal Processing*, 21(2): 706–723.

Snyman, J.A. (2005) *Practical Mathematical Optimization: An Introduction to Basic Optimization Theory and Classical and New Gradient-based Algorithms*. Springer Science & Business Media, Inc.

Soh, C.K., Chiew, S.P. and Dong, Y.X. (1999) Damage model for concrete-steel interface. *Journal of Engineering Mechanics ASCE*, 125(8): 979–983.

Soh, C.K., Liu, Y., Dong, Y.X. and Lu, X.Z. (2003) Damage model based reinforced-concrete element. *Journal of Materials in Civil Engineering ASCE*, 15(4): 371–380.

Sohn, H. and Law, K.H. (1997) A Bayesian probabilistic approach for structural damage detection. *Earthquake Engineering and Structural Dynamics*, 26(12): 1259–1281.

Spacone, E. and Limkatanyu, S. (2000) Responses of reinforced concrete members including bond-slip effects. *ACI Structural Journal*, 97(6): 831–839.

Staszewski, W.J. (1998) Structural and mechanical damage detection using wavelets. *The Shock and Vibration Digest*, 30(6): 457–472.

Stone, M. (1974) Cross-validatory choice and assessment of statistical prediction. *Royal Statistics Society*, 36: 111–147.

Stubbs, N., Park, S., Sikorsky, C. and Choi, S. (1998) A methodology to nondestructively evaluate the safety of offshore platforms. *Proc. of the 18th Int. Offshore and Polar Engrg Conf.*, Montreal, Canada.

Suarez, L.E. and Singh, M.P. (1992) Dynamic condensation method for structural eigenvalue analysis. *AAIA Journal*, 30(4): 1046–1054.

Sun, Z. and Chang, C.C. (2002a) Wavelet packet signature: a novel structure condition index. *China-Japan Workshop on Vibration Control and Health Monitoring of Structures and Third Chinese Symposium on Structural Vibration Control*, Shanghai, China.

Sun, Z. and Chang, C.C. (2002b) Structural damage assessment based on wavelet packet transform. *Journal of Structural Engineering ASCE*, 128(10): 1354–1361.

Tanaka M. and Bui H.D. (1992) *Inverse Problems in Engineering Mechanics: IUTAM symposium, Tokyo*. Springer-Verlag.

Tanaka M. and Dulikravich G.S. (1998) *Inverse Problems in Engineering Mechanics: International Symposium on Inverse Problems in Engineering Mechanics 1998 (ISIP '98), Nagano, Japan*. Elsevier Science Ltd.

Tanaka M. and Dulikravich G.S. (2000) *Inverse Problems in Engineering Mechanics II: International Symposium on Inverse Problems in Engineering Mechanics 2000 (ISIP 2000), Nagano, Japan*. Elsevier Science Ltd.

Teng, J.G., Chen, J.F., Smith, S.T. and Lam, L. (2002) *FRP Strengthened RC Structures*, Chichester, UK: John Wiley and Sons.

Terrell, M.J., Friswell, M.I. and Lieven, N.A.J. (2007) Constrained generic substructure transformations in finite element model updating. *Journal of Sound and Vibration*, 300(1–2): 264–279.

Tikhonov, A.N. (1963) On the solution of ill-posed problems and the method of regularization. *Soviet Mathematics*, 4: 1035–1038.

Tong, L. and Steven, G.P. (1999) *Analysis and Design of Structural Bonded Joints*. Dordrecht: Kluwer.

Tong, L., Sun, D.C. and Atluri, S.N. (2001) Sensing and actuating behaviours of piezoelectric layers with debonding in smart beams. *Smart Materials and Structures*, 10(4): 713–723.

Tracy, J.J. and Pardoen, G.C. (1989) Effect of delamination on the natural frequencies of composite laminates. *Journal of Composite Materials*, 23(12): 1200–1215.

Trujillo, D.M. and Busby, H.R. (1997) *Practical Inverse Analysis in Engineering*. CRC Press.

Vanik, M.W., Beck, J.L. and Au, S.K. (2000) Bayesian probabilistic approach to structural health monitoring. *Journal of Engineering Mechanics ASCE*, 126(7): 738–745.

Van Laarhoven, P.J.M. and Aarts, E.H.L. (1987) *Simulates Annealing: Theory and Applications*. Dordrecht, Holland: Reidel.

Vanlanduit, S., Parloo, E., Cauberghe, B., Gulllaume, P. and Verboven, P. (2005) A robust singular value decomposition for damage detection under changing operating conditions and structural uncertainties. *Journal of Sound and Vibration*, 284(3): 1033–1050.

Visser, R. (2001) Regularization in nearfield acoustic source identification. *Proceedings of the 8th International Congress on Sound and Vibration*, 2–6 July 2001, Hong Kong, China: 1637–1644.

Voyiadjis, G.Z. and Kattan, P.I. (2005) *Damage Mechanics*. Taylor and Francis.

Wahab, M.M., De Roeck, G. and Peeters, B. (1999) Parameterization of damage in reinforced concrete structures using model updating. *Journal of Sound and Vibration*, 228(4): 717–730.

Wang, B.P. (1985) Improved approximate method for computing eigenvector derivatives in structural dynamics. *AIAA Journal*, 29: 1018–1020.

Wang, J.T.S., Liu, Y.Y. and Gibby, J.A. (1982) Vibration of split beams. *Journal of Sound and Vibration*, 84(4): 491–502.

Wang, Q. and Deng, X.M. (1999) Damage detection with spatial wavelets. *International Journal of Solids and Structures*, 36(23): 3443–3468.

Wang, X., Hu, N., Fukunage, H. and Yao, Z.H. (2001) Structural damage identification using static test data and changes in frequencies. *Engineering Structures*, 23(6): 610–621.

Williams, D. and Jones, R.P.N. (1948) *Dynamic Loads in Aeroplanes under given Impulsive Loads with Particular Reference to Landing and Gust Loads on a Large Flying Boat*. Aeronautic Research Council, TR#2221.

Wu, C.G., Liang, Y.C., Lin, W.Z., Lee, H.P. and Lim, S.P. (2003) A note on equivalence of proper orthogonal decomposition methods. *Journal of Sound and Vibration*, 265(5): 1103–1110.

Wu, D. and Law, S.S. (2004a) Damage localization in plate structures from uniform load surface curvature. *Journal of Sound and Vibration*, 276(1–2): 227–244.

Wu, D. and Law, S.S. (2004b) Model error correction from truncated modal flexibility sensitivity and generic parameters. I: Simulation. *Mechanical Systems and Signal Processing*, 18(6): 1381–1399.

Wu, D. and Law, S.S. (2004c) Model error correction from truncated modal flexibility sensitivity and generic parameters. II: Experimental Verification. *Mechanical Systems and Signal Processing*, 18(6): 1401–1419.

Wu, D. and Law, S.S. (2005a) Sensitivity of uniform load surface curvature for damage identification in plate structures. *Journal of Vibration and Acoustics ASME*, 127(1): 84–92.

Wu, D. and Law, S.S. (2005b) Crack identification in thin plates with anisotropic damage model and vibration measurements. *Journal of Applied Mechanics ASME*, 72(6): 852–861.

Wu, D. and Law, S.S. (2007) Delamination detection-oriented finite element model for a fiber reinforced polymer bonded concrete plate and its application with vibration measurements. *Journal of Applied Mechanics ASME*, 74(2): 240–248.

Wu, J.J. (2007) Use of moving distributed mass element for the dynamic analysis of a flat plate undergoing a moving distributed load. *International Journal for Numerical Methods in Engineering*, 71(3): 347–362.

Wu, S.Q. and Law, S.S. (2009) Stochastic bridge-vehicle interaction problem with Gaussian uncertainties. *The 10th International Conference on Structural Safety and Reliability (ICOSSAR2009)*, Osaka, Japan. September 13–17, 2009.

Xia, Y., Hao, H., Brownjohn, J.M.W. and Xia, P.Q. (2002) Damage identification of structures with uncertain frequency and mode shape data. *Earthquake Engineering and Structural Dynamics*, 31(5): 1053–1066.

Yang, J.N. and Huang, H.W. (2006) Substructure damage identification using sequential nonlinear LSE method. *Proceedings of the 4th International Conference on Earthquake Engineering*, Taipei, Taiwan, October 12–13, 2006. Paper No. 119.

Yang, Y.B. and Yau, J.D. (1997) Vehicle-bridge interaction element for dynamic analysis. *Journal of Structural Engineering ASCE*, 123(11): 1512–1528.

Yuen, K.V. and Katafygiotis, L.S. (2006) Substructure identification and health monitoring using noisy response measurements only. *Computer-aided Civil and Infrastructure Engineering*, 21: 280–291.

Yun, C.B. and Bahng, E.Y. (2000) Substructural identification using neural networks. *Computer & Structures*, 77: 41–52.

Yun, C.B. and Lee, H.J. (1997) Substructural identification for damage estimation of structures. *Structural Safety*, 19(1): 121–140.

Zabaras, N., Woodbury, K.A. and Raynaud, M. (1993) *Inverse Problems in Engineering: Theory and Practice*. New York: Published on behalf of the Engineering Foundation by the American Society of Mechanical Engineers.

Zastrau, B. (1985) Vibration of cracked structures. *Archives of Mechanics*, 37(6): 731–743.

Zhang, Z. and Aktan, A.E. (1998) Application of modal flexibility and its derivatives in structural identification. *Research in Nondestructive Evaluation*, 10(1): 43–61.

Zhu, X.Q. and Law, S.S. (2001) Orthogonal function in moving loads identification on a multi-span bridge. *Journal of Sound and Vibration*, 245(2): 329–345.

Zhu, X.Q. and Law, S.S. (2002) Dynamic load on continuous multi-lane bridge deck from moving vehicles. *Journal of Sound and Vibration*, 251(4): 697–716.

Zhu, X.Q. and Law, S.S. (2002) Moving loads identification through regularization. *Journal of Engineering Mechanics ASCE*, 128(9): 989–1000.

Zhu, X.Q. and Law, S.S. (2006) Wavelet-based crack identification of bridge beam from operational deflection time history. *International Journal of Solids and Structures*, 43(7–8): 2299–2317.

Zhu, X.Q. and Law, S.S. (2007a) Damage detection in simply supported concrete bridge structure under moving vehicular loads. *Journal of Vibration and Acoustics ASME*, 129(1): 58–65.

Zhu, X.Q. and Law, S.S. (2007b) A concrete-steel interface element for damage detection of reinforced concrete structures. *Engineering Structures*, 29(12): 3513–3524

Zhu, X.Q., Law, S.S. and Hao, H. (under review) Damage assessment of reinforced concrete beams including the load environment. *International Journal of Mechanical Sciences*.

Subject Index

A

AASHTO 228
Admissibility condition 232
Algorithm
  for damage detection 219, 224
  for optimization 15
  for system identification 209
Ambient excitation 271, 275
Ambient vibration test 283–284
Assumed mode shape 89
Auto-covariance 284, 285

B

Bayesian theorem 291–294
Beams
  concrete 78–86, 154–164, 243
  steel 207, 208, 211
Bond
  damage index 82
  interface 82
  slip 81
  stress distribution function 79–80
Boundary conditions 44–45
Bridge-vehicle system 227–230, 236

C

Cantilever plate 192
Chebyshev polynomial 182–184
Concrete
  pre-stressed 92
  reinforced 78–79, 80, 83, 157, 160, 243
Condensation
  static 39, 131–132
  dynamic 132–135
  iterative 135
Connection spring element 32–33, 43–44
Constrained minimization 145–151
Covariance 271
Covariance sensitivity 271–275

Crack
  anisotropic 100, 107
  elliptical 97, 98
  flexural 78, 79
Crack model 237–239
Crack zone 244
Criteria of convergence 25
Cross-correlation function 271, 272, 275
Curvatures
  modal flexibility 181
  mode shape 181
  uniform load surface 106, 107, 185–188
  unit load surface 181–182
Cyclic loading 29–32

D

Damage assessment 46, 154
Damage model 27, 163, 198, 271
Damage detection 145, 167, 219, 224, 283, 285, 308
Damage-detection-oriented-model 28
Damage distribution function 152
Damage index 82
Daubechies wavelet 266
Dilation parameter 232
(DLS) Damped least-squares method 207
Damper 41
Delamination 113, 116, 127–128
Dini-Lipschitz condition 183
Dirac delta function 87, 237
Dynamic characteristic equivalence 100

E

Earthquake loading 275
Eigen-decomposition 52
Eigenpairs 169, 170
Eigenvalues 168, 170
  close 170–172
  distinct 169
  repeated 170–172

Eigenvectors 168
Elemental damage indicator 82–83
Elemental stiffness matrix 33, 34, 46
Elements
    beam 28, 70
    frame 49–50, 54–55
    generic 47–52
    hybrid 33, 38, 39
    rectangular 58–60, 61
    rod 50
    shell 56–58
    spring 32–33, 43–46
    super 69–78
Euler-Bernoulli beam 236
Euler-Bernoulli beam element 50
European Space Agency Structure 49–50

F
False positive 222
Filter factor 13, 15
Finite element method 90–91
Flexibility
    flexural 42–46
    shear 29, 42–46
Force identification 205–207
Frequency-domain 199
FRP (Fibre-reinforced-plastic) 113, 116

G
GA (Genetic algorithm) 17–18
Gap-smoothing 184
GCV (Generalized cross-validation) 21–23
Generic element 47–50
Geometric stiffness matrix 35–37
Gradient-based optimization 15
Gradient vector 15, 16
Griffith theory 99
Guyan/Iron method 133

H
Hysteretic damping 29

I
Ill-conditioned 10, 14
Ill-posedness 10
Ill-posed problems 10
Impulsive excitation 203, 205
Impulse response function 264–271
Interface 116–117
Inverse problems 9
Inverse wavelet transform 232
IRF (impulse response function) 264–271
IRS (Improved Reduction System)
    method 133

K
Kirchhoff's theory 101
Karhunen-Loéve expansion 294

L
Laplacian operator 181
L-curve 23–25
Linear least-square problem 10
Load-carrying capacity 93
Load environment 275
London Millennium Footbridge 37

M
Macro stiffness 59
Mass topology matrix 54
Matrix transformation 39
Measurement noise 154, 195, 256, 264
    see also white noise
Mexican Hat wavelet 240
Micro slip element 29
Modal flexibility 176–177
Modal flexibility sensitivity 177–179
Mode shape changes 281
Model error 256, 260–262
Model updating 46–47, 172–173, 218–219
Monte-Carlo simulation 291
Mother wavelet 232
Moving loads 154, 223, 236
Moving vehicular loads 236
MSE (Modal Strain Energy) 173
MSECR (MSE Change Ratio) 173

N
Nelson's method 169
Nonlinear stiffness 29
Null space 12

O
OPENFEM 106, 113
Operational deflection 236
Operational loads 223–230
Orthogonal basis 233
Output error function
    displacement 146
    strain 147–148
Orthogonal function expansion 226–227

P
Perturbation method 291–292
Picard condition 13, 14, 21
Plate
    thin 97, 100 3–76, 3–80
    thick 107, 113, 116
Pre-stress 27, 91
    force 92
    tendon 92–93
    identification 205
Proper orthogonal decomposition 294
Probability density function 293

Q
Quadratic programming problem 149

R
Random excitations
  single 272–273
  multiple 273–274
Random support excitation 280
Rank deficient 12, 20
Rayleigh damping 90, 201
Regularization parameter 10, 11, 24
Reinforced concrete beam element 80
Reissner-Mindlin plate 188
Residual pre-stress identification 229
Response sensitivity 199, 201, 207, 245,
  258–260, 277
Richard-Abbott model 42

S
SA (Simulated annealing) 19
Sampling frequency 211, 243
Scordelis-Lo roof 63
Semi-rigid joints 70, 76
Sensitivity 245
  of cross-correlation function 271
  of covariance 271
  of eigenvalues 168
  of eigenvectors 168
  of impulse response function 264–265
  of MSEC 174
  of modal flexibility 176
  of ULS curvature 185
  of wavelet 246
  of WPT coefficients 234, 276, 277
  of WPT energy 285
Shape function 44–46
Shear-slip model 43
Simply supported
  beam 211
  plate 190
Singular values 11, 13
Sinusoidal excitation 203, 208
SBCE (slotted bolted connection element) 29
Spatial distribution system 167
State-space formulation 200
Static response changes 148–150
Stiffness
  anti-symmetric torsional 61
  saddle and symmetric bending 61
  symmetric twisting 61
  topology matrix 54
Strain energy equivalence 97–98
Super element 69

SVD (Singular value decomposition)
  11–12, 187
  generalized SVD 12–13
  truncated SVD 20–21

T
Tangent stiffness matrix 37
Taylor series 133, 172
Temperature effect 212–215
Tendon
  bonded 92
  pre-stressing 95
  unbonded 93
Tikhonov regularization 10, 15, 19–25
Time domain 199
Timoshenko beam element 50
TMF (truncated modal flexibility) 178
Translation parameter 232
Tsing Ma Bridge 70, 73

U
ULS (unit load surface) 180
Uniform load surface 106, 180
Uniform load surface curvature 185
Uncertainty
  environment 289
  measurement 289
  modelling 288
  parameter 289
Uncertainty analysis 287

V
Variance of static deflections 152
Vibration mitigation 38

W
Wavelet
  bandwidth 252
  coefficient 245, 248, 249, 250, 257, 258
  continuous 232, 236
  discrete 233, 265
WBZ (weak bonding zone) 116
Weights 64
WPD (wavelet packet decomposition) 234
WPT (wavelet packet transform) 234,
  246, 251
WPT component energy 247
WPT component energy sensitivity 246
WPT sensitivity 277
WT (wavelet transform) 154, 232, 234,
  239, 267
  continuous 232, 236
  discrete 233, 265, 266
  spatial 154
White noise 160, 240, 263

Structures and Infrastructures Series

Book Series Editor: Dan M. Frangopol

ISSN: 1747–7735

Publisher: CRC/Balkema, Taylor & Francis Group

1. Structural Design Optimization Considering Uncertainties
   Editors: Yiannis Tsompanakis, Nikos D. Lagaros & Manolis Papadrakakis
   2008
   ISBN: 978-0-415-45260-1 (Hb)

2. Computational Structural Dynamics and Earthquake Engineering
   Editors: Manolis Papadrakakis, Dimos C. Charmpis,
   Nikos D. Lagaros & Yiannis Tsompanakis
   2008
   ISBN: 978-0-415-45261-8 (Hb)

3. Computational Analysis of Randomness in Structural Mechanics
   Christian Bucher
   2009
   ISBN: 978-0-415-40354-2 (Hb)

4. Frontier Technologies for Infrastructures Engineering
   Editors: Shi-Shuenn Chen & Alfredo H-S. Ang
   2009
   ISBN: 978-0-415-49875-3 (Hb)

5. Damage Models and Algorithms for Assessment of Structures
   under Operating Conditions
   Siu-Seong Law and Xin-Qun Zhu
   ISBN: 978-0-415-42195-9 (Hb)

6. Structural Identification and Damage Detection using Genetic Algorithms
   Chan Gee Koh and Michael. J. Perry
   ISBN: 978-0-415-46102-3 (Hb)